Vascular Technology

AN ILLUSTRATED REVIEW

Vascular Technology

AN ILLUSTRATED REVIEW

New Fourth Edition

CLAUDIA RUMWELL, RN, RVT, FSVU

MICHALENE McPHARLIN, RN, RVT, FSVU

DAVIES PUBLISHING

DAVIES
Registry Reviews & Study Aids

Library of Congress Cataloging-in-Publication Data

Rumwell, Claudia, 1946–
Vascular technology : an illustrated review / Claudia Rumwell, Michalene McPharlin.—New 4th ed.
p. ; cm.
Includes bibliographical references and index.
ISBN 0-941022-73-0
1. Angiography. 2. Angiography—Examinations, questions, etc. I.
McPharlin, Michalene, 1949– II. Title.
[DNLM: 1. Vascular Diseases—diagnosis—Examination Questions. 2.
Diagnostic Techniques, Cardiovascular—Examination Questions. WG 18.2 R938v 2009]
RC691.6.A53R86 2009
616.1'30076–dc22

2008038522

Davies Publishing, Inc.
Publishers in Medicine and Surgery
32 South Raymond Avenue
Pasadena, California 91105-1935
Phone 626.792.3046
Facsimile 626.792.5308
E-mail info@DaviesPublishing.com
Website www.DaviesPublishing.com

Cover, interior design, and art direction by Bill Murawski
Production editing by Christian Jones
Illustrations by Stephen Beebe, Donald Ridgway, Ted Bloodhart, Jay Knipstein, and The Left Coast Group, Inc.
Index by Bruce Tracy, PhD
Production by The Left Coast Group, Inc.

Printed and bound in the United States of America

ISBN 0-941022-73-0

Preface

Now in its fourth edition, this book remains concise, practical, and clinically oriented. Its purposes also remain the same:

- A text for sonographers and technologists in training (and cross training).
- A reference and guide for practicing technologists and sonographers.
- A resource for interpreting physicians and RPVI candidates.
- A very efficient and powerful means of preparing for the national registry examinations in vascular technology, which indeed was a primary goal of the first, second, and third editions.

In this fourth edition we have greatly expanded the content of our illustrated review. The text has been reorganized into chapters, some of them new and all of them updated and revised. There are many more figures as well. We have added and revised hundreds of images and illustrations to provide examples of normality, abnormality, and hemodynamics as they relate to physiologic Doppler and plethysmographic waveforms, duplex scanning, and color flow studies. We selected, annotated, and explain these new images to promote an understanding of not only the data acquisition process, but also the interpretation of findings.

Vascular diagnosis depends on a broad spectrum of diagnostic modalities and techniques. Like the first three editions, this fourth edition continues to explain these modalities and techniques in a concise, orderly manner, integrating a variety of elements—text, illustrations, photographs, radiographic findings, and self-assessment case-study questions. In so doing it addresses the topics on the ARDMS outline for the specialty examination in vascular technology, and also the CCI exam. In addition, it explains vascular-specific physical principles and also provides an excellent resource for those physicians taking the ARDMS physician interpretation (RPVI) exam.

The text clearly and systematically covers anatomy and hemodynamics, patient history, clinical manifestations, testing capabilities, limitations, physical principles, techniques, current diagnostic criteria, interpretation, therapeutic interventions, test validation, and related subtopics. For those whose familiarity with the nonimaging tests (physiologic studies) may be limited, it explains and illustrates them in great detail.

A complete questionnaire and exam at the end of the book make it possible to earn continuing medical education credit toward satisfaction of ARDMS, ICAVL, and other requirements for professional registration and facility accreditation.

We are very happy to offer this fourth edition to you. It is designed to provide a strong educational foundation upon which those with less experience can build and those with more experience can expand. *Vascular Technology* describes and explains how, why, and when imaging and nonimaging techniques are applied to the noninvasive diagnosis of vascular disease. It is our hope that this new edition will become a valuable tool for you in the years ahead.

Claudia Rumwell, RN, RVT, FSVU

Portland, Oregon

Michalene McPharlin, RN, RVT, FSVU

Detroit, Michigan

Acknowledgments

We are very thankful for the overwhelmingly positive response to our first, second, and third editions—from students, educators, physicians, cross-training sonographers and echocardiographers, and our own colleagues in the field of vascular technology. Your response provided the impetus for this expanded fourth edition.

A project like this cannot be completed without help. Once again, we gratefully acknowledge the expert opinion and direction of our publisher, Michael Davies. In addition, our appreciation goes to Don Ridgway for his editing and comments. We also want to make sure that the technologists and sonographers in the vascular labs at Henry Ford Hospital in Detroit, Michigan, supervised by Micky, and at Oregon Health & Science University Hospital, Providence St. Vincent Medical Center and Providence Portland Medical Center in Portland, Oregon, previously supervised by Claudia, know how much we appreciate them and their contributions to this and all previous editions. In addition, we would like to thank technologists from other laboratories across the country who made contributions to this book.

We could not have undertaken this endeavor without the unwavering support of our husbands, Mel Rumwell and Dan McPharlin. Without their encouragement, patience, and ongoing support, we would not have been able to start, let alone complete, this huge undertaking.

We are grateful to have had the opportunity to work with so many outstanding physicians, technologists, and sonographers, all friends and colleagues. You keep us motivated and continue to serve as role models. We feel privileged to share what we have learned with you, and we are very thankful for what we have learned from you.

Our ultimate goal is to provide the very best service and care to those patients with whom we are entrusted. The fact that you have this book is evidence that you share this same goal, and we thank you for caring enough to participate actively in your own professional development. Our best wishes to you.

Color Plate 1. Longitudinal image of the carotid bifurcation in systole. The inset shows the same image in diastole. Note that flow separation (in blue) is present during systole, absent in diastole. See figure 2-12 and the discussion of pressure/velocity relationships (Bernoulli) in chapter 2.

Color Plate 2. Blue toe syndrome. Note the bluish discoloration of the great toe. See the discussion of embolism under *Mechanisms of Disease* in chapter 3.

Color Plate 3. Pseudoaneurysm of the common femoral artery (CFA). The "neck" and unique Doppler waveforms (to and fro) are diagnostic hallmarks of a pseudoaneurysm. See figures 3-5, 13-3, and 17-13. See also the discussion of aneurysms under *Mechanisms of Disease* in chapter 3, the evaluation of technique for bypass grafts in chapter 13, and manual compression under *Pseudoaneurysm Treatment Options* in chapter 17.

Color Plate 4. Patient with Raynaud's phenomenon. The patient's finger tips are cyanotic. See the discussion of vasospastic disorders/cold sensitivity under *Mechanisms of Disease* in chapter 3.

Color Plate 5. Ulcerations on the tibia and lateral ankle. See chapter 28 for a discussion of venous vs. arterial ulcers. See also figure 3-8 and the discussion of lesions under *Physical Examination* in chapter 3.

Color Plate 6. Dependent rubor in both lower extremities. See the discussion of elevation/dependency changes under *Physical Examination* in chapter 3.

Color Plate 7. Arrow #1 Flow moving in a vein bypass graft (from left to right on the image) toward the patient's foot. **Arrow #2** Flow moving in the same direction as #1, but in the native dorsalis pedis artery. **Arrow #3** The native anterior tibial artery. **Arrow #4** Flow moving from right to left on the image as it moves proximally in the native anterior tibial artery to fill any branches that may be providing collateral flow around a proximal occlusion or high-grade stenosis. See figure 13-4 and the discussion of Doppler signals in chapter 13.

Color Plate 8. Transverse view of the abdominal aorta also showing the SMA, left renal vein, and splenic vein. This image is a good reminder to watch the color bars for flow direction. In this example, the veins are not blue. See figure 14-4 and the discussion of renal artery scanning techniques in chapter 14.

Color Plate 9. Longitudinal (sagittal) view of the abdominal aorta with duplicate right renal arteries and a single left renal artery. This image is often obtained by using an oblique approach from the right side of the abdomen. Because of the configuration of the image, it is often referred to as the "banana peel" image. See figure 14-7 and the discussion of renal artery scanning techniques in chapter 14.

Color Plate 10. The thin layer of lipid material on the arterial wall can be difficult to visualize because of its low-level echoes. Color flow imaging is helpful in determining that the vessel does not fill completely. See figure 19-1 and the discussion of atherosclerosis under *Mechanisms of Disease* in chapter 19.

Color Plate 11. Sagittal view of a carotid body tumor. Carotid body tumors are highly vascular structures that develop between the internal carotid artery and external carotid artery (ECA) and are usually fed by the ECA. See figure 19-8 and the discussion of nonatherosclerotic lesions under *Mechanisms of Disease* in chapter 19.

Color Plate 12. Longitudinal view of a common carotid artery dissection. The inset shows the wall defect in transverse view. Flow direction and velocities differ in each lumen. A distinguishing B-mode feature is the thin membrane dividing the main arterial lumen from the false lumen. See figure 19-9A and the discussion of nonatherosclerotic lesions under *Mechanisms of Disease* in chapter 19.

Color Plate 13. A sagittal view of the distal CCA and proximal ICA obtained one year post-endarterectomy. The thickened arterial wall with very low-level echoes is consistent with intimal hyperplasia. This differs from a recurrence of atherosclerosis, which usually is not evident until two years following carotid endarterectomy. See figure 19-12 and the discussion of nonatherosclerotic lesions under *Mechanisms of Disease* in chapter 19.

Color Plate 14. B-mode image of the vertebral artery. See figure 22-1A and the discussion of longitudinal technique in chapter 22.

Color Plate 15. Longitudinal view of an occluded common carotid artery. The retrograde blood flow detected in the external carotid artery represents the collateral pathway maintaining blood flow to the internal carotid artery (ICA). The inset documents flow in the ICA. See figure 22-14 and the discussion of interrelated, corroborative findings in chapter 22.

Color Plate 16. Lipodermatosclerosis in a patient with chronic venous insufficiency. The tissue becomes hard and over time can change the contour of the ankle area, resulting in a "bottle-neck deficiency." See figure 28-1 and the discussion of skin changes under *Physical Examination* in chapter 28.

Color Plate 17. Phlegmasia ceruleans dolens, a potentially limb-threatening complication of acute deep venous thrombosis. The condition is an unusual finding in the lower extremity; upper extremity involvement is rare and in this case did result in amputation of the affected limb. See figure 28-2 and the discussion of skin changes under *Physical Examination* in chapter 28.

Color Plate 18. Acute partial thrombosis of this portal vein is evident by lack of color-filling and dilatation of the vein wall. Although there is no accompanying artery for comparison, dilatation is apparent in the contour of the vein wall. See figure 34-16 and the discussion of acute thrombosis in chapter 34.

Color Plate 19. Longitudinal and transverse (inset) view of the axillary vein documenting acute thrombosis. The vein is dilated, and very low-level echoes are evident in sagittal view. Color flow imaging reveals partial filling of the vessel consistent with partial deep venous thrombosis. See figure 34-17 and the discussion of acute thrombosis in chapter 34.

Color Plate 20. Sagittal view of proximal great saphenous vein. With patient at rest, blood is moving in its normal direction (cephalad). See figure 34-22 and the discussion of chronic changes following acute DVT in chapter 34.

Color Plate 21. During the Valsalva maneuver, which is comparable to a proximal compression, reversed flow is evident. This flow reversal may also be seen following release of distal compression and is consistent with venous reflux. See figure 34-22 and the discussion of chronic changes following acute DVT in chapter 34.

Color Plate 22. Sagittal view of the infrainguinal ligament area demonstrating a vascular mass. Any mass, vascular or avascular, must be evaluated. Color flow imaging reveals a vascularized lymph node consistent with this patient's history of lymphoma. The B-mode image (inset) confirms a mass. See figure 34-24 and the discussion of miscellaneous findings in the upper extremities in chapter 34.

Color Plate 23. Transverse view of the brachial vessels, the basilic vein, and an incompressible circular structure. See figure 34-28 and the discussion of miscellaneous findings in the upper extremities in chapter 34.

With sincere gratitude we give special thanks to the following laboratories and manufacturers for various images: Henry Ford Hospital Vascular Lab (Detroit, Michigan), Oregon Health & Science University Hospital Vascular Lab (Portland, Oregon), Ochsner Clinic Vascular Lab (New Orleans, Louisiana), Providence Portland Medical Center Noninvasive Vascular Lab (Portland, Oregon), Providence St. Vincent Medical Center Noninvasive Vascular Lab (Portland, Oregon), GE Diasonics, Philips Medical Systems, Inc., and Siemens Medical Systems, Inc.

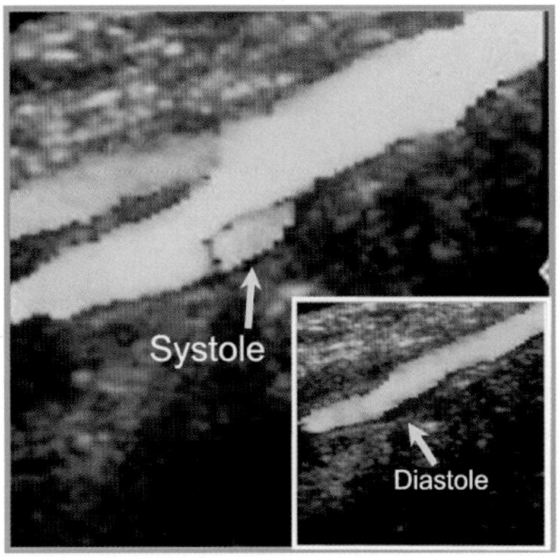

1

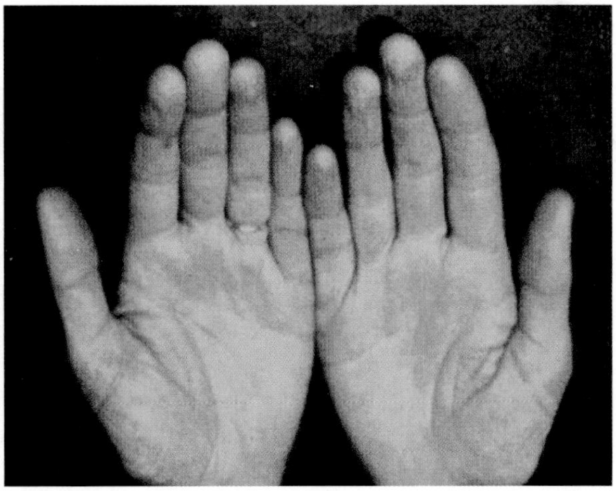
4

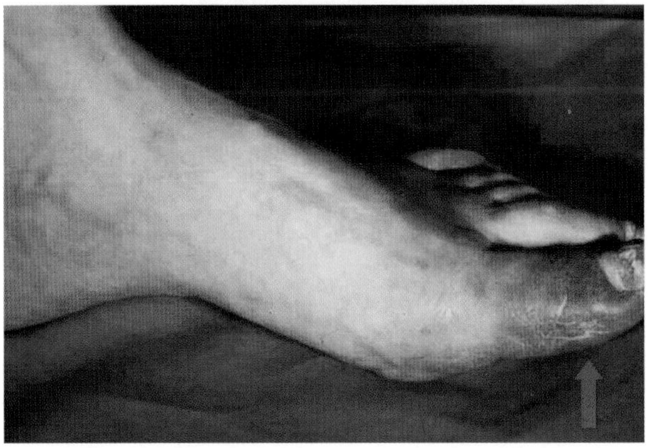
2

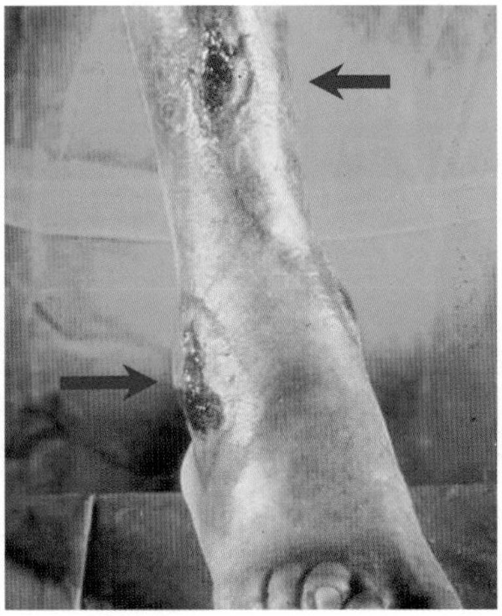
5

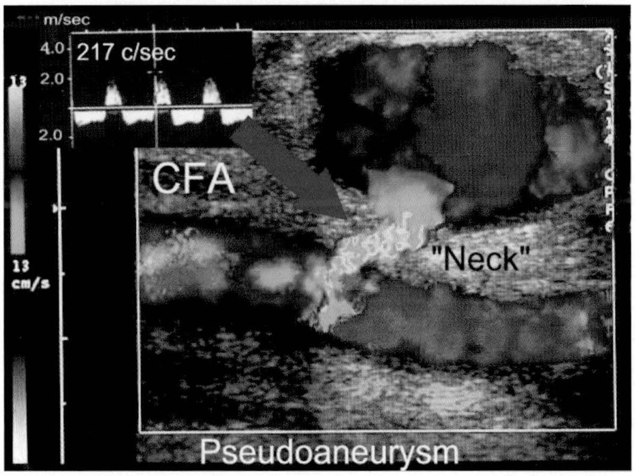

3

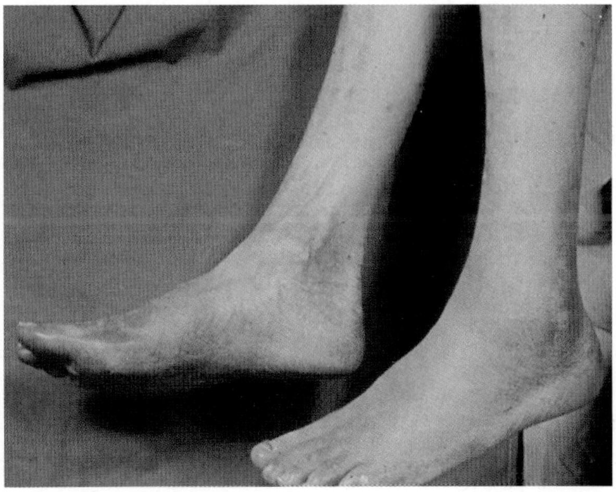
6

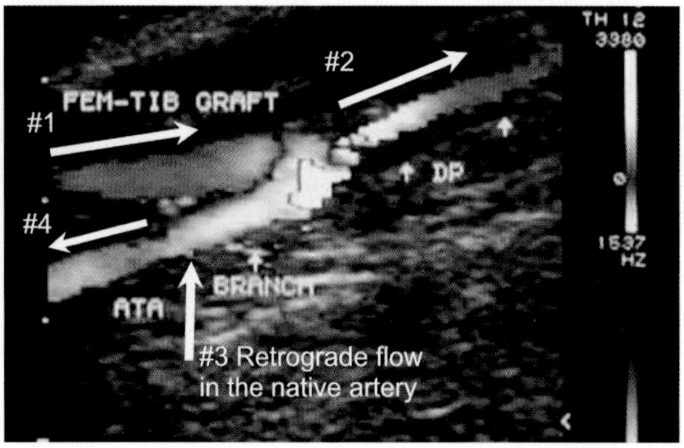
TH 12
3980
#1
FEM-TIB GRAFT
#2
DP
#4
1537
HZ
BRANCH
ATA
#3 Retrograde flow
in the native artery
7

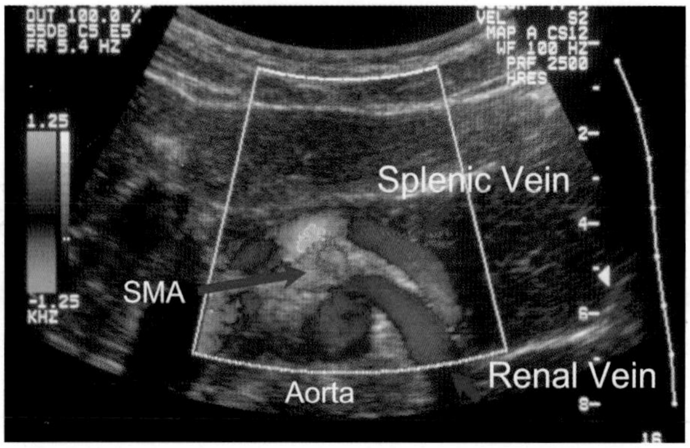
1.25
-1.25
KHZ
Splenic Vein
SMA
Aorta
Renal Vein
8

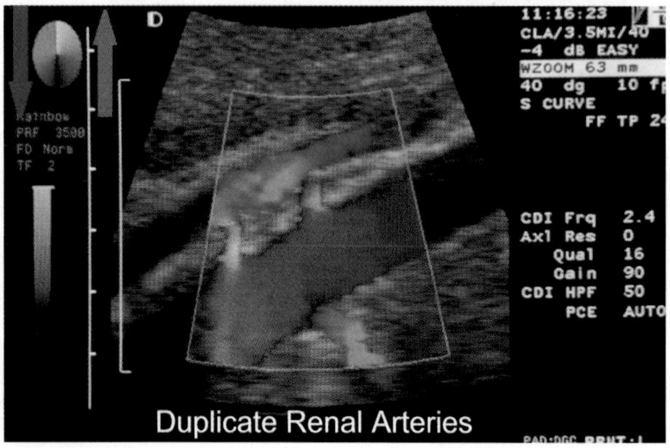
11:16:23
-4 dB EASY
WZOOM 63 mm
S CURVE
CDI Frq 2.4
Axl Res 0
Qual 16
Gain 90
CDI HPF 50
PCE AUTO
Duplicate Renal Arteries
9

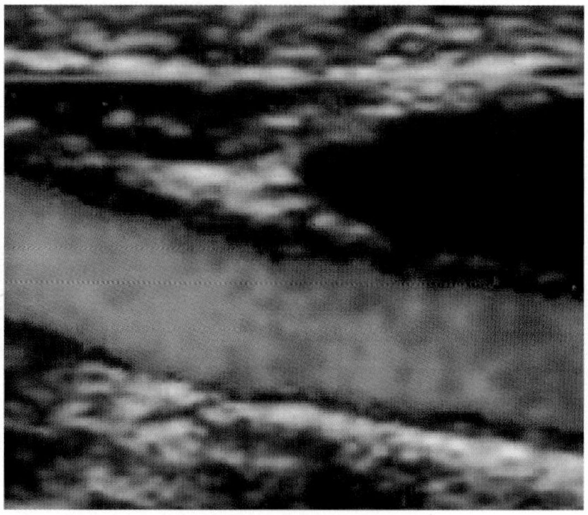
10

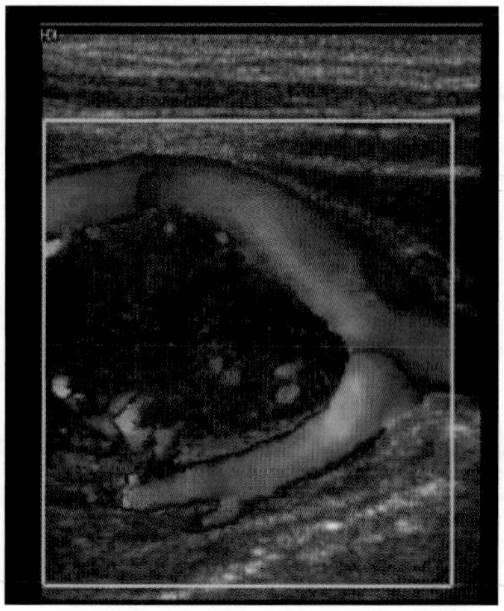
11

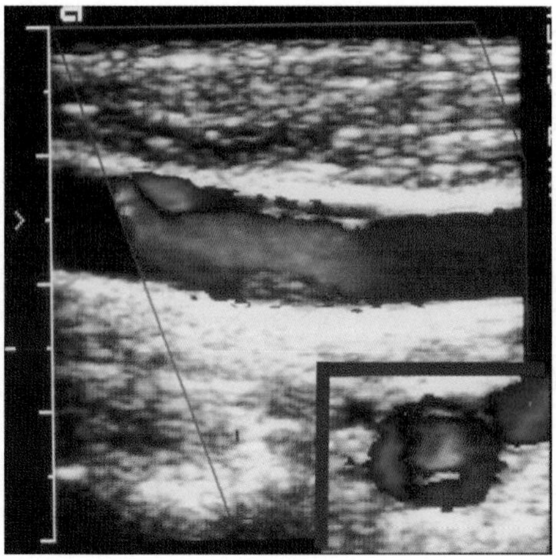
12

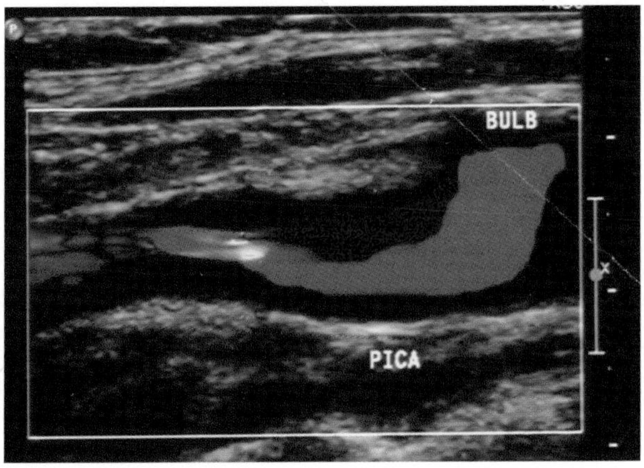

13

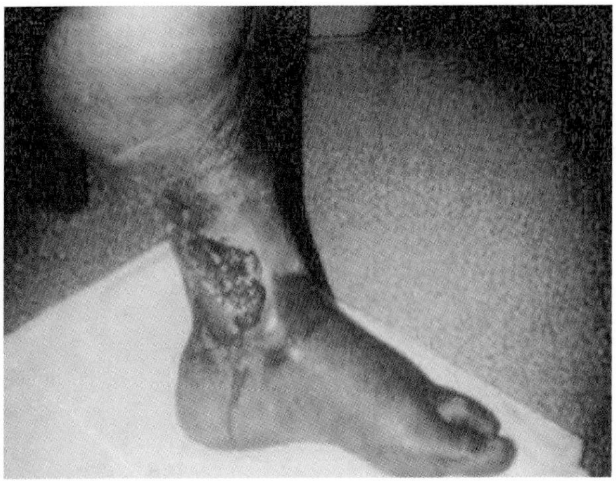

16

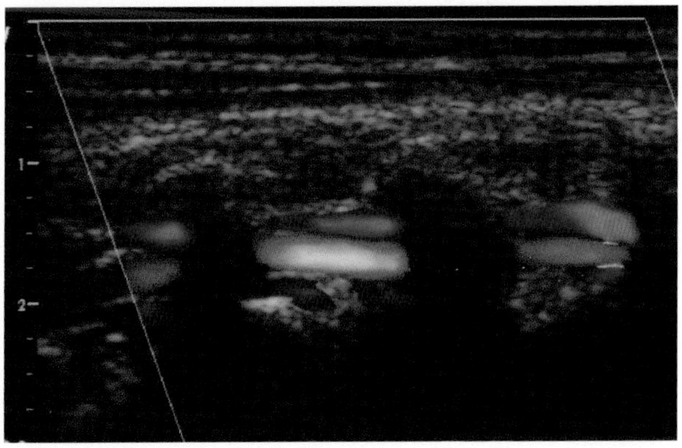

14

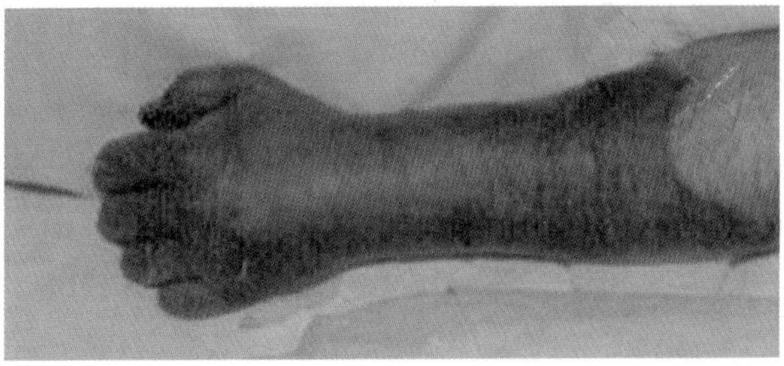

17

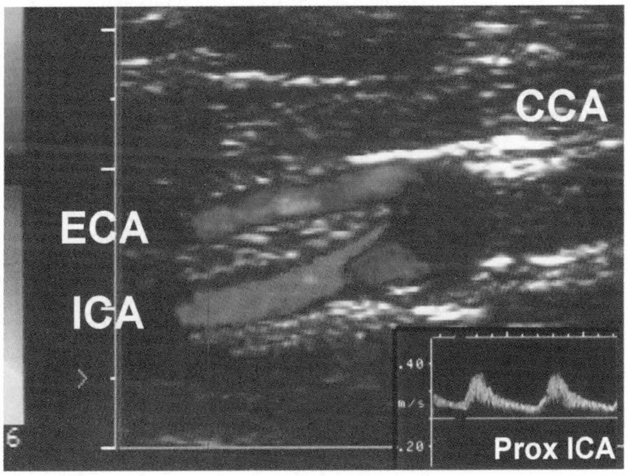

15

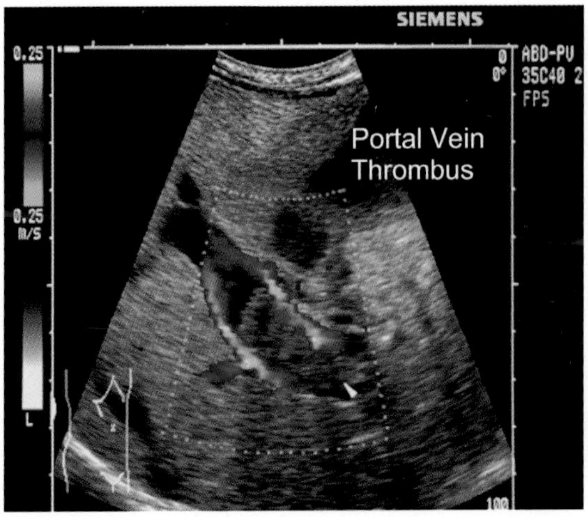

18

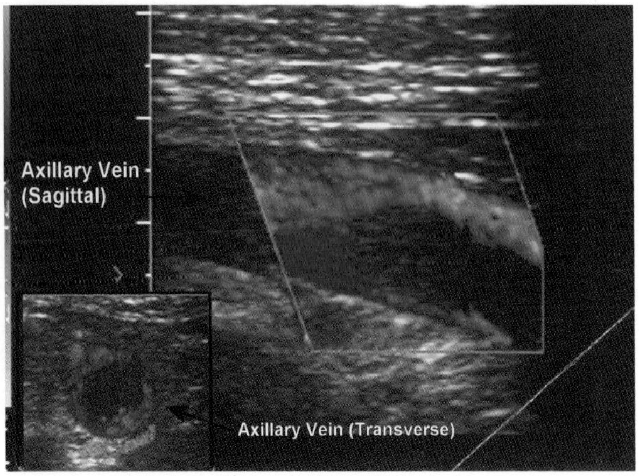
Axillary Vein (Sagittal)
Axillary Vein (Transverse)

19

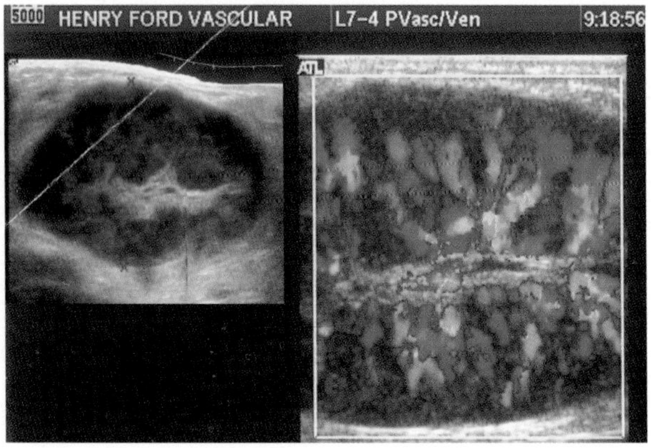
HENRY FORD VASCULAR
L7-4 PVasc/Ven
9:18:56

22

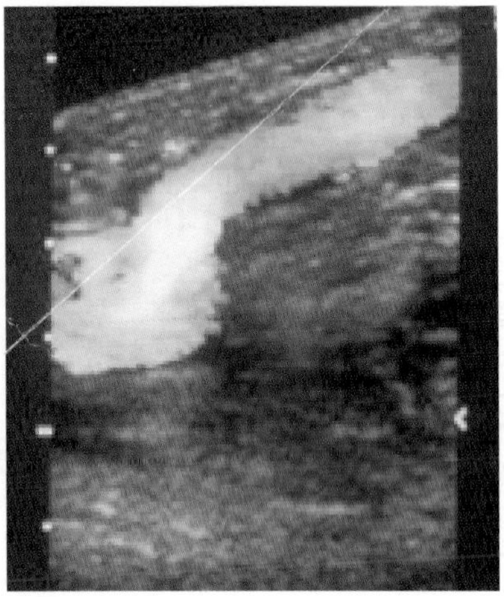

20

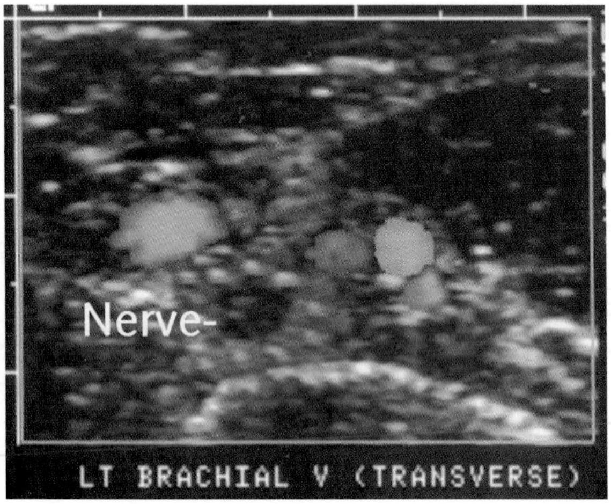
Nerve-
LT BRACHIAL V (TRANSVERSE)

23

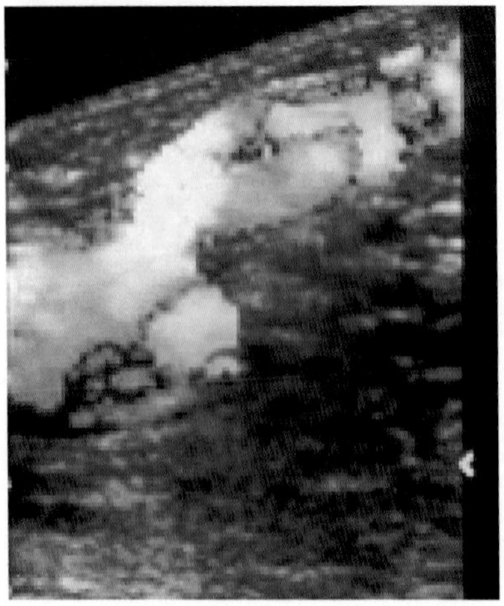

21

Contents

PART I

Arterial Evaluation

Gross Anatomy of the Central and Peripheral Arterial System

CHAPTER 1

The Aortic Arch

The *aortic arch* has several branching vessels, which are shown in figure 1-1 on the following page and listed below:

1 The *innominate/brachiocephalic artery:*

- Arises on the right only; it is the first branch off of the aortic arch.
- Divides into the right common carotid and subclavian arteries.

2 The *left common carotid artery:*

- Is the second branch off of the aortic arch.
- Terminates at the carotid bifurcation.

3 The *left subclavian artery:*

- Is the third branch off of the aortic arch.
- Terminates at the thoracic outlet.

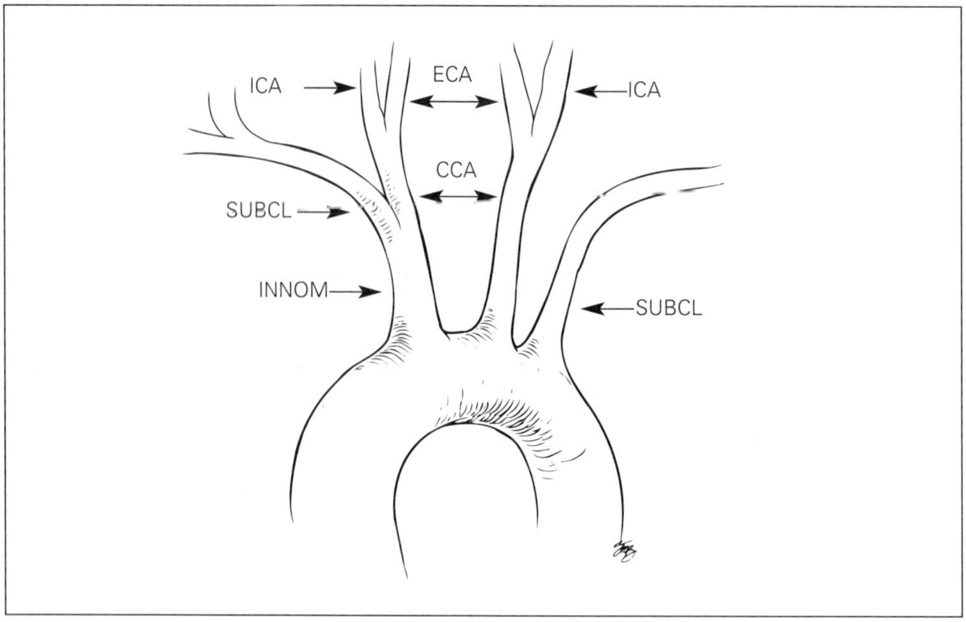

Figure 1-1.
The aortic arch and its branches. From Belanger AC: *Vascular Anatomy and Physiology: An Introductory Text*. Pasadena, Davies Publishing, 1999.

Upper Extremity Arteries

1 The *subclavian artery* becomes the axillary as it travels along the shoulder to the upper arm:

- It arches above the clavicle, in front of the apex of the lung and behind the scalenus anterior muscle.
- It runs laterally and downward to the outer border of the first rib; there it becomes the axillary artery.
- Its most important branches are the vertebral, thyrocervical, internal thoracic, and costocervical arteries.

2 The *axillary artery* becomes the brachial after giving off seven branches:

- Superior artery
- Thoracic artery
- Thoracoacromial artery
- Lateral thoracic artery
- Subscapular artery
- Anterior and posterior humeral artery
- Thoracodorsal artery

3 The *brachial artery* (figure 1-2) courses down the upper arm, ending about 1 cm beyond the bend of the elbow where it divides into the radial and ulnar arteries.

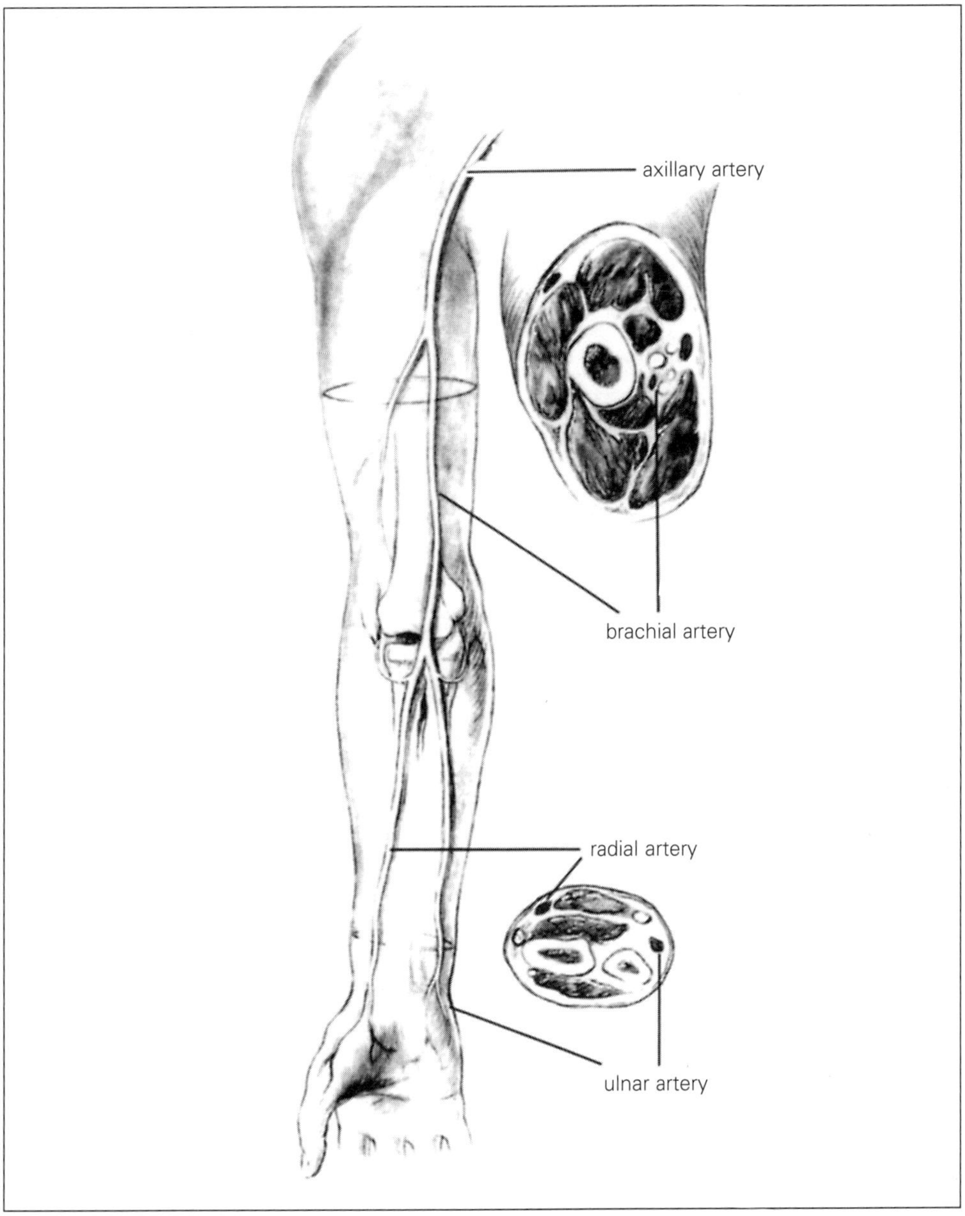

Figure 1-2. Major arteries of the upper extremity. From Salles-Cunha SX, Andros G: *Atlas of Duplex Ultrasonography: Essential Images of the Vascular System.* Pasadena, Appleton Davies, 1988.

Note: The *antecubital fossa* is a triangular region located anterior to and below the elbow.

4 The *radial artery* (figure 1-2):

- Originates from the brachial artery and travels down the lateral side of the forearm into the hand.
- Gives off a branch in the hand to form the superficial palmar arch and terminates in the deep palmar arch of the hand by joining the deep branch of the ulnar artery.

Figure 1-3.
The superficial palmar arch (A) and the deep palmar arch (B).

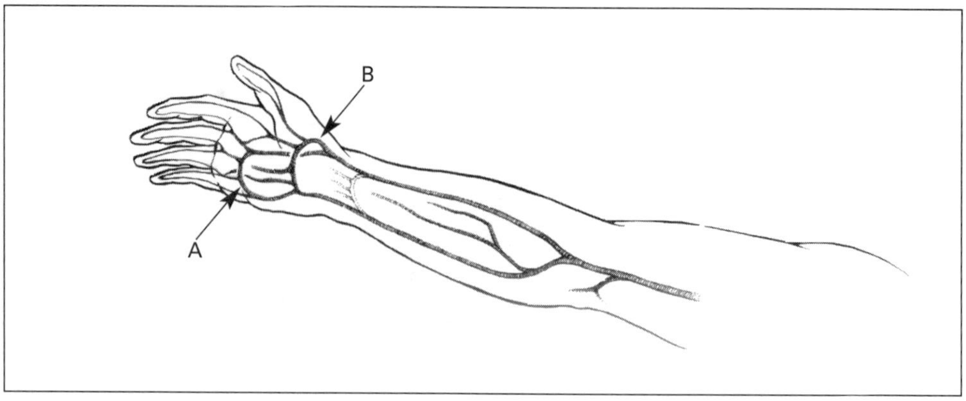

5 The *ulnar artery* (figure 1-2):

- Originates from the brachial artery and travels down the medial side of the forearm into the hand.
- Gives off a deep palmar branch and then terminates in the superficial palmar arch.

6 The *superficial palmar (volar) arch* (figure 1-3) consists of the distal portion of the ulnar artery, as it continues into the hand, and a branch of the radial artery.

7 The *deep palmar (volar) arch* (figure 1-3) consists of the deep palmar branch of the ulnar artery and the distal portion of the radial artery.

8 The *digital arteries:*

- Arise from the palmar arches.
- Extend into the fingers/toes.
- Divide into lateral and medial branches.

The Thoracic and Abdominal Arteries

1 The *ascending aorta:*

- Arises from the left ventricle.
- Has two branches, the right and left coronary arteries.

2 The *aortic arch,* formed by the ascending aorta, gives off the three branches described above (the innominate, left common carotid, and left subclavian arteries).

3 The *descending thoracic aorta* extends downward from the aortic arch to just above the diaphragm (figure 1-4).

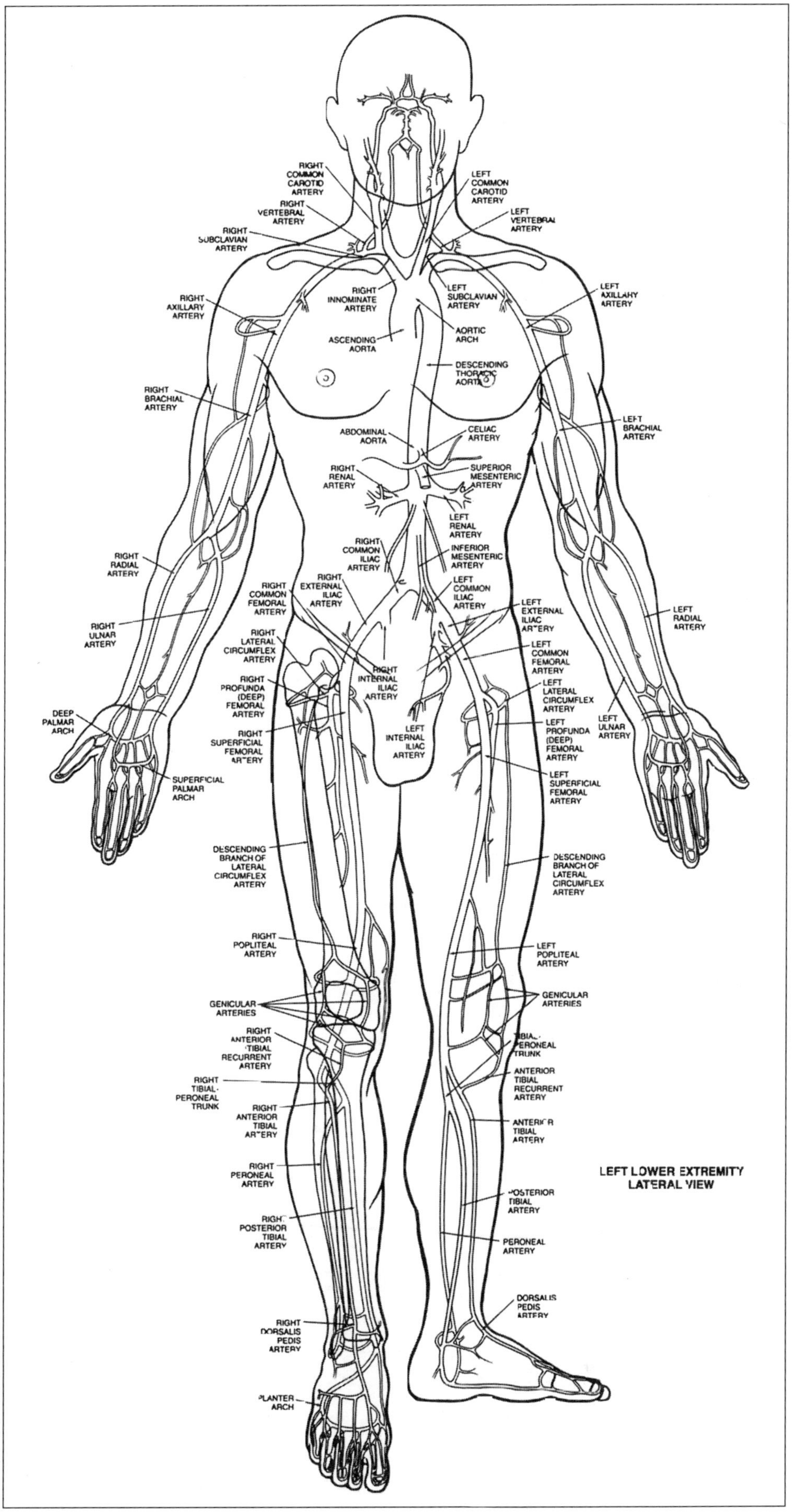

Figure 1-4. Arterial circulatory system. Courtesy of MedaSonics.

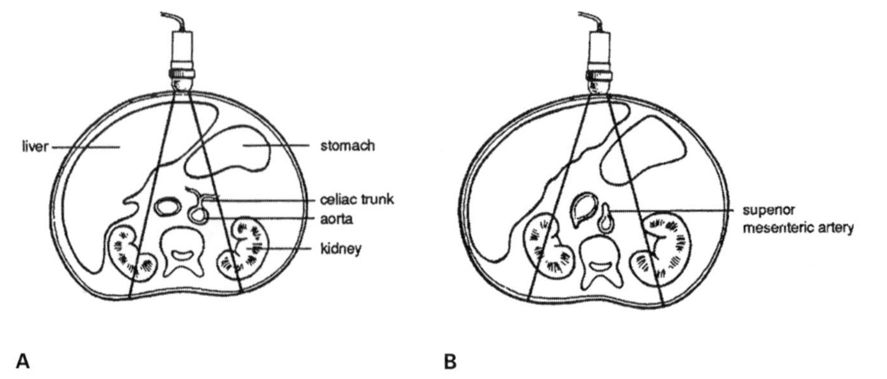

Figure 1-5. Transverse views of (**A**) the proximal aorta at the level of the celiac trunk and (**B**) the proximal aorta at the level of the superior mesenteric artery. From Ridgway DP: *Introduction to Vascular Scanning,* 3rd edition. Pasadena, Davies Publishing, 2004.

4 The major *visceral branches of the abdominal aorta* are:

- The *celiac artery* (figures 1-4, 1-5A, and 1-6), which feeds the stomach, liver, pancreas, duodenum, and spleen, and which branches into the *left gastric, splenic,* and *common hepatic arteries.*
- The *superior mesenteric artery* (figures 1-4, 1-5B, and 1-6), which feeds the small intestine, cecum, ascending colon, and part of the transverse colon. Commonly abbreviated "SMA," the superior mesenteric artery is located approximately 1 cm distal to the celiac artery. The SMA and celiac artery occasionally share a common trunk.
- The *renal arteries* (figures 1-4 and 1-6), which supply blood to the kidneys, suprarenal glands, and ureters. Multiple renal arteries are not uncommon bilaterally. A landmark for locating the left renal artery is the left renal vein, which crosses the aorta anteriorly and is positioned superior to the artery. After the right renal artery branches off the aorta, it courses underneath the inferior vena cava (IVC).
- The *inferior mesenteric artery* (figures 1-4 and 1-6), which supplies the left half of the transverse colon, the descending, iliac, and sigmoid colon, and part of the rectum. It arises from the distal abdominal aorta approximately 3–4 cm above the aortic bifurcation and can act as a collateral connection.

5 The major *parietal branches of the abdominal aorta* are:

- The *inferior phrenic artery*
- The *lumbar arteries*
- The *middle sacral artery*

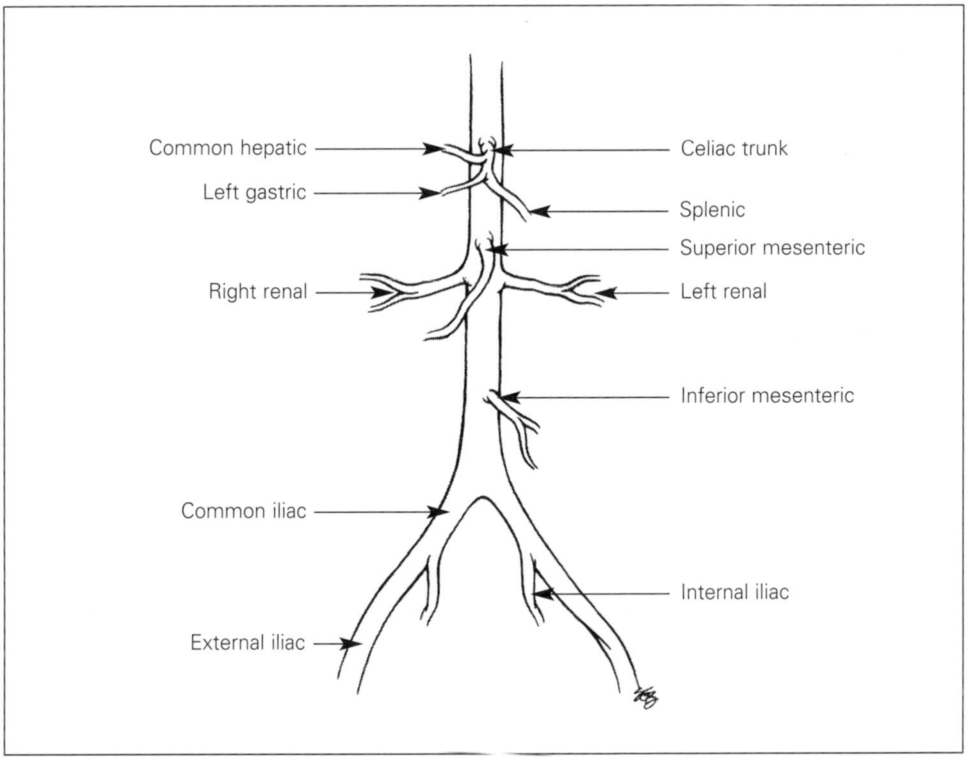

Figure 1-6. The abdominal aorta and branch vessels. From Belanger AC: *Vascular Anatomy and Physiology: An Introductory Text.* Pasadena, Davies Publishing, 1999.

6 The *terminal branches of the abdominal aorta* are the right and left common iliac arteries—the distalmost branches of the aorta carrying blood to the pelvis, abdominal wall, and lower limbs (figures 1-4 and 1-6).

The Lower Extremity Arteries

1 The common iliac arteries (figures 1-4 and 1-6) divide into the internal (hypogastric) and external iliac arteries at the level of the lumbosacral junction about 5 cm from their origin:

- The *internal iliac (hypogastric) arteries* are 3–4 cm in length. They descend into the pelvis, each dividing into two vessels—anterior and posterior—at the upper margin of the greater sciatic foramen. The internal iliac artery and its branches are highly variable. Specific branches provide arterial inflow to regions such as the pelvic wall, gluteal muscle, pelvic viscera, thigh, and perineum. Some of these branches anastomose with other arterial branches and can provide collateral flow when necessary.

- The *external iliac arteries* are essentially continuations of the corresponding common iliac artery. In the adult they are larger than the internal iliac artery. The external iliac artery travels in a lateral and inferior direction along the medial side of the psoas major muscle. When it passes underneath the inguinal ligament, the external iliac becomes the common femoral artery. The inferior epigastric artery arises from the external iliac just above the inguinal ligament, supplying vessels to the abdominal muscles and skin. The deep circumflex iliac artery arises from the lateral aspect of the external iliac near the inguinal ligament and supplies the abdominal muscles. Both of the external iliac branches anastomose with other arterial branches and can serve as collateral connections when necessary.

2 The *common femoral artery* divides into:

- The *superficial femoral artery,* which runs the length of the thigh and enters the popliteal fossa behind the knee. (See figures 1-4 and 1-7.) A well-known landmark for the point at which the superficial femoral artery becomes the popliteal artery is Hunter's canal (also known as the adductor canal). Created by the confluence of the quadriceps and adductor muscles in the mid to distal thigh, Hunter's canal is the channel for the femoral vessels. Although many people refer to this as the adductor hiatus, the adductor hiatus is the gap in the adductor magnus muscle.

- The *deep femoral artery* (profunda femoris) is a large branch and arises approximately 5 cm from the inguinal ligament on the lateral side. It can act as a collateral connection. (See figures 1-4 and 1-7.)

3 The *popliteal artery* (figures 1-4 and 1-7) is a distal continuation of the superficial femoral artery:

- Once the superficial femoral artery enters the popliteal fossa (Hunter's canal/adductor canal), it becomes the popliteal artery and gives off a number of genicular branches to supply the muscles, knee joint, and skin. These arteries also can act as collateral connections.

- At the interval between the tibia and fibula at the lower portion of the popliteus muscle, the popliteal artery branches.

4 The *anterior tibial artery* is the first branch of the distal popliteal artery:

- After arising from the distal popliteal artery, the anterior tibial artery (ATA) passes superficial to the interosseous membrane and runs deep in the front of the leg along the anterior surface of the interosseous membrane. (See figures 1-4 and 1-7.)

- At its distal end, the anterior tibial artery courses to the anterior aspect of the tibia. As it passes in front of the ankle joint it is more superficial and becomes the dorsalis pedis artery (DPA), traversing the dorsum of the foot toward the base of the first toe. On the dorsum of the foot the dorsalis pedis artery forms two branches, the first dorsal metatarsal and

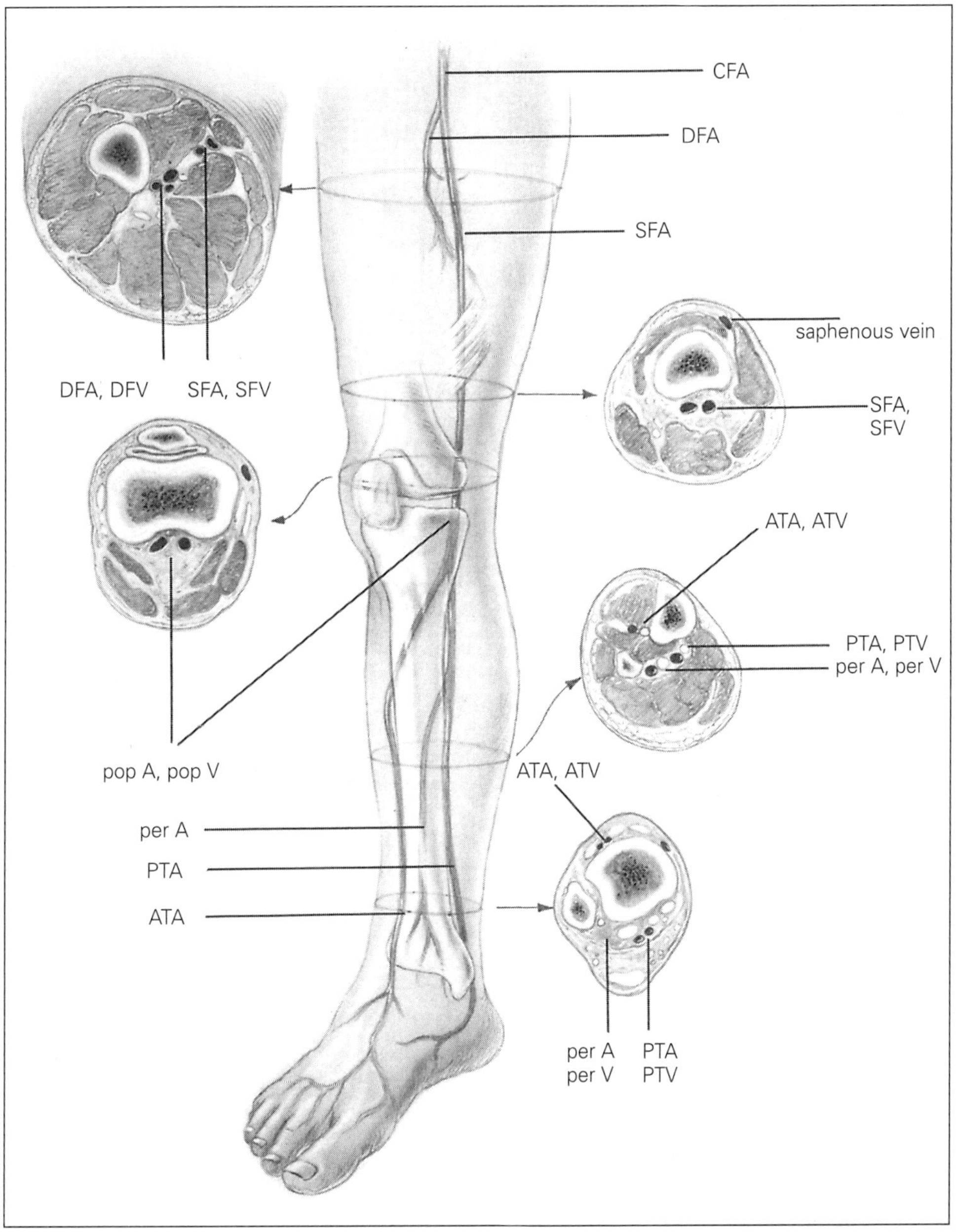

Figure 1-7. Arteries of the lower extremities: common femoral (CFA), superficial femoral (SFA), deep femoral (DFA), popliteal (pop A), anterior tibial (ATA), posterior tibial (PTA), and peroneal arteries (per A). Adjacent veins (SFV, DFV, pop V, ATV, PTV, per V) are noted in cross section. From Salles-Cunha SX, Andros G: *Atlas of Duplex Ultrasonography*. Pasadena, Appleton Davies, 1988.

the deep plantar arteries. The deep plantar artery penetrates into the sole of the foot, uniting with the lateral plantar artery to complete the plantar arch of the foot. (See figures 1-4 and 1-8.)

5 The *tibioperoneal trunk:*

- The second branch of the distal popliteal artery, the tibioperoneal trunk quickly gives rise to the posterior tibial and peroneal arteries.

- This short segment has also been called the proximal portion of the posterior tibial artery.

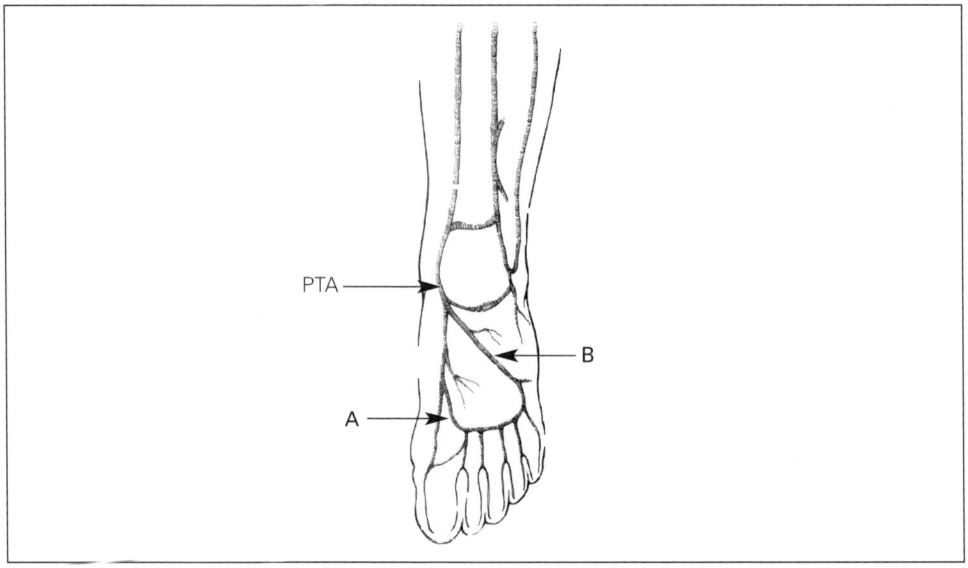

Figure 1-8.
Plantar arch (**A**) and the lateral plantar artery (**B**).

6 The *posterior tibial artery:*

- Extends obliquely down the posterior and medial side of the leg. It is commonly referred to as one of two branches of the tibioperoneal trunk (a short segment between the anterior tibial branch and this region where the posterior tibial [PTA] and peroneal arteries develop) as shown in figures 1-4 and 1-7.

- Divides into the medial and lateral plantar arteries in the foot, below the medial malleolus, to supply the sole of the foot.

7 The *peroneal artery:*

- Arises at the distal end of the tibioperoneal trunk, along with the posterior tibial artery.

- Passes toward the fibula and travels down the medial side of that bone to supply structures in the lateral side of the leg and in the calcaneal region of the foot. (See figures 1-4 and 1-7.)

8 The *digital arteries:*

The *plantar arch*, which consists of the deep plantar artery (branch of the dorsalis pedis artery) and the lateral plantar artery (branch of the posterior tibial artery), and the dorsal metatarsal arteries distribute blood into the digits. (See figures 1-4 and 1-8.)

9 The *capillaries:*

- Are vessels of the microcirculation.

- Are not much more than a millimeter long.

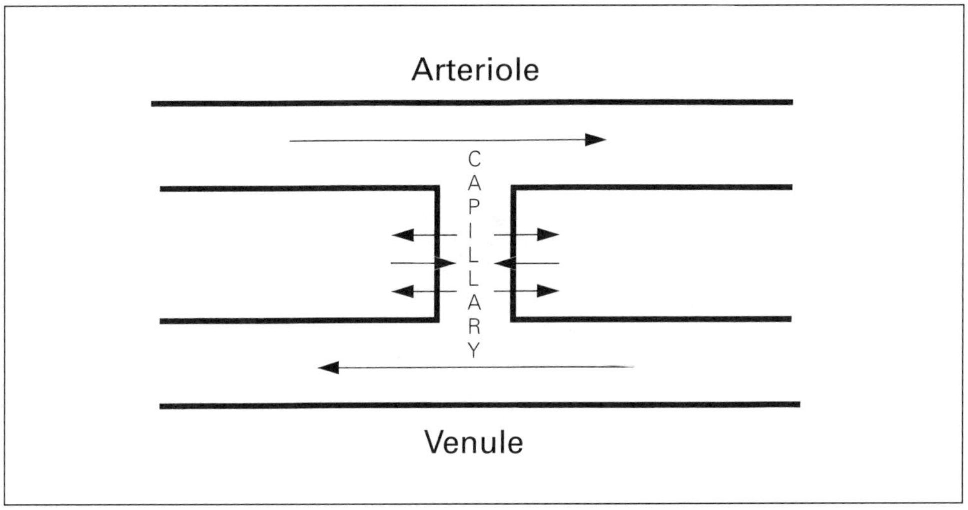

Figure 1-9. Vessels of the microcirculation, where nutrients and waste products are exchanged between blood and tissue.

- Are 8–10 microns in diameter (about the same as a red blood cell).
- Flow quality is steady with low flow velocity.
- Have walls that consist of endothelial cells, forming a layer one cell thick.
- Form vast networks with a total surface area of about 1.5 acres or 6,000 square meters.
- Are supplied by the arteries, which transport gases, nutrients, and other essential substances to the capillary beds. Arteries progressively decease in size from the aorta (largest) to the arterioles (smallest). Arterioles, which are considered to be resistance vessels, help to regulate blood flow by contracting and relaxing.
- Constitute a most vital part of the circulatory system. It is through the walls of the capillaries that nutrients and waste products are exchanged between tissue and blood to maintain the constancy of the internal environment (figure 1-9). It has been said that all other circulatory system organs exist only to serve the capillary beds.

Microscopic Anatomy of the Arterial Wall

1 The function of the artery is to transport blood—and the gases, nutrients, and other essential substances it contains—away from the heart and out to the tissues.

2 Anatomy:

- The *tunica intima,* the innermost layer of the arterial wall, is thin and consists of a surface layer of smooth endothelium over a base membrane and connective tissue.

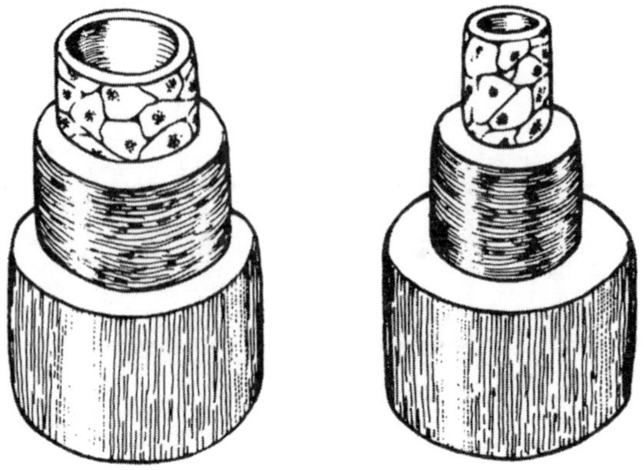

Figure 1-10. The three layers of the arterial wall. From Belanger AC: *Vascular Anatomy and Physiology: An Introductory Text.* Pasadena, Davies Publishing, 1999.

- The *tunica media,* the intermediate layer of the arterial wall, is thicker and composed of smooth muscle and connective tissue, largely of the elastic type. These components are usually arranged in a circular pattern.
- The *tunica externa* (or *adventitia*) is the outer layer of the arterial wall. It is somewhat thinner than the media and contains white fibrous connective tissue and, at times, a few smooth muscle fibers, all arranged longitudinally. The adventitial layer usually contains the vasa vasorum, tiny vessels that carry blood to the walls of the larger arteries.
- These three coats of the arteries may be separated by internal and external elastic membranes. (See figure 1-10.)

CHAPTER 2

Physiology and Fluid Dynamics

The Arterial System

1 The arterial system is a multibranched elastic conduit that carries blood away from the heart and outward to the most distant tissues.

- The arterial tree oscillates with every beat of the heart, each one of which pumps approximately 70 milliliters of blood into the aorta and causes a blood pressure pulse.

- At the beginning of the cardiac contraction, the pressure in the left ventricle rises rapidly, quickly exceeding that in the aorta so that the aortic valve opens, blood is ejected, and the blood pressure rises. The amount of blood ejected is called the *stroke volume.*

- Increased heart rate delivers an increased blood volume that supplies more nutrients. Conversely, the lower the heart rate, the smaller the volume of pumped blood.

- The patient's cardiac status plays an important role in the movement of blood throughout the vascular system.

2 The heart pump generates the pressure *(potential energy)* to move the blood. The stroke volume of blood produced by each heartbeat creates a pressure (or *energy*) wave that travels rapidly throughout the arterial system (figure 2-1):

- The propagation speed, shape, and strength of the pressure wave change as the wave moves through the arterial system. *Example:* As the arterial pressure wave moves distally, away from the heart and out toward the periphery, the propagation speed—the pulse wave velocity—increases with the growing stiffness of the arterial walls.

- Variations in the characteristics of the vessels influence these alterations in blood flow. Velocity and flow direction also vary with each heartbeat.

- As the pressure wave moves from the large arteries through the high-resistance vessels, capillaries, and then into the venous side, the *mean* pressure gradually declines because of losses in total fluid energy.

3 The pumping action of the heart maintains a high volume of blood in the arterial side of the system that in turn sustains a high pressure gradient between the arterial and venous sides of the circulation. This pressure gradient is necessary to maintain flow.

Figure 2-1.
Once the heart generates the pressure to move the blood, the energy wave produced travels rapidly throughout the system beginning with (A) to the capillary bed (E) back to the heart via the venous system (F–I).

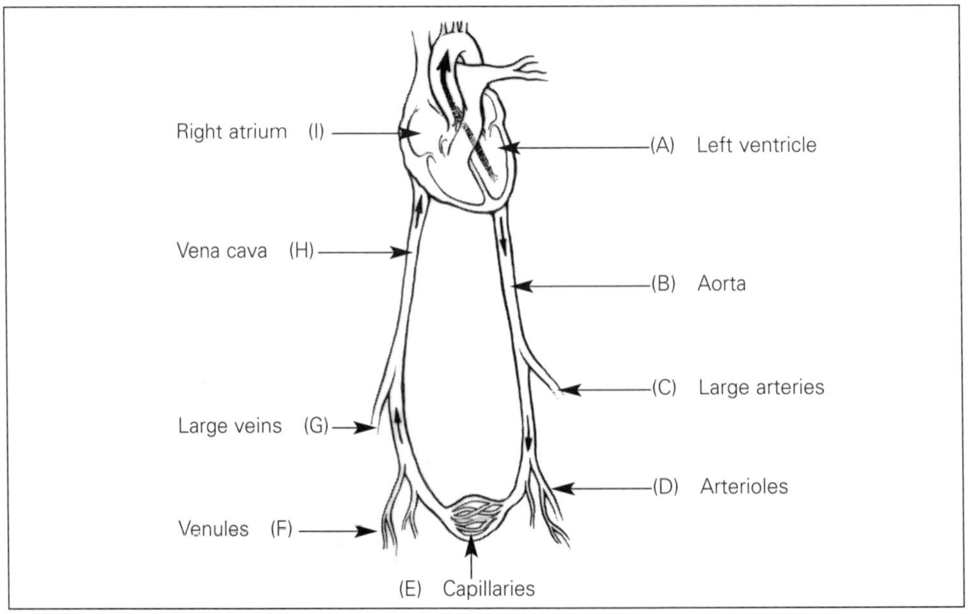

4 Cardiac output governs the amount of blood that enters the arterial system, while arterial pressure and total peripheral resistance (which is controlled by the level of vasoconstriction in the microcirculation) determine the volume of blood that leaves it.

5 A large portion of the energy created with each left ventricular contraction results in distention of the arteries, producing an arterial "reservoir" that stores some of the blood volume and the potential energy supplied to the system.

It is this store of energy and volume that promotes the flow of blood into the tissues during diastole. That is, potential energy is stored in the distended arterial wall and is released when the wall recoils.

6 Pressure is greatest at the heart and gradually decreases as the blood moves distally. This pressure difference (or *gradient*) is necessary to maintain blood flow.

Energy

The movement of any fluid medium between two points requires two things: (1) a route along which the fluid can flow and (2) a difference in energy (pressure) levels between the two points. The volume of flow depends on the net energy difference between these two points, a factor that is affected by losses resulting from the movement of the fluid—i.e., friction—and any resistance within the pathway that opposes such movement.

1 The greater the energy difference (or the lower the resistance), the greater the flow, as illustrated in figure 2-2.

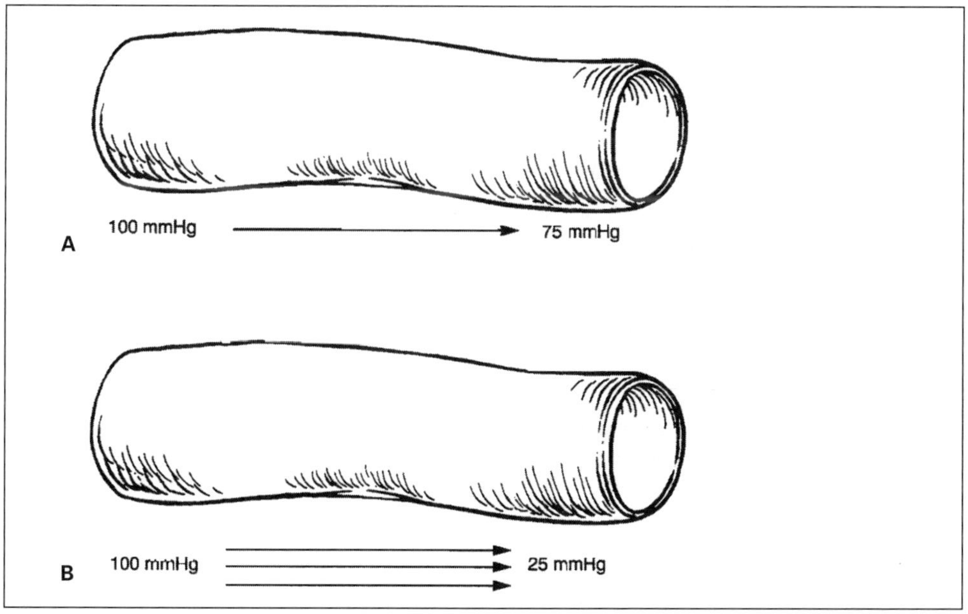

Figure 2-2. The greater the pressure gradient, the greater the flow. Example (**A**) has less pressure difference and therefore less flow than that of (**B**) which has a larger pressure gradient and greater flow. From Belanger AC: *Vascular Anatomy and Physiology: An Introductory Text.* Pasadena, Davies Publishing, 1999.

2 The greater the pathway resistance and/or energy losses, the lower the flow.

lower resistance = higher flow rate higher resistance = lower flow rate

3 The total energy contained in moving fluid is the sum of pressure (potential) energy, kinetic (movement) energy, and gravitational energy, as explained below:

- *Pressure energy* is the main form of energy present in flowing blood. It is created by the pumping action of the heart, which subsequently distends the arterial vessels. In this instance, pressure energy is also referred to as potential energy. Potential energy has several components. The dominant source is the pumping action of the heart muscle. Another source is hydrostatic pressure.

- *Kinetic energy* is the ability of flowing blood to do work as a result of its velocity; it is the energy of something in motion. With regard to blood flow, the kinetic energy portion is small compared to the pressure energy. Kinetic energy is also proportional to the density of blood (which is normally stable) and to the square of its velocity. The everyday example of a dam illustrates the difference between potential and kinetic energies: The water behind the dam has *potential* energy (with the height of the dam providing a form of *gravitational* energy), while the water flowing through the dam has *kinetic* energy.

- *Gravitational energy (hydrostatic pressure)* is expressed in millimeters of mercury (mmHg). Changes in the height of the fluid column introduce the element of gravitational energy, which is hydrostatic pressure. In the circulatory system, hydrostatic pressure is equivalent to the weight of the column of blood extending from the heart, where the right atrium is considered the 0 pressure reference point (i.e., atmospheric pressure) to the level where the pressure is being measured.

In a supine, medium-sized patient, for example, in whom the arteries and veins are at approximately the same level as the heart, there is negligible difference in the hydrostatic pressure effect on the arteries and veins. Nevertheless, at the level of the ankle the mean venous pressure is about 0 mmHg, while the mean arterial pressure for this example is 95 mmHg. When the same individual stands, the hydrostatic pressure increases, with the venous side more noticeably affected because of its normally lower mean pressure. (The reference point for 0 pressure is the right atrium of the heart.) In this example, our now standing patient has a hydrostatic pressure of approximately 100 mmHg added to the mean venous pressure of about 0 mmHg so that the total venous pressure at the ankle increases to about 100 mmHg. Similarly, the arterial pressure at the ankle increases to 195 mmHg in this standing patient, reflecting the addition of the hydrostatic pressure of 100 mmHg to the previous supine pressure of 95 mmHg.

See also Chapter 27 for further discussion of hydrostatic pressure in the venous system.

4 An energy gradient (pressure difference) is necessary for blood or any other fluid to move from one point to another (see figure 2-2). In other words, movement will occur only from a high-pressure state to a low-pressure state. *Inertia*—the tendency of objects to maintain their status quo—helps to maintain flow. As blood moves farther out to the periphery, energy dissipates, largely in the form of heat, but the pumping action of the heart continually restores this energy.

5 Flow velocity and the cardiac cycle:

- During the acceleration phase of the pulse *(early systole* to *peak systole)*, pressure is evenly distributed in all directions (figure 2-3).

- The greatest amount of energy is produced at peak systole, where the greatest velocities are observed.

- During the deceleration phase of the pulse *(late systole/early diastole)*, cardiac output decreases to the point at which outflow through the high-resistance peripheral vessels exceeds the volume ejected from the heart and the pressure declines.

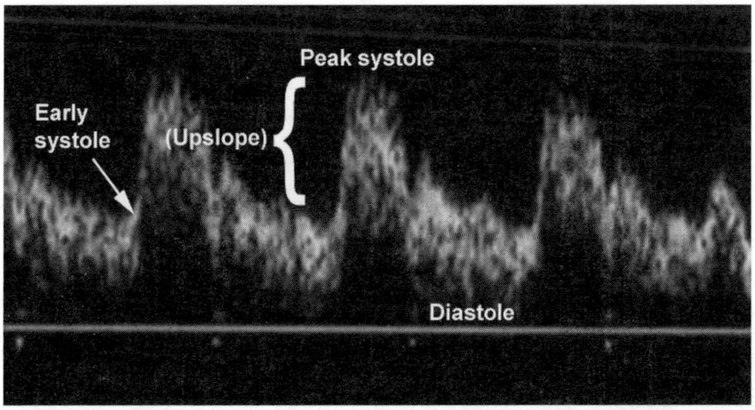

Figure 2-3.
The upslope (acceleration phase) in this spectral waveform represents the arrival of the pulse wave where the pressure is evenly distributed (early systole).

Blood Flow Characteristics

1 *Laminar flow* is the type of flow thought to exist in many vessels. In this type of flow, blood moves in concentric layers, or laminae. Each very thin layer flows at a different velocity, slowest at the vessel wall and fastest at the center of the vessel. It is thought that a thin layer of blood is held stationary at the blood vessel wall. The next concentric layer moves at a slightly higher velocity, but not as fast as it might if there were no friction with the stationary layer of blood. This pattern of thin, concentric layers of blood flowing faster as they approach the center stream—where the highest velocities are thought to exist—typifies laminar flow (see figure 2-4). Laminar flow is stable, with the streamline formations staying intact. As discussed below, friction and energy losses increase to the extent that laminar flow is disturbed.

2 There are two forms of laminar flow, both of which are associated with normal physiology. *Parabolic flow* has a profile that resembles the shape of a bullet. Velocity is highest in the center of the lumen, gradually decreasing at the vessel wall. *Plug flow* occurs when all of the blood cells and layers travel at the same velocity, as is usually seen at the origin of vessels.

3 The movement of fluid (blood) through a conduit (artery) depends on both the physical properties of the fluid and the interaction between the conduit and the fluid.

4 The longer the conduit, the longer the fluid is in contact with the conduit and the higher the pressure required to maintain flow.

5 *Viscosity* can be described as the thickness of a fluid. If a fluid is thin (e.g., gasoline), it moves quite freely through a conduit. If the fluid is thick, it moves less freely through the conduit; there are usually more internal forces between the molecules of a thick fluid, and it takes much more energy to move the fluid. The forces in this description are called the *viscous forces.*

> increased viscosity = decreased velocity
> decreased viscosity = increased velocity

- In the circulatory system, energy is lost in the form of heat as the layers of red blood cells rub against each other, creating friction. Friction is generated by the viscous properties of the fluid. The thicker the fluid, the greater the molecular attraction and the more energy required to move the fluid.

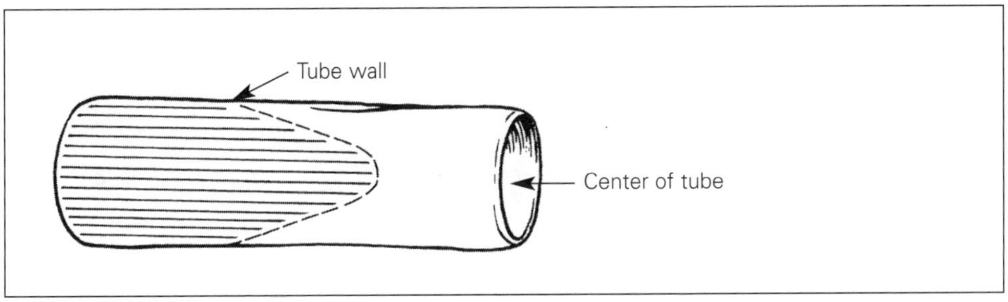

Figure 2-4. Parabolic profile of laminar flow. From Belanger AC: *Vascular Anatomy and Physiology: An Introductory Text.* Pasadena, Davies Publishing, 1999.

- Energy loss during blood flow occurs because of friction. In part, the dimensions of the vessels determine the amount of friction and energy loss. The smaller the vessel, the greater the friction and resistance. In the microcirculation, the blood flow layers are relatively close to the vessel walls and therefore are delayed considerably, resulting in a resistance to flow.

- Example of high viscosity in the circulatory system: elevated hematocrit, which means there are more red blood cells per volume of plasma. Low viscosity: low hematocrit; severe anemia.

6 Although both viscosity and vessel length affect resistance, a change in vessel diameter has an even more dramatic effect on resistance. The following equation illustrates this relationship:

$$R = 8\eta L/\pi r^4$$

where R = the overall flow resistance, η = is the viscosity of the fluid, L = vessel length, π = 3.1416, and r = the radius of the vessel lumen. (See figure 2-5.) Note: R is directly proportional to variables in the numerator (viscosity and length). R is inversely proportional to the variable in the denominator (radius).

7 Energy loss in a fluid system can also be inertial in nature. *Inertial losses* are caused by changes in direction and/or velocity and increase with deviations from laminar flow. The parabolic flow profile (figure 2-4) becomes flattened and flow moves in a disorganized fashion. This type of energy loss can occur at the exit of a stenosis.

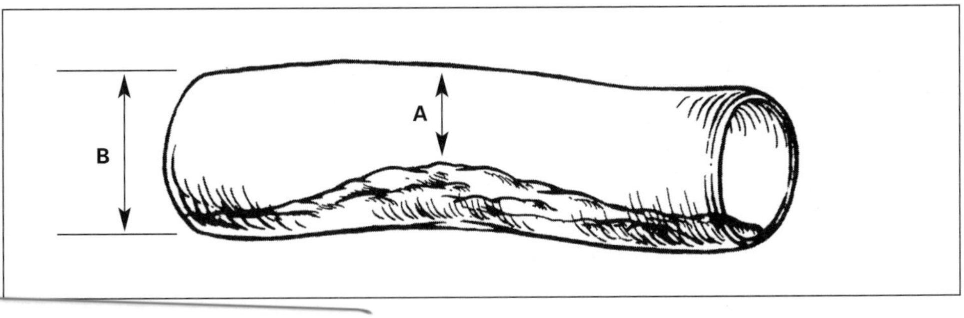

Figure 2-5. Representation of a stenotic arterial segment. The radius of the vessel is decreased, which increases the resistance and, additionally, the peak systolic velocities. A: Reduced diameter. B: Original diameter.

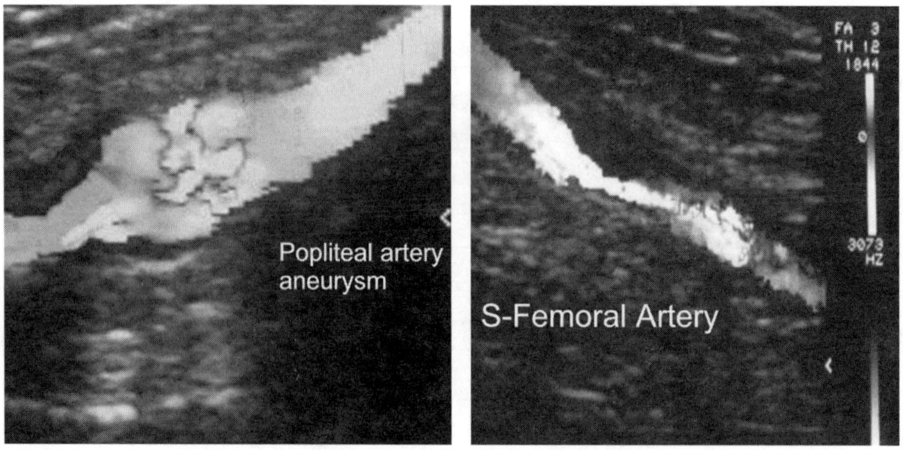

Figure 2-6.
A In this popliteal artery aneurysm, resistance to flow is reduced due to the larger-than-normal diameter. Although the volume of blood flow through this vessel remains constant, velocities are decreased compared to normal. **B** In this superficial femoral artery (SFA) stenosis, resistance to flow is increased because of the stenotic lumen. Although the volume flow remains constant, velocities are elevated as a result of the stenosis.

Poiseuille's Law

1 Poiseuille's law defines the relationship between pressure, volume flow, and resistance. It helps to answer the question of how much fluid is moving through a vessel.

- A change in the diameter of a vessel affects resistance more dramatically than either viscosity or vessel length (see figures 2-6A and 2-6B).
- The radius of a vessel is directly proportional to the volume flow. Small changes in radius may result in large changes in flow, as demonstrated in figure 2-7 on the page opposite.
- As vessel radius decreases, resistance increases. The volume of blood flow through the vessel nevertheless remains constant. To maintain volume flow as vessel size decreases, velocity must increase. As illustrated in figure 2-8 and figures 2-6A and B, the size of a vessel is inversely proportional to the velocity of blood flow. According to the basic laws of fluid dynamics, most notably the law of conservation of mass (i.e., what goes in must come out), the relationship among velocity, volume flow, and cross-sectional area of the vessel is

$$V = Q/A$$

where V = velocity (cm/sec), Q = volume flow (cm^3), and A = cross-sectional area (cm^2).

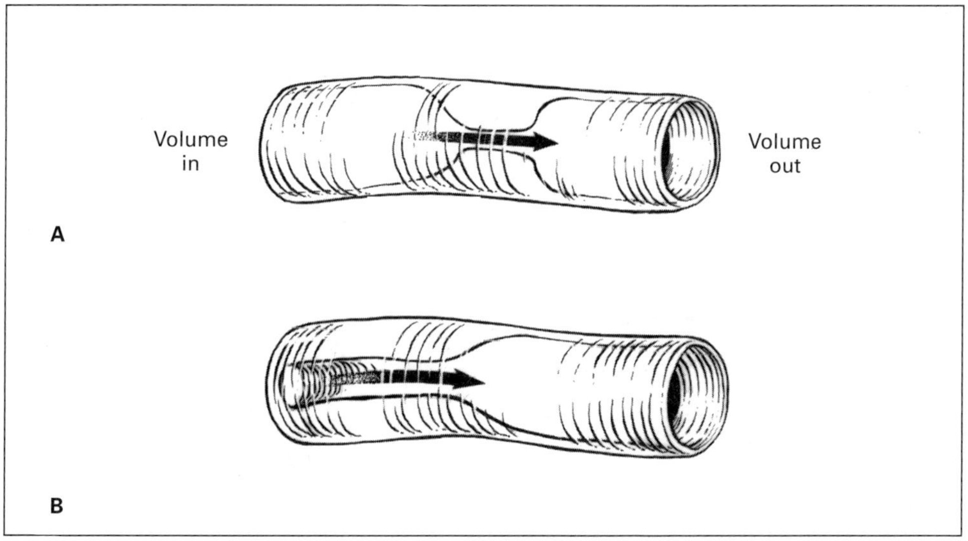

Figure 2-7.
Parts A and B both illustrate narrowed segments. The reduction in volume flow through segment A will be greater than that through segment B, where the radius of the stenotic segment is only slightly larger.

● In the cardiovascular system, the length of the vessels and the viscosity of the blood usually do not change much. This means that changes in blood flow occur mainly as a result of changes in the radius of a vessel and in the pressure energy gradient that makes flow possible.

2 Poiseuille's equation may be stated as follows:

$$Q = \frac{(P_1 - P_2)\,\pi r^4}{8\eta L}$$

where Q = volume flow, P_1 = pressure at the proximal end of the vessel, P_2 = pressure at the distal end of the vessel, r = radius of the vessel, L = length of the vessel, $\pi = 3.1416$, and η = viscosity of the fluid.

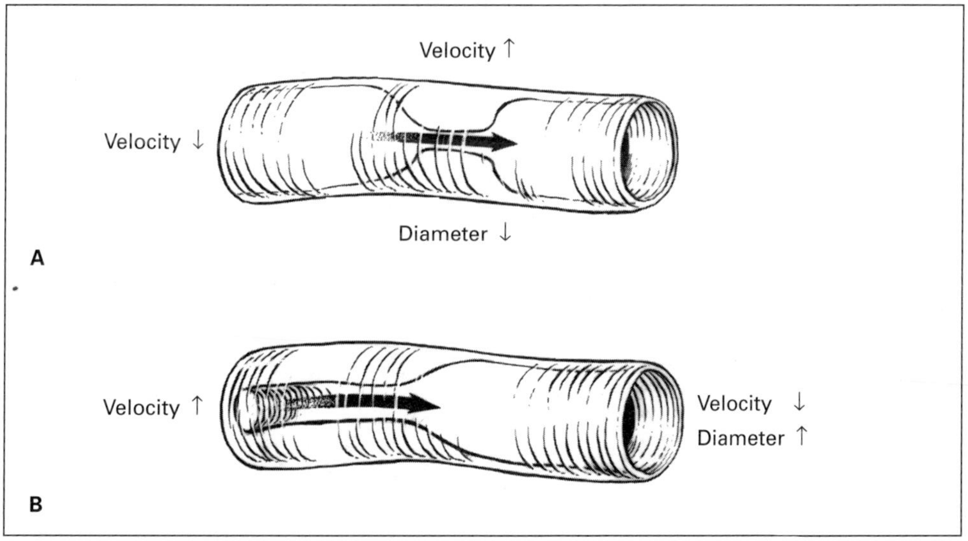

Figure 2-8.
A As the radius decreases, the velocities increase.
B As the radius increases, the velocities decrease.

3 An abbreviated pressure/volume flow relationship is expressed in the equation Q = P/R. This equation is quite similar to one used in electronics to understand the flow of electricity—Ohm's law—commonly expressed as I (flow of electrons) = E/R. Please note that it may be written as I=V/R in some references. Here is how the two equations compare:

Q = P/R		**I = E/R**
Flow volume (Q)	similar to	Current (I), flow volume of electrons
Pressure (P)	similar to	Voltage (E)
Resistance(R)	similar to	Resistance (R)

Pressure/Flow Relationships (Reynolds Number)

1 After the initial acceleration in systole, blood movement continues and develops into distinct streamline formations. Where P = pressure, note in figure 2-9A that the streamlines are evenly distributed. When the flow pattern becomes unstable, these continuous streamlines break up and form small circular currents called eddy currents and vortices (swirling patterns of rotational flow) (figure 2-9B).

2 Osborne Reynolds sought to determine how viscosity, vessel radius, and the pressure/volume relationship influence the stability of flow through a vessel. Although most of his work applied to straight, rigid tubes, it still provides insight into the physics of blood flow.

3 Flow volume increases as pressure increases, but only to a point. As flow changes from stable to disturbed, Reynolds found that pressure increases no longer increased flow volume. Instead, it increased flow disturbance, contributing to the formation of eddy currents.

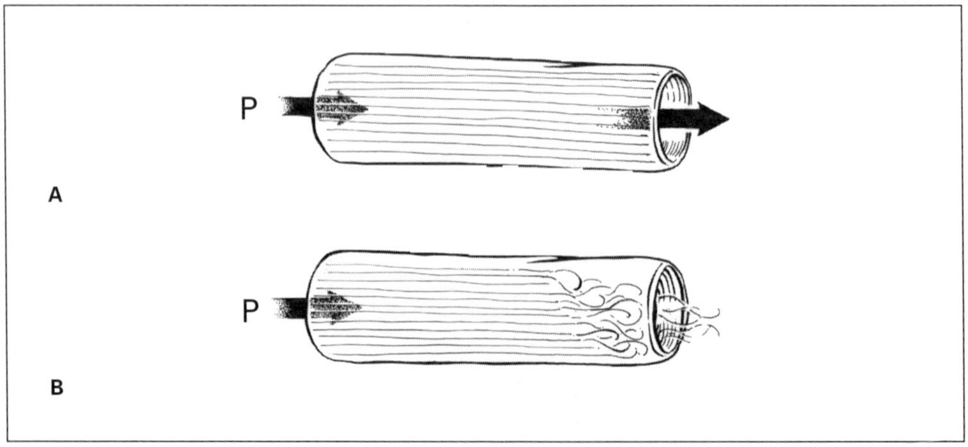

Figure 2-9. Arterial flow streamlines. **A** Evenly distributed. **B** Disrupted (turbulent).

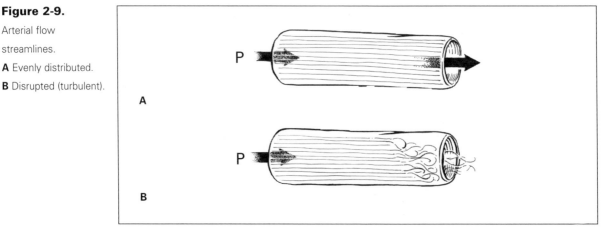

4 The elements that affect the development of turbulent flow are expressed by a "dimensionless" number called *Reynolds number* (Re). The factors that affect the development of turbulence are expressed by this number according to the following equation:

$$Re = \frac{Vq2r}{\eta}$$

where Re = Reynolds number, V = velocity, q = the density of the fluid, r = the radius of the tube, and η = the viscosity of the fluid. Because the density and viscosity of the blood are fairly constant, the development of turbulence depends mainly on the size of the vessels and the velocity of flow. When the Reynolds number exceeds 2,000, laminar flow tends to become disturbed. Flow disturbances also can occur at lower values because of other factors such as body movement, pulsatility of blood flow, and irregularities of the vessel wall and plaque.

5 Turbulent flow may cause vessel walls to vibrate. The harmonics of this vibration produce vascular bruits.

Pressure/Velocity Relationships (Bernoulli)

1 As previously described in the section on energy, the total energy contained in moving fluid is the sum of potential (i.e., pressure), kinetic, and gravitational energies. If one of these variables changes, the others also must change to maintain total fluid energy at the same level. *Example:* If gravitational energy remains unchanged (that is, there is no change in the height of the fluid) but kinetic energy (velocity) increases, then potential (pressure) energy must decrease to maintain the same total fluid energy.

2 The Bernoulli equation shows that velocity and pressure are inversely related. Where there is high velocity, there is low pressure; where there is low velocity, there is high pressure. This inverse relationship between pressure and velocity explains why pressure decreases where fluid velocity increases (within the stenotic segment of an artery, for example) and why pressure distal to a stenosis (the region of poststenotic turbulence, where velocity decreases) is higher than that within the lesion itself. (See figure 2-10.)

3 In other words, this is the law of conservation of energy: (1) In the region proximal to the stenosis (prestenosis), the pressure energy is higher and the kinetic energy lower. This region has the highest total energy sum. (2) As blood flows into the area of the stenosis, the pressure energy decreases and the kinetic energy increases (higher). However, total

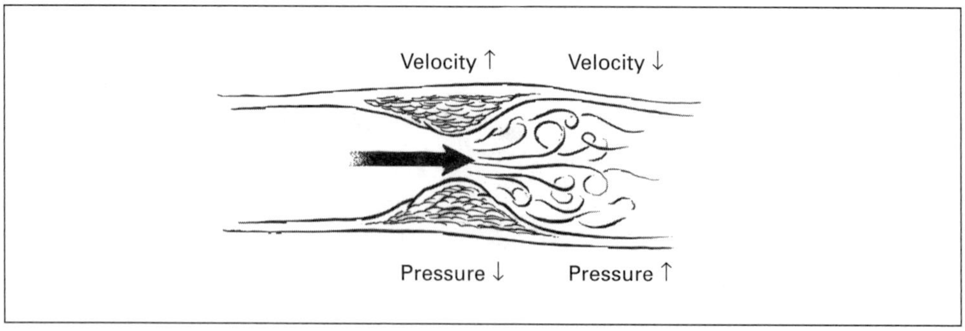

Figure 2-10. Velocities are elevated and pressure diminished within a stenosis, while velocities decrease and pressure increases distal to a stenosis.

energy in this stenotic segment is less than that in the prestenotic segment because energy is lost (i.e., converted into heat) as it moves through the narrowing. (3) Lastly, distal to the stenosis (poststenosis), the kinetic energy decreases and the pressure energy increases. At this point the total energy sum is lowest.

4 Pressure gradients—the difference in pressure between two points in a vessel—are described as flow separations. Flow separations within a vessel may be caused by changes in the geometry of the vessel (with or without intraluminal disease) or the direction of the vessel, as depicted in figures 2-11A and B. See figure 2-12 and color plate 1.

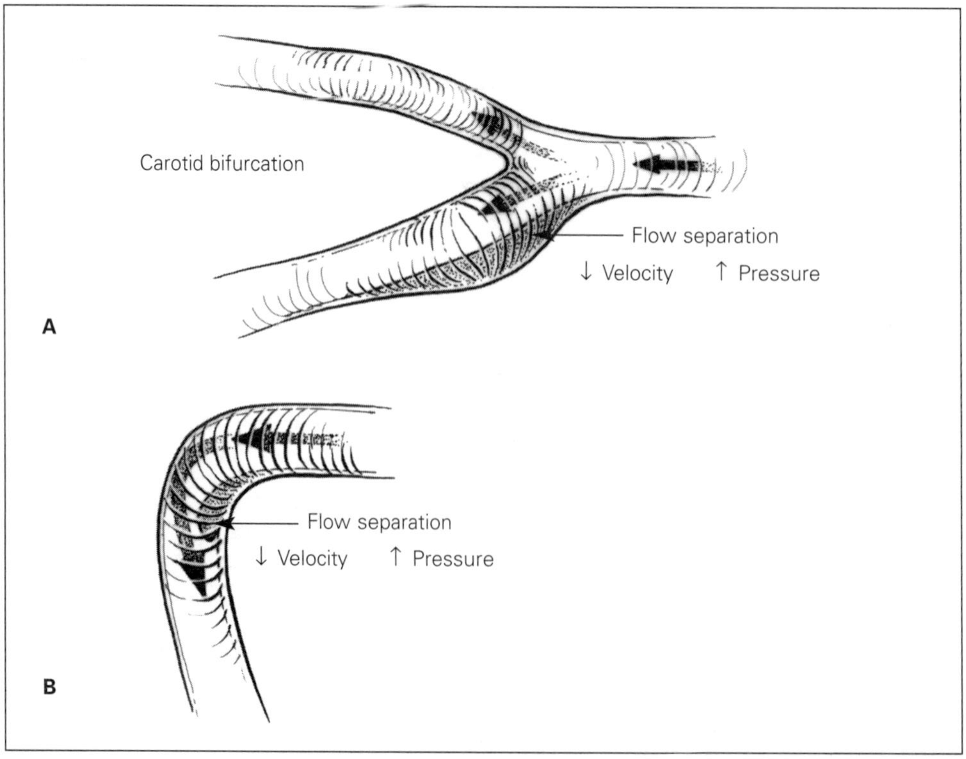

Figure 2-11. Flow separation patterns. In both examples A and B, pressure energy is higher and velocity energy lower in the area of the pressure gradient (flow separation). This causes flow direction to move to the area of lower pressure energy (large arrows). **A** Flow separation in the carotid bulb. **B** Flow separation at a curve (note that it is located on the inside of the vessel curve).

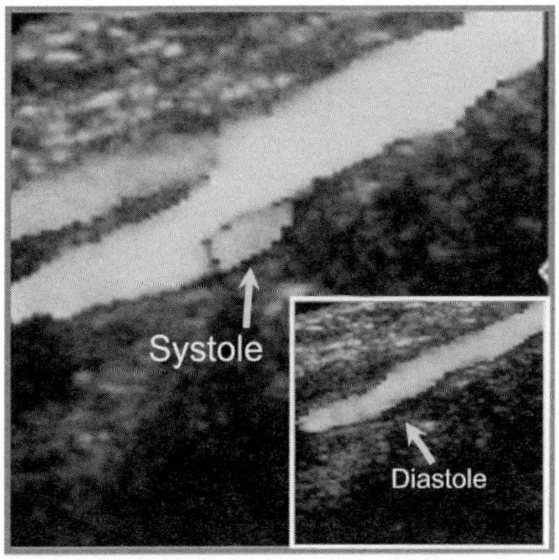

Figure 2-12. Longitudinal image of the carotid bifurcation in systole. The inset shows the same image in diastole. Note that flow separation (blue in color plate 1) is present during systole, absent in diastole.

5 Flow separations leave behind regions of stagnant or little movement. Figure 2-13 demonstrates two different regions within a bypass graft where these flow separations can occur.

6 Because flow moves from high to low pressure (described as a *pressure gradient),* the direction of flow in the region of flow separation (e.g., carotid bulb, bypass graft anastomosis) changes with respect to the transducer, causing a visible color change in the color flow image at systole.

During diastole, when flow at the vessel wall is stagnant, there is no movement of blood and therefore no color in the color flow image. The flow separation pattern is an ideal one to use to help define whether an image is in systole or diastole.

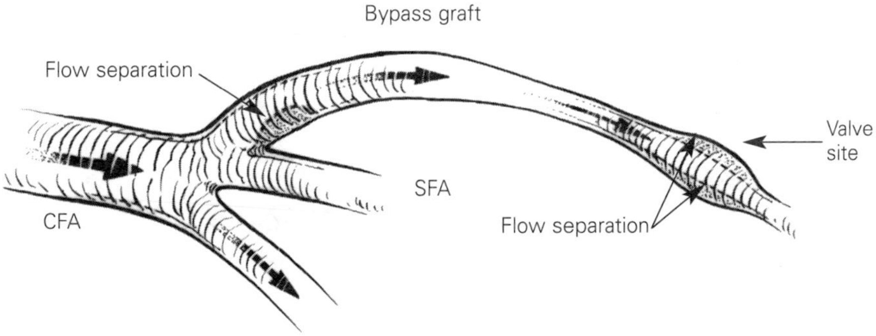

Figure 2-13. In this reversed saphenous vein graft at the end-to-side proximal anastomosis there is an area of flow stagnation (flow separation) on the inside wall. Distally, there are areas of flow separation at the site of a valve cusp. Note the variability of vessel dimension in a bypass graft.

Steady versus Pulsatile Flow

1 Steady flow is the result of a steady driving pressure that initiates movement of a viscous fluid through a conduit. If there is no decrease in pressure downstream, there will be no flow away from the driving pressure.

- Steady pressure states are relatively easy to analyze and describe because fluid dynamics are more easily predictable.

- Where there is steady flow in a rigid tube, energy losses are mainly viscous and can be described by Poiseuille's equation (see above).

2 Pulsatile flow reflects changes in both the driving pressure conditions and the response of the vascular system. Nonsteady conditions include fluid acceleration, deceleration, and rest, and they affect fluid behavior over space and time.

- During systole there is forward flow throughout the periphery. Figure 2-14A depicts early systole; figure 2-14B depicts peak systole. In addition, the walls of the aorta distend storing potential energy.

- At the end of systole, the aortic valve closes and maximum ventricular contraction is obtained.

- During late systole there is temporary flow reversal because of a phase-shifted negative pressure gradient and peripheral resistance, which cause a reflection of the wave proximally (figure 2-14C). The negative pressure gradients are related to different arrival times of the pressure wave at various sites in the arterial system. In comparing the common femoral artery to the dorsalis pedis artery, the pressure gradient is greatest during the first half of systole when the peak of the wave arrives at the femoral site. By the time the peak of the pressure wave arrives at the dorsalis pedis artery, the femoral pressure has fallen and the pressure at the dorsalis pedis artery is higher, causing the change in flow direction (phase-shifted).

- During early diastole (figure 2-14D), flow moves forward again as the reflective wave hits the proximal resistance of the next oncoming wave and reverses direction again.

- As the pressure wave quickly moves through the system, the vessels recoil, converting potential energy to kinetic energy, maintaining flow more distally (figure 2-14E represents late diastole).

- In the biological system, flow always moves along the path of least resistance. *Analogy:* If you drop a golf ball and there is no floor (no resistance), it keeps moving downward just as blood flows forward during systole. If you drop a golf ball onto a cement floor (resistance), it reverses

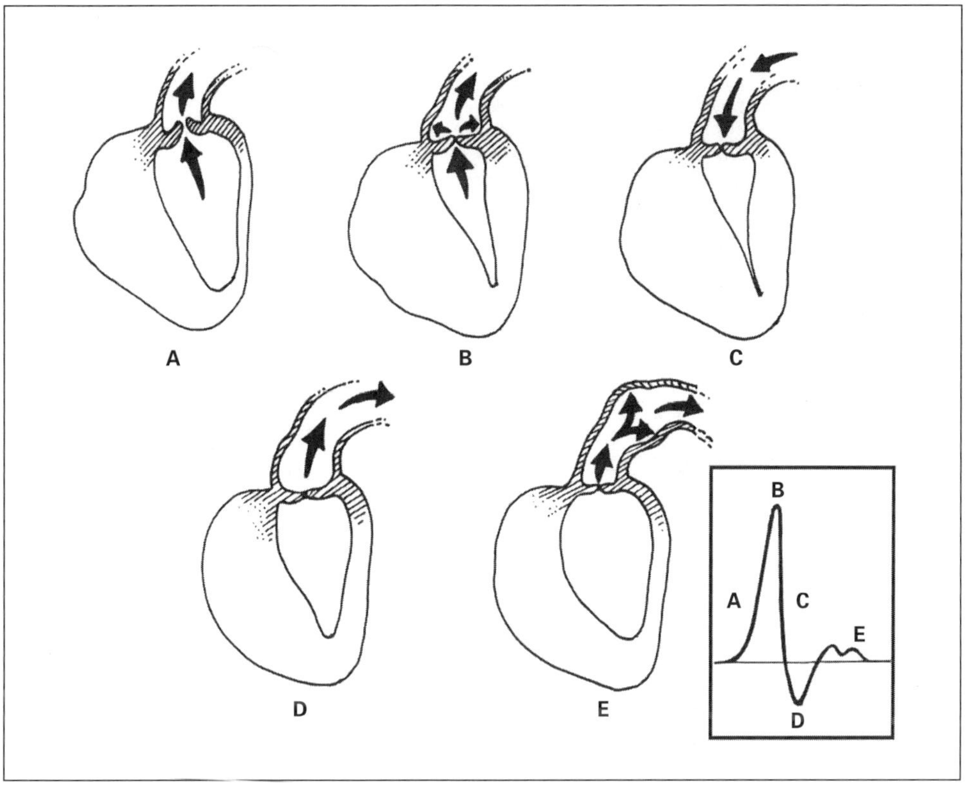

Figure 2-14. Pulsatile arterial flow and the relationship between cardiac activity and peripheral arterial waveforms. **A** Early systole. **B** Peak systole. **C** Late systole. **D** Early diastole. **E** Late diastole.

direction just as blood does during diastole. If you put your hand out as the golf ball travels back to you (resistance again), it will reverse direction once again and travel downward toward the cement floor just as blood flows forward again during late diastole.

- Diastolic flow reversal is a hallmark of vessels that supply high-resistance peripheral vascular beds.
- Flow reversal decreases or is absent with vasodilatation, as illustrated in figure 2-15A. Vasodilation can be produced by body heating, exercise, and stenosis. Additionally, flow reversal is not evident in the presence of an arteriovenous fistula.
- Flow reversal increases with vasoconstriction, as demonstrated in figure 2-15B.

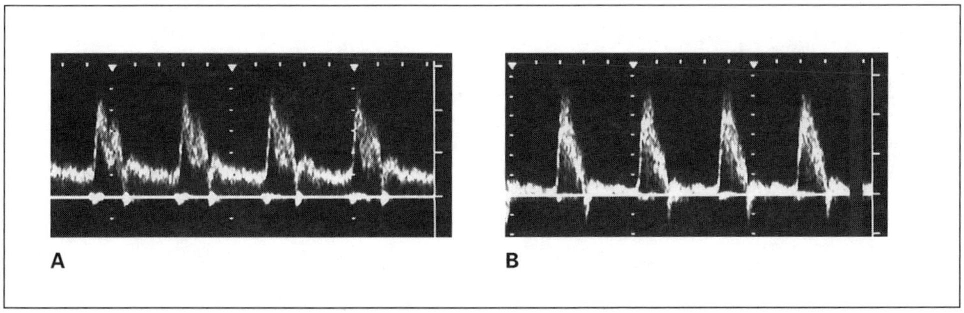

Figure 2-15. Spectral analytic Doppler waveforms. **A** This signal is from an artery taking blood to a vasodilated arterial bed (i.e., resistance decreases). **B** Spectral analytic Doppler waveform from an artery taking blood to a vasoconstricted arterial bed (i.e., resistance increases).

Figure 2-16. Low-resistance (steady, continuous flow pattern) spectral Doppler waveform.

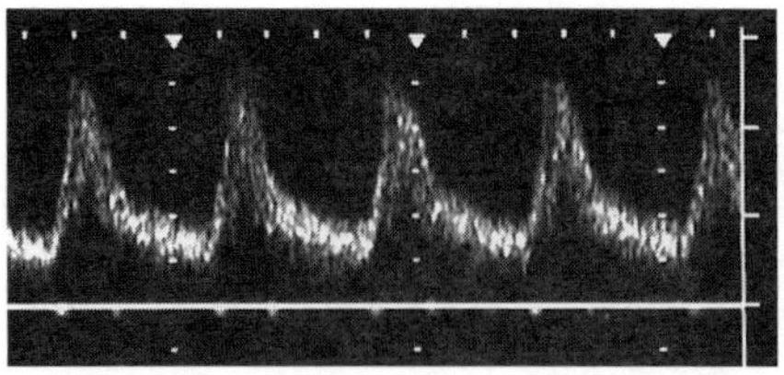

Peripheral Resistance

1 Low-resistance flow is described as flow of a continuous (steady) nature throughout systole and diastole (figure 2-16) feeding a dilated vascular bed. The internal carotid, vertebral, renal, celiac, splenic, and hepatic arteries are characterized by low-resistance flow.

2 High-resistance (pulsatile) flow is described as flow of a pulsatile nature. Between incident pulses, hydraulic reflections travel back up the vessel from the periphery (under normal circumstances) producing flow reversals in the whole vascular compartment. Figure 2-17A describes a normal, triphasic Doppler signal display, while figure 2-17B presents a biphasic Doppler signal display. The external carotid, subclavian, aorta, iliac, extremity arteries, and fasting superior mesenteric arteries are characterized by high-resistance flow.

3 The characteristic reversal of flow in a biphasic or triphasic high-resistance signal disappears distal to a stenosis for a number of reasons, including decreased peripheral resistance as a result of relative ischemia.

4 It is not unusual for a normally high-resistance biphasic or triphasic Doppler signal to become monophasic (losing the reversal component as shown in figure 2-18) proximal to a significant stenosis due to the increased resistance in the stenotic vessel segment.

Figure 2-17. Analog peripheral arterial Doppler signals. **A** Triphasic (normal). There is forward flow, reversed flow, and then forward flow again. **B** Biphasic Doppler signal. There is a forward flow component and a reversed flow component only.

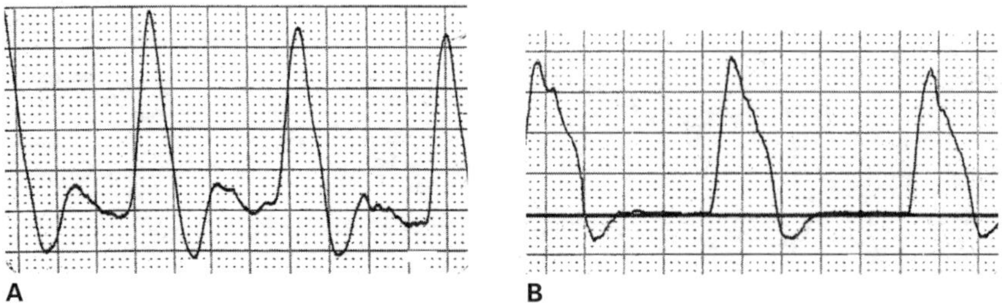

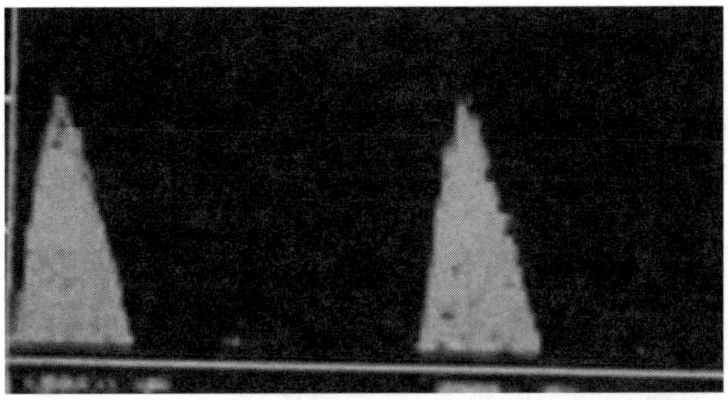

Figure 2-18. Monophasic analog Doppler signal. This type of signal is often seen proximal to a significant obstruction.

5 In response to vasoconstriction, the pulsatility of flow in medium and small-size arteries of the limbs increases (see figure 2-19) but decreases in minute arteries, arterioles, and capillaries.

6 In response to vasodilatation, the pulsatility of flow in medium and small-size arteries of the limbs decreases (see figure 2-20) but increases in minute arteries, arterioles, and capillaries. As the perfusion pressure (inflow pressure) diminishes as a result of a stenosis, the natural response of the high-resistance peripheral vessels is to vasodilate to maintain flow.

7 Because such compensatory responses of the microcirculation tend to maintain pressures and flow at rest, it is important to obtain information about pressures and flow patterns both at rest and after exercise.

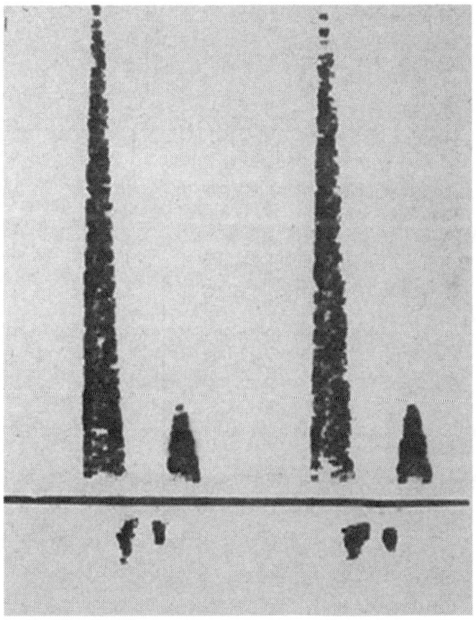

Figure 2-19. As flow becomes more pulsatile, the Doppler signal becomes much more resistive in quality.

Figure 2-20. Decreased pulsatility produces a more continuous, low-resistance Doppler signal.

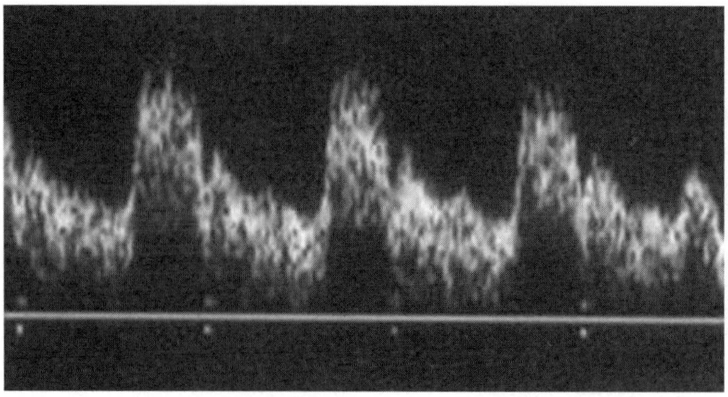

Collateral Effects

1 In an extremity at rest, total blood flow may be fairly normal even in the presence of severe stenosis or complete occlusion of the main artery because of the development of a collateral network, as well as the aforementioned compensatory decrease in peripheral resistance.

2 To evaluate the approximate location of the obstructed artery, Doppler segmental pressures would be necessary.

3 Arterial obstruction may alter flow in nearby or more distant collateral channels, increasing volume flow, reversing flow direction, increasing velocity, and/or altering the pulsatility of the waveform. Note in figure 2-21 the low-resistance quality of the spectral waveform.

4 The location of collateral vessels helps to provide a tentative indication of the obstruction level, as demonstrated in figure 2-22.

Figure 2-21. With a proximal superficial femoral artery (SFA) occlusion, flow in this popliteal artery is reconstituted via collaterals. Because flow is collateral-based and flowing into a vasodilated vascular bed, its quality is of low resistance.

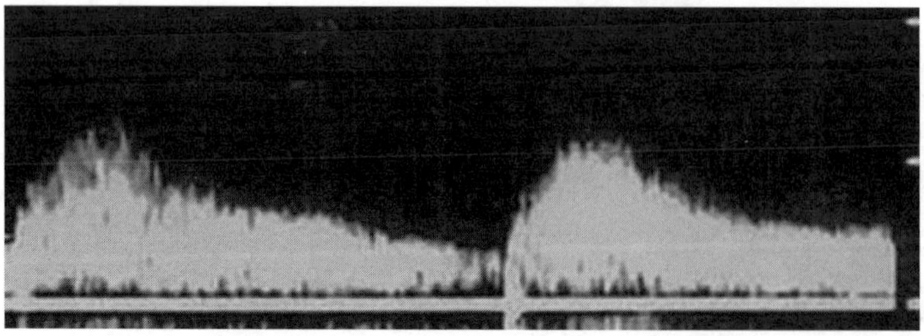

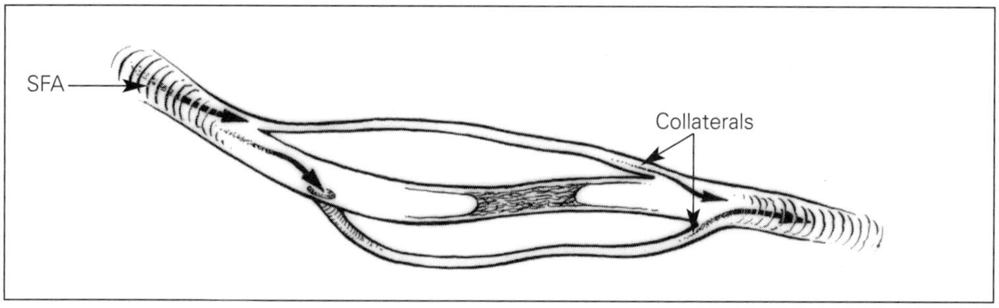

Figure 2-22. Representation of an occluded superficial femoral artery with reconstituted flow distally due to collateralization. Arterial branches that were small are now taking a larger portion of the arterial flow, actually bypassing the obstruction in order to provide needed arterial flow to the foot.

5 Secondary collateral changes (such as evidence of tissue healing or granulation, increased capillary refill, and decreased symptomology) also provide some limited information regarding the adequacy of a collateral system that has evolved in response to arterial obstruction.

Effects of Exercise

1 Exercise should induce peripheral vasodilation in the microcirculation so that distal peripheral resistance diminishes and blood flow markedly increases.

2 Peripheral resistance—the resistance to blood flow caused by the ever-decreasing size of the vessels, especially in the microcirculation—changes in response to a variety of stimuli such as heat, cold, tobacco use, and emotional stress.

3 Vasoconstriction and vasodilation of the blood vessels within skeletal muscles are also influenced by sympathetic innervation fibers that function primarily to regulate body temperature.

4 Exercise is probably the best single vasodilator of high-resistance vessels within skeletal muscle.

5 Autoregulation also controls vasoconstriction and vasodilatation. Autoregulation accounts for the ability of most vascular beds to maintain a constant level of blood flow over a wide range of perfusion pressures.

- Autoregulation does not function normally when perfusion pressure drops below a critical level.
- High-resistance vessels constrict in response to increased blood pressure and dilate in response to decreased blood pressure.

6 By decreasing resistance in the working muscle, exercise normally decreases reflection (flow reversal) of the Doppler flow signal in the exercising extremity. *Example:* A low-resistance, monophasic Doppler flow signal (figure 2-23) may be present normally in an extremity artery after vigorous exercise because the exercise causes peripheral dilatation and

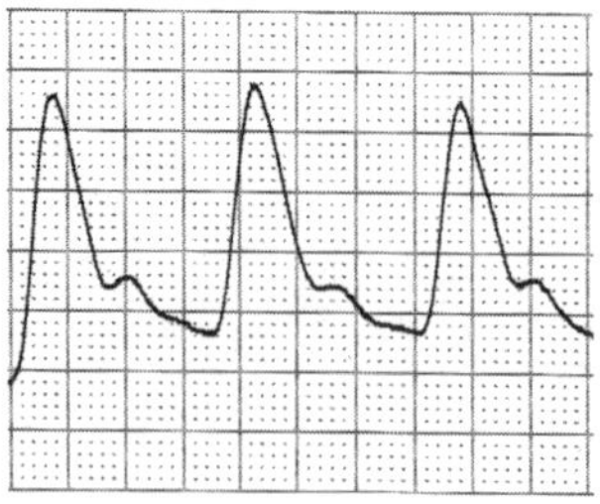

Figure 2-23. Because exercise produces a demand for blood to the muscles, a normally high-resistance arterial Doppler signal becomes low-resistance. In this analog waveform, the usual reversal below the baseline is seen as a foward reflection instead. This finding is quite normal after exercise.

reduced flow resistance. This same monophasic pattern is also seen pathologically when peripheral dilatation occurs in response to proximal arterial obstruction.

7 On the other hand, a high-resistance signal (figure 2-24) may occur from normal (physiologic) vasoconstriction at the arteriolar level *or* from distal arterial obstruction.

8 Additional notes:

- Proper characterization of velocity waveforms requires an understanding of both the normal flow characteristics of a particular artery and the physiologic status of the circulation supplied by the vessel.
- Questions need to be asked: Was the extremity cooled or warmed? Has it been exercised prior to the exam? Flow to a cool, vasoconstricted extremity will have pulsatile signals. Flow to a warm, vasodilated extremity will have continuous, steady signals.
- Proximal and distal pulsatility changes do not precisely differentiate between occlusion and severe stenosis.
- If good collateralization is present, proximal or distal Doppler velocity waveform qualities may not be altered.
- The distal effects of obstructive disease may be detectable in the presence of exercise or hyperemic evaluation.

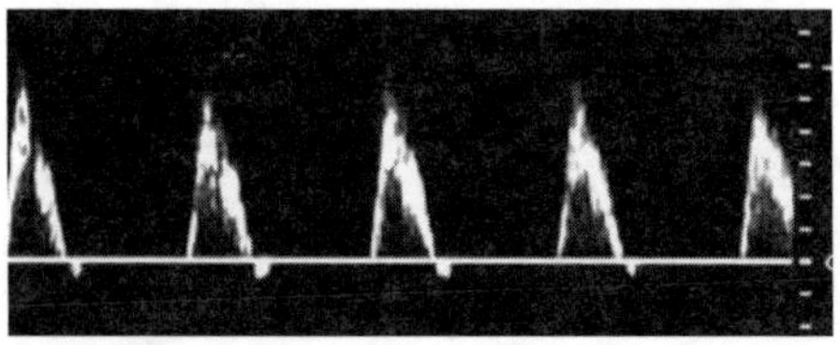

Figure 2-24. This high-resistance Doppler signal can occur with normal vasoconstriction at the arteriolar level. It can also occur proximal to distal arterial obstruction.

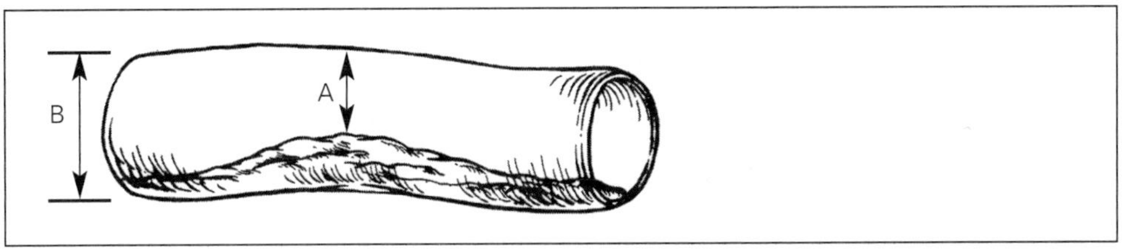

Figure 2-25. The nondiseased arterial segment is measured from wall to wall in a longitudinal approach (B) and then compared to the residual flow channel at the area of stenosis (A). A percentage diameter reduction is calculated.

Effects of Stenosis on Flow

1 Laminar flow review:

- Laminar flow has an even distribution of frequencies at systole, with the lower frequencies distributed at the walls (the boundary layer) and the higher frequencies in center stream.
- Stable flow through a relatively straight vessel is usually laminar, the layers of fluid slipping over one another with minimal friction normally.

2 A critical, hemodynamically significant stenosis causes a major reduction in volume flow and pressure. A stenosis usually becomes hemodynamically significant when the cross-sectional area of the arterial lumen is reduced 75%, which corresponds to a diameter reduction of 50%.

- Diameter reduction is a one-dimensional measurement (figure 2-25).
- Area reduction is a two-dimensional measurement (figure 2-26).
- In critical stenoses, both pressure and flow volume decrease.
- Two or more stenotic lesions that occur in series have a greater effect on volume flow and distal pressure than a single lesion of equal total length. The difference is due to the large loss of energy at the entrance and especially at the exit of the lesion, which results from very disturbed flow patterns such as jet effects, turbulence, and eddy formation. It is because of this that the energy losses in tandem lesions far exceed those that result from the increased resistance in a single stenosis (see *Poiseuille's Law* on page 22 above).

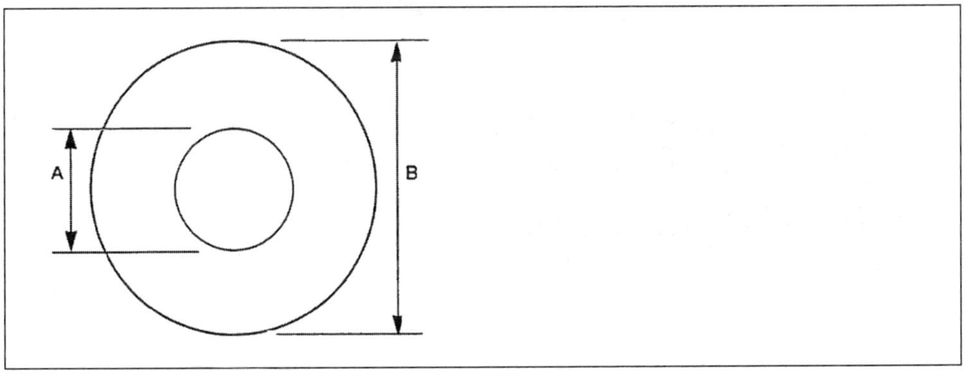

Figure 2-26. The original lumen (B) is measured and compared to the residual lumen (A) when the vessel is in a transverse approach. A percentage area reduction is calculated.

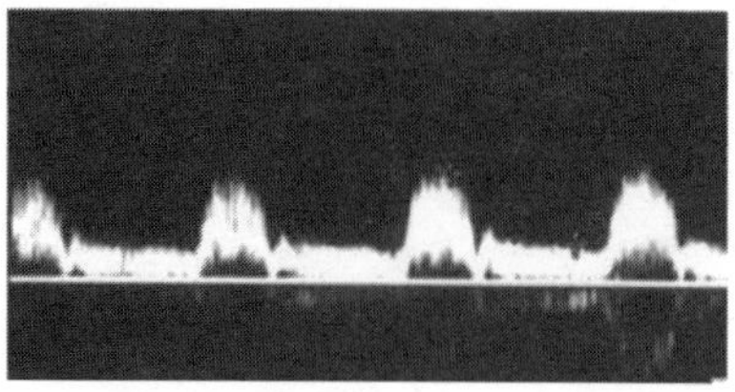

Figure 2-27. Dampened, attenuated peak systolic velocities proximal to a significant stenosis.

3 The occurrence and degree of hemodynamic abnormality produced by a stenosis depend on many factors, including these:

- Length of narrowing
- Diameter of narrowing
- Roughness of endothelial surface
- Shape and degree of narrowing
- Arteriovenous pressure gradient
- Peripheral resistance distal to the stenosis
- Collateral circulation

4 Critical arterial stenosis produces certain identifiable hemodynamic phenomena that are described in the literature as the *stenosis profile:*

- Proximal to a stenosis, flow frequencies—or velocities, the term we use in this book—are usually dampened, with or without disturbance (figure 2-27).

- As flow enters, passes through, and/or exits the stenosis, there is an increase in Doppler shift frequencies and velocities. The increased range of Doppler shift frequencies are displayed as spectral broadening in the Fourier display. (See figure 2-28.) Flow also becomes disorganized and

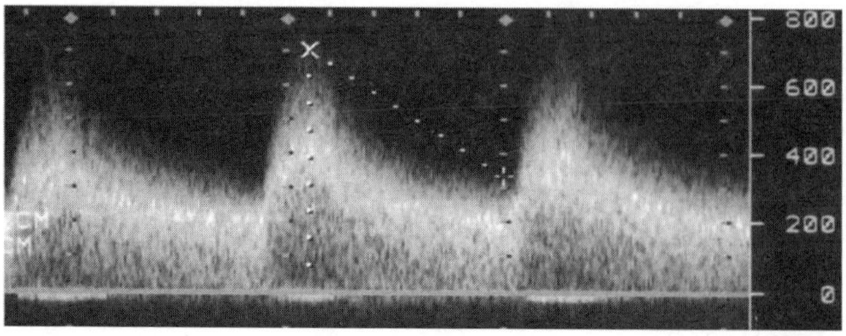

Figure 2-28. Increased peak systolic velocities with severe spectral broadening from the site of a significant stenotic arterial segment.

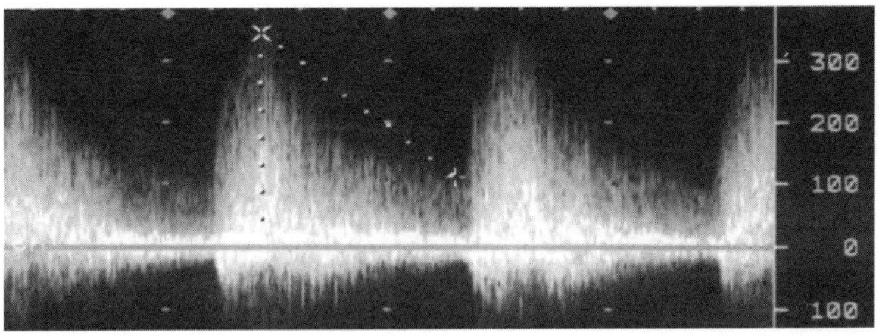

Figure 2-29. Poststenotic turbulence (forward and reversed flow directions due to eddy currents and vortices distal to the area of stenosis).

unstable as elevated frequencies/velocities and eddy currents disrupt normal flow. **Note:** The abnormal stenotic jet with its characteristically elevated velocities is not isolated to just the stenotic segment; it is not uncommon to find it just proximal to and/or just distal to the stenosis.

- Poststenotic turbulence is typically seen at the exit from the stenosis. Such flow is characterized by flow reversals, flow separations, vortices, and eddy currents, which occur toward the outer boundaries of the higher velocity flow, ultimately producing the bidirectional, disturbed flow pattern seen distal to the stenosis and illustrated in figure 2-29. When the poststenotic arterial lumen widens rather quickly, large flow separation patterns are observed at the walls as a result of the velocity gradients described by Bernoulli.

- Energy is expended largely in the form of heat as the eddy currents and vortices work against the viscosity of the blood. The spectral display becomes quite broad in response to the multiplicity of frequencies found in the poststenotic vessel.

5 Reductions in the diameter of a vessel increase the velocity of flow. As noted previously, the size of a vessel is inversely proportional to the velocity of blood flow. According to the basic laws of fluid dynamics, most notably the law of conservation of mass (i.e., what goes in must come out), the relationship among velocity, volume flow, and cross-sectional area of the vessel is

$$V = Q/A$$

where V = velocity (cm/sec), Q = volume flow (cm^3), and A = cross-sectional area (cm^2).

Patient History, Mechanisms of Disease, and Physical Examination

CHAPTER 3

It is very important to obtain a complete history and to conduct a limited physical examination. The history and physical examination help to define the clinical context in which tests are ordered (i.e., the appropriate and medically necessary indications for the tests), performed, and interpreted.

Patient History: Signs and Symptoms

1 Chronic occlusive disease:

- *Claudication:* Pain in muscles occurring during exercise but subsiding with rest. The patient notes muscle fatigue with work, and the discomfort is usually predictable, occurring with the same amount of work and disappearing within minutes of activity cessation. True claudication results from inadequate blood supply to the exercising muscle, which may be caused by arterial spasm, atherosclerosis, arteriosclerosis, or an occlusion. Differential diagnosis includes neurogenic (nerve involvement) and musculoskeletal (called pseudo- [false] claudication) causes that mimic the symptoms produced by vascular claudication. Various types of claudication include:

- Buttock claudication, which strongly suggests aortoiliac disease. If the symptoms are unilateral, this suggests iliofemoral disease.
- Thigh claudication, which suggests distal external iliac/common femoral disease.
- Calf claudication, which suggests femoral/popliteal disease.

For example, a patient history may include the note, "2 block claudication," which indicates that the patient complains of pain in a specific part of the leg after walking 2 city blocks.

- *Ischemic rest pain:* A more severe symptom of diminished blood flow to the most distal portion of the extremity. Pain at rest usually occurs when the limb is not in a dependent position and the patient's blood pressure is decreased (e.g., when sleeping). Symptoms occur in the forefoot, heel, and toes, but not in the calf.
- *Tissue loss:* Necrosis (tissue death), usually due to a deficient or absent blood supply. Necrosis is the most severe symptom of arterial insufficiency.

2 Acute arterial occlusion:

- Symptoms include the five Ps: pain, pallor, pulselessness, paresthesia, paralysis. Some authors include a sixth "P" for "polar," i.e., cold. Some even include a seventh "P" for "purplish," the darkly purple color of cyanosis seen in patients presenting with a cold, blue foot.
- Occlusion may result from thrombus, embolism, or trauma.
- This is an emergency situation since the abrupt onset does not provide for the development of collateral channels.

3 Cold sensitivity:

- Symptoms include changes in skin color such as pallor (paleness), cyanosis (bluish discoloration), or rubor (dark red coloration), and the patient often experiences paresthesia and pain. In more severe cases, trophic changes may be evident.
- *Raynaud's phenomenon* is a condition that exists when symptoms of intermittent ischemia of the fingers or toes occur in response to cold exposure as well as emotional stress.

4 Mesenteric ischemia and renovascular hypertension: The signs and symptoms of these conditions are described below and in chapter 14.

Patient History: Risk Factors and Contributing Diseases

1 Diabetes:

- Diabetes mellitus is the most common form.
- Atherosclerosis is more common among diabetics and occurs at a younger age.
- Diabetics have a higher incidence of occlusive disease of the distal popliteal artery and tibial vessels.
- Medial calcification develops in the lower extremity arteries.
- There is a higher incidence of gangrenous changes and, ultimately, amputations.
- Poor sensation as a result of neuropathy leads to increased likelihood of trauma.

2 Hypertension:

- It remains unclear whether high blood pressure is a causative factor or enhances and complicates the development of the atherosclerotic process.
- Systemic hypertension is associated with a greater incidence of coronary atherosclerosis and also increases an individual's susceptibility to peripheral and cerebrovascular involvement (e.g., atherosclerosis).

3 Hyperlipidemia:

- Because they are insoluble in water, elevated plasma lipids are closely associated with the development of atherosclerosis. Although increased lipids may result from metabolic problems associated with heredity, a diet high in certain types of fat is also associated with hyperlipidemia.

4 Smoking:

- Studies have demonstrated that the chemicals in cigarettes irritate the endothelial lining of the arteries in addition to causing vasoconstriction.

5 Other risk factors:

- Age
- Family history
- Male gender

Mechanisms of Disease

1 Atherosclerosis:

- The most common arterial pathology. The term "atherosclerosis" (also known as *arteriosclerosis obliterans*) is applied to a number of pathological conditions in which there is thickening, hardening, and loss of elasticity of the walls of the arteries. These changes occur in the intima and media layers of the vessel.

- Major risk factors include smoking, hyperlipidemia, and family history. Less important factors are hypertension, diabetes mellitus, and sedentary lifestyle.

- Atherosclerosis most often occurs at the carotid bifurcation, origins of the brachiocephalic vessel, origins of the visceral vessels, the infrarenal aortoiliac system, common femoral bifurcation, superficial femoral artery at the adductor canal, and the popliteal trifurcation.

- Example: Leriche syndrome, caused by obstruction of the terminal aorta. It usually occurs in males and is characterized by fatigue in the hips, thighs, or calves on exercising, absence of pulsation in the femoral arteries, impotence, and often coldness/pallor of the lower limbs.

2 Embolism:

- Obstruction of a blood vessel by a foreign substance or blood clot. Emboli may be solid, liquid, or gaseous and may arise from the body or may enter from without. The most frequent cause of embolism, however, is plaque breaking loose and traveling distally until it lodges in a small vessel.

- Emboli move distally and become stuck in the vessel(s) of the smallest caliber (i.e., the digital arteries). Toe ischemia results, which often improves mainly as the result of blood flow from smaller, collateral arterial branches. (See color plate 2.)

- The blue toe syndrome may be caused by ulcerated and/or atherosclerotic lesions, embolization, the inflammatory process of arteritis (which can produce thrombosis), and some angiographic procedures.

3 Aneurysm:

- Pathology-related aneurysm types include true aneurysm, dissecting aneurysm, and pseudoaneurysm.

- A *true aneurysm* is a dilatation (bulging) of all three layers of the arterial wall (figure 3-1), differentiating it from a *false aneurysm*, which does not contain all arterial wall layers.

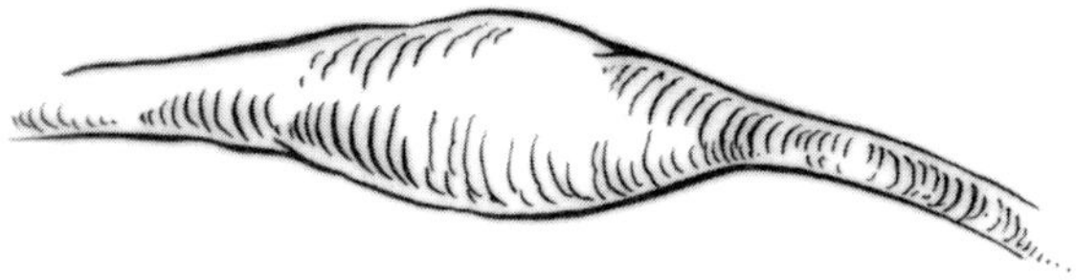

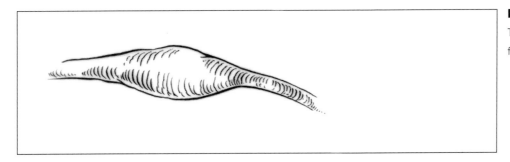

Figure 3-1.
True aneurysm, fusiform in shape.

- The most common location for arterial aneurysms is the infrarenal aorta, but they can occur in nearly any artery of the body. Patients with one aneurysm have a higher chance of having a second aneurysm. Other locations include: thoracic aorta, femoral, popliteal, carotid, renal, and splenic arteries. Interestingly, popliteal aneurysms are often bilateral.
- The cause of aneurysms is unknown but may include poor arterial nutrition, congenital defects, infection, or atherosclerosis, trauma, and iatrogenic injury.
- Some terms that relate to aneurysm shape include focal aneurysm, fusiform aneurysm, saccular aneurysm, and concentric aneurysm. *Fusiform aneurysm* is a diffuse, circumferential dilatation of an arterial segment (see figure 3-1). *Saccular aneurysm* is a localized out-pouching of an artery, resulting from wall thinning and stretching (figure 3-2).
- The main complications of aneurysm include rupture of aortic aneurysms and distal embolization of peripheral aneurysms. Any aneurysmal formation has the propensity to form thrombotic material at the walls. This is likely due to the wall dilatation in combination with the normally lower velocities found at the wall, which become stagnant in diastole.

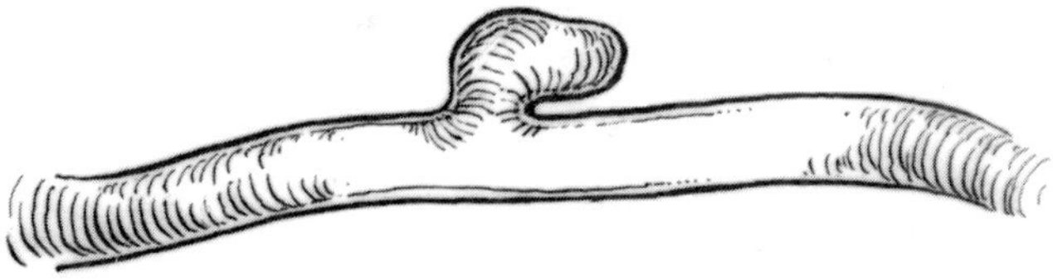

Figure 3-2.
Saccular aneurysm.

Figure 3-3.
Dissecting aneurysm.

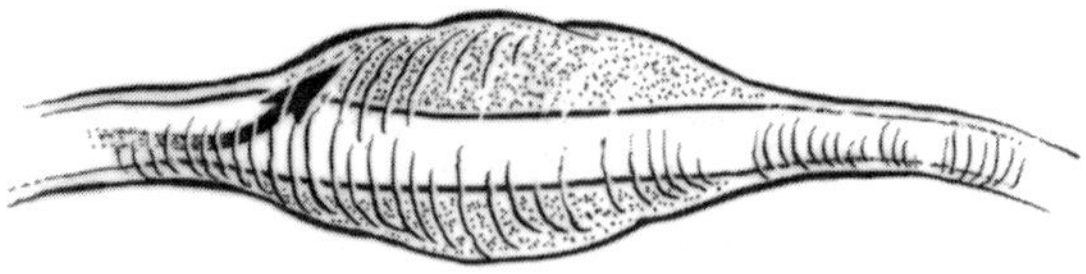

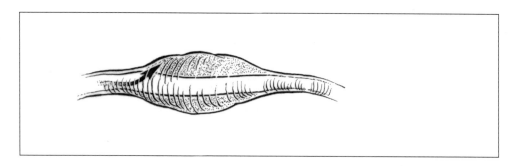

- A *dissecting aneurysm* occurs when a small tear of the intima allows blood to form a cavity between two wall layers (figure 3-3). A new lumen, the false lumen, is formed, and blood may flow through this lumen as well as through the original vessel lumen to supply branch vessels. Two conditions must usually be met for a dissecting aneurysm to form: weakening of the media of the vessel and development of an intimal tear through which blood then leaks into the media. Atherosclerosis is not generally considered to be a causative factor. A dissection most often occurs in the thoracic aorta.

- A pseudoaneurysm is essentially a pulsating hematoma. A hole in the arterial wall permits blood to escape under pressure, into a contained area in the adjacent tissue. A hematoma forms in the tissue and, if confined by the surrounding structures and if there is continuous blood flowing from the artery to the nonthrombotic region of the hematoma, the pseudoaneurysm is created. Such an aneurysm usually results from a defect in the main arterial wall following the insertion of a catheter (e.g., for angiography or an endovascular procedure, or after dialysis). To be considered a pseudoaneurysm, there must be a communication (channel) from the main artery to the pulsatile structure in the tissue. (See figure 3-4; see also figure 3-5 and color plate 3.)

4 Arteritis:

Arteritis—inflammation of the arterial wall—often results in thrombosis of the vessel and can affect tibial and peroneal arteries as well as the smaller and more distal arterioles and nutrient vessels. There are several

Figure 3-4.
Pseudoaneurysm.

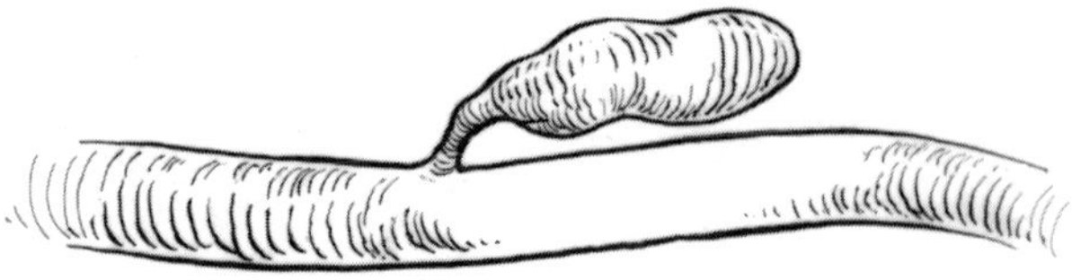

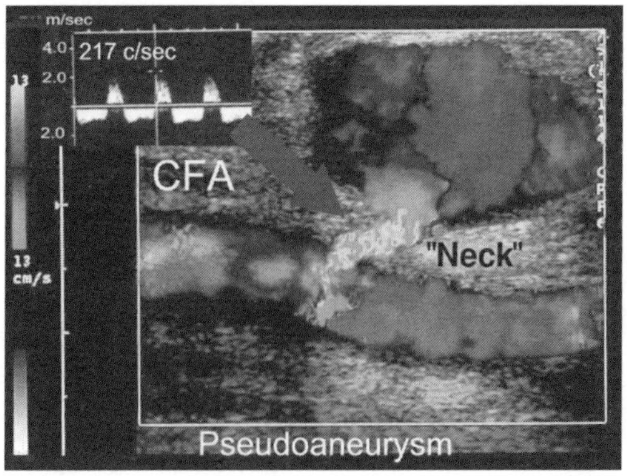

Figure 3-5. Pseudoaneurysm of the common femoral artery (CFA). The "neck" and unique Doppler waveforms (to and fro) are diagnostic hallmarks of a pseudoaneurysm.

types of arteritis, including Takayasu's, temporal, and polyarteritis diseases. The most common form of arteritis is *Buerger's disease*, also known as *thromboangiitis obliterans*, the characteristics of which include the following:

Associated with heavy cigarette smoking.

Occurs primarily in men younger than 40 years of age.

Patients present with occlusions of the distal arteries.

Rest pain and ischemic ulceration occur early in the course of the disease.

Inflammation of the arterial wall occurs and often results in thrombosis of the vessel.

Superficial thrombophlebitis is a secondary result.

Common clinical presentation includes patchy areas of ulceration or gangrene.

It may be coincidentally associated with atherosclerotic occlusive disease of the larger arteries.

Most often associated with collagen vascular syndromes such as rheumatoid arthritis and lupus.

5 Coarctation of the aorta:

One of several congenital anomalies of the arterial system, coarctation of the aorta is a congenital narrowing or stricture of the thoracic aorta that may affect the abdominal aorta as well (figure 3-6). Clinical findings may include hypertension due to decreased kidney perfusion or manifestations of lower extremity ischemia (e.g., decreased pulses, decreased segmental Doppler pressures).

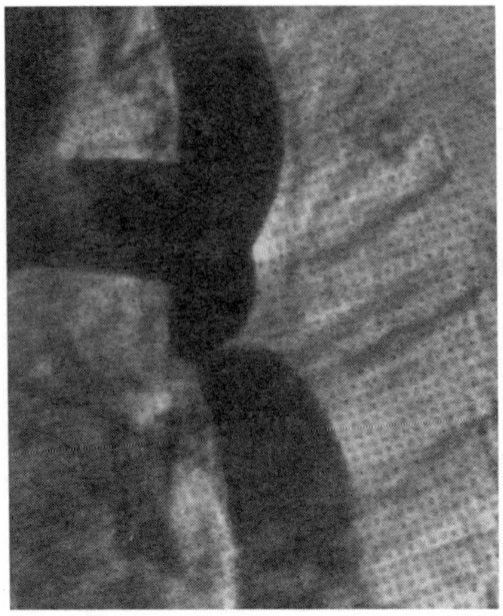

Figure 3-6. Coarctation of the aorta. Note the stricture in the aorta.

6 Dissection

- The media is weakened; the intima develops a tear through which blood leaks into the false lumen that has developed between the intima and the media, as illustrated in figure 3-7.
- Flow velocities differ in each lumen (main and false).
- Dissection can affect the aorta and peripheral arteries.
- The distinguishing ultrasonographic feature is the thin membrane that divides the arterial lumen into two compartments.
- Aortic dissections, which can also extend to the iliac arteries, may occur as a consequence of hypertension or severe chest trauma.

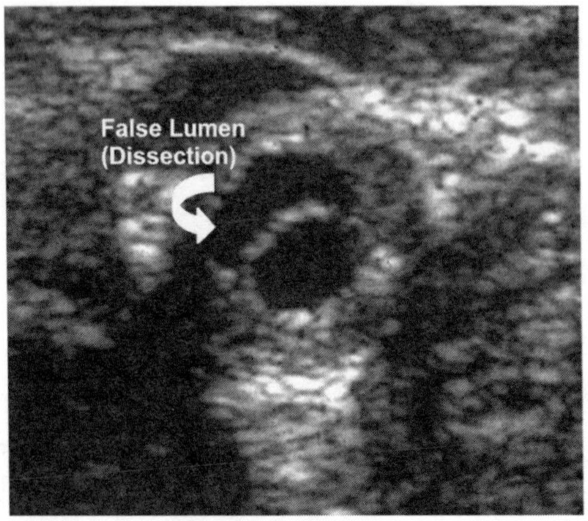

Figure 3-7. Transverse image of a common carotid artery dissection. Note the two lumens.

- Complications: As the dissection enlarges, there is the risk of significant stenosis and/or occlusion of the main artery or, at the very least, arterial branches.

7 Vasospastic disorders/cold sensitivity:

- Symptoms include changes in skin color such as pallor (paleness), cyanosis (bluish discoloration), or rubor (dark red coloration), and the patient often experiences paresthesia and pain. In more severe cases, trophic changes may be evident. See color plate 4.

- *Raynaud's phenomenon* is a condition that exists when symptoms of intermittent ischemia of the fingers or toes occur in response to cold exposure as well as emotional stress. Raynaud's phenomenon may be idiopathic, or it may be the result of an underlying disease or anatomic abnormality.

- Primary (idiopathic) Raynaud's phenomenon, also called *Raynaud's disease* and *spastic Raynaud's syndrome,* is intermittent digital ischemia caused by digital arterial spasm. Common in young women, primary Raynaud's disease may be hereditary. It is usually bilateral, and the patient commonly presents with a long-term history of symptoms without progression or evidence of underlying cause. The condition is rather benign and has an excellent prognosis.

- Secondary Raynaud's, also known as *secondary Raynaud's phenomenon* and *obstructive Raynaud's syndrome*, consists of normal vasoconstrictive responses of the arterioles superimposed on a *fixed* arterial obstruction. Ischemia is constantly present. Secondary Raynaud's may be the first manifestation of a collagen disorder, Buerger's disease, anatomic abnormality, or other underlying causes.

- Waveform pulse contours (described on pages 99–101) provide a fairly good method of distinguishing among the various causes of digital ischemic obstruction from vasospasm.

8 Entrapment syndromes:

Entrapment of the popliteal artery is the most written about entrapment syndrome. It is thought to be caused by compression of the popliteal artery by the medial head of the gastrocnemius muscle or fibrous bands. It is commonly found in young men and is bilateral in about a third of the cases. See pages 169–170 for additional information about symptoms and test procedures.

Physical Examination

1 Skin changes:

Color

- *Pallor* is a result of a deficient blood supply. Skin color is pale.
- *Rubor*, a dark reddish discoloration, suggests damaged, dilated vessels or vessels dilated as a result of reactive hyperemia or infection. *Dependent rubor* is a specific type of rubor that is explained on page 49.
- *Cyanosis*, a bluish discoloration of the skin and mucous membranes, occurs when there is a concentration of deoxygenated hemoglobin.
- *Livedo reticularis*, in which there are purple patches (similar to bruising) on the skin of the dorsum of the foot, is usually the result of dilated capillary and venule filling, not arterial obstruction.

Temperature

The patient's skin should be warm to the touch. It is essential to touch and feel the patient's skin to determine whether it is warm or cold.

Lesions

- Ulcerations as a result of arterial insufficiency are usually deep and regular in shape (figure 3-8; see also color plate 5), often located over the tibial area, and quite painful compared to venous ulcerations. It is important to elicit from the patient the length of time the ulceration(s) has been present and also to examine and record observations of the foot and toes.
- Gangrene is the death of tissue, usually caused by deficient or absent blood supply.

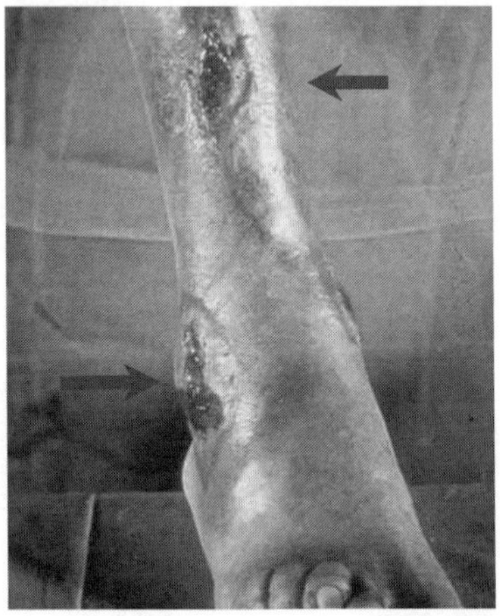

Figure 3-8. Ulcerations on the tibia and lateral ankle. See table 28-2 and the discussion of arterial vs. venous ulcers in chapter 28.

Trophic Changes

- Loss of hair on the extremity may reflect a poor nutritional state caused by decreased circulation, although hair loss alone is a poor indicator of peripheral vascular occlusive disease. The skin can also have a shiny, scaly appearance.
- Thickened toe nails are commonly seen.

Capillary Filling

The healthy flesh color blanches in appearance as superficial vessels are constricted by manual pressure. Normal skin color should return immediately upon the release of the pressure. An increase in the capillary refill time denotes decreased arterial perfusion.

Elevation/Dependency Changes

- Elevating the extremity with impaired circulation produces a cadaveric pallor because of very poor arterial perfusion. When tissue is starved for arterial blood, the arterial vessels dilate to provide as much flow as possible. But with the extremity elevated (or even at the level of the heart), only minimal arterial blood flow can make its way distally. There is no hydrostatic pressure to assist the flow because of the extremity's position.
- Returning the extremity to a dependent position causes a slow return to normality followed by the red discoloration called *dependent rubor* (see color plate 6). The ruborous coloration is caused by the large amount of blood flowing into small vasodilated superficial arteries. Hydrostatic pressure also assists blood flow to the dependent extremity.

2 Palpation (pulses, aneurysms):

- The rhythmic throbbing caused by the regular contraction and alternate expansion of an artery in time with the heartbeat usually signifies an adequate circulatory status. A diminished or absent pulse suggests arterial insufficiency.
- Grading pulses on a scale of 0 to 4+ is fairly standard. Always compare the pulse palpated on one side to the pulse at the same site on the other side.

  0 = None

  1+ = Weak

  2+ = Good

  3+ = Strong

  4+ = Bounding

- Aneurysms, if present, are easily palpated and their pulses quite bounding in response to the pressure of palpation.
- Palpable "vibration" or "thrill" over a pulse site may indicate a fistula, poststenotic turbulence, or even a patent dialysis graft.
- When palpating a pulse, the examiner should not place his or her thumb on the patient's skin since it is quite probable that his/her own pulse will be felt.
- Palpable pulses include the aorta, femoral, popliteal, dorsalis pedis, and posterior tibial arteries. The peroneal artery cannot be palpated.

3 Auscultation (bruits):

- Normal flow patterns evident when listening through a stethoscope are often described as a lub-dub sound that corresponds with the closure of the atrioventricular valves followed by the closure of the aortic and pulmonary valves.
- *Bruits* are abnormal, low-frequency sounds heard on auscultation. They can be caused by significant stenosis that sets up a vibratory response in the tissue distal to the stenosis. Because bruits are low frequency, and depending on the examiner's hearing range, they may or may not always be heard.
- Bruits may be graded on the basis of their strength and duration as:

  1+ = Mild

  2+ = Moderate

  3+ = Severe
- A bruit that extends throughout diastole (sometimes referred to as *pan-diastolic*) is usually associated with more severe arterial disease.
- Sites for auscultation include: carotid, abdominal aorta, femoral, and popliteal arteries.
- The absence of a bruit suggests normality but cannot rule out disease. In cases of severe stenosis, usually greater than 90% diameter reduction, the bruit disappears.
- Poor cardiac output can decrease the strength of the bruit, making bruit auscultation difficult or impossible.

Doppler Waveform Analysis in the Upper and Lower Extremities

CHAPTER 4

It is essential to maintain a warm environment for the patient, especially during the physiologic studies. If a patient arrives for a test when the outside temperature is low, it is important to permit warming (and peripheral dilation) to occur.

An adequate amount of acoustic coupling gel is applied to the skin whenever direct, indirect Doppler, or imaging information is to be obtained. This instruction will not be repeated below under *Technique.*

1 Capabilities:

- Helps to confirm the diagnosis of arterial occlusive disease.
- Differentiates the severity of the occlusive process from pre- to postexercise.
- Indicates the approximate location of the significant obstruction.
- Provides follow-up information about progression of disease, results of medical therapy, or postoperative status.

2 Limitations:

- Patients with casts or extensive bandages on the lower extremities that cannot be removed.
- Waveforms may be affected by ambient temperature.
- Uncompensated congestive heart failure may result in dampened waveforms.
- Unable to discriminate stenosis from occlusion and cannot precisely localize the obstruction.
- The test is technologist-dependent. It is essential that the technologist understand the importance of maintaining an adequate Doppler angle.

3 Patient positioning:

- The patient is positioned supine with the extremities at the same level as the heart. The head of the bed can be elevated slightly, and the patient's head can rest on a pillow.
- The patient's hip is externally rotated with the knee slightly bent to help facilitate the lower extremity evaluation.
- Alternative positions for lower extremity examinations include right or left lateral decubitus (patient on his or her side) or prone for access to the popliteal space.
- When evaluating the upper extremities, position the arms at the patient's side in a relaxed state. The patient's head can be slightly elevated.

4 Physical principles:

- The *Doppler effect:* When a wave is reflected from a moving target, the frequency of the wave received differs from that of the transmitted wave. This difference is known as the *Doppler shift.* There is a Doppler effect whenever there is relative motion between the source and the receiver of the sound. The frequency of the sound emitted from a source moving toward an observer seems higher than the actual transmitted frequency. If the source is moving away from the observer, the perceived frequency is lower than the transmitted frequency. A classic example of this change in frequency or Doppler shift is the changing pitch of the ambulance siren, which is higher as the ambulance approaches, and lower as the ambulance moves away.
- A continuous wave (CW) Doppler is used for this study. Blood is the moving target and the transducer is the stationary source. Depending on the direction of flow relative to the Doppler beam, the reflected frequency is higher or lower than the transmitted frequency (Doppler shift). There are two crystals in a CW Doppler, one that transmits the signal and one that receives the reflected frequencies. See figure 4-1.
- The Doppler probe (transducer) must be positioned on the long axis of the vessel. An angle of insonation of 45–60 degrees to the skin surface is usually appropriate for this study. This angle usually works fairly well since most peripheral major arteries are located parallel to the skin.

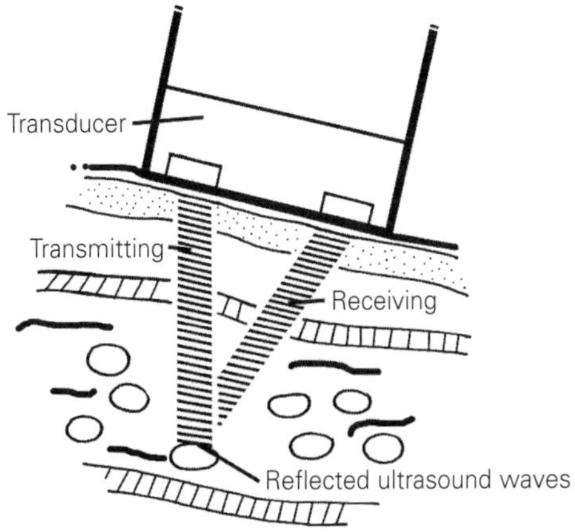

Figure 4-1. Continuous wave Doppler has two crystals. One transmits the signals; the second crystal receives the reflected signals. The transducer is the stationary source, and the blood is the moving target.

5 Technique:

- Types of Doppler velocimetry:

  Auditory: Processes the Doppler signal as sound. It has the advantage of containing all Doppler frequencies with the exception of those extreme frequencies removed by filtering. A trained observer can easily distinguish normal signals from those received proximal to, within, or distal to a stenosis.

  Note: A higher-pitched signal can mean that the probe angle is very acute to the vessel angle, or it can indicate a significant arterial obstruction.

  Analog: This method employs a zero-crossing frequency meter to display the signals graphically on a strip-chart recorder. The circuitry counts every time the input signal crosses the zero baseline within a specific time span. Because the number of times the sound waves oscillate each second varies (e.g., high-frequency waves have many oscillations while low-frequency waves have few) and because the direction of blood flow varies during the cardiac cycle, the machine *estimates* the frequency of the reflected signal and displays it. The horizontal axis of the tracing represents time while the vertical axis represents the amplitude of the Doppler-shifted frequencies.

  Although analog recording has acceptable accuracy, it is not as sensitive as spectral analysis. Drawbacks include noise, less sensitivity, and a tendency to underestimate high velocities and to overestimate low velocities. One example of a Doppler analog waveform appears in figure 4-2. Most equipment self-calibrates when the system is activated.

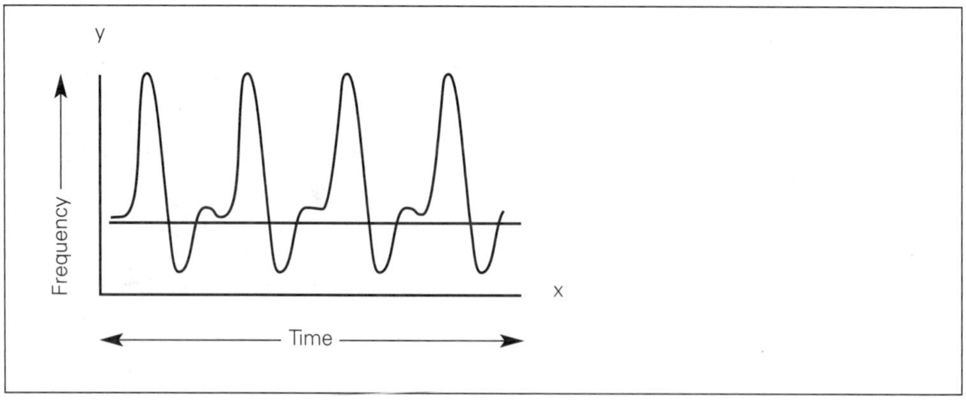

Figure 4-2. Analog Doppler waveform display. This is a normal triphasic signal. X-axis = time, Y-axis = frequency.

Spectral analysis: This method displays frequency (or velocity) on the vertical axis, time on the horizontal axis, and the amplitude of backscattered signals at any given frequency and time. It has the advantage of showing the amplitudes at all frequencies, but is free of many of the drawbacks previously described. It is more sensitive to displaying multiple frequencies. An example of a Doppler signal using spectral analysis appears in figure 4-3. Figure 4-4 further demonstrates the differences between an analog Doppler waveform and a spectral waveform.

- Acoustic gel is applied to the site to be evaluated.
- An 8–10 MHz Doppler probe is utilized.
- For the upper extremities, Doppler velocity waveforms are recorded from the following arteries bilaterally:

Subclavian

Axillary (axilla)

Brachial (at elbow—antecubital fossa)

Radial (lateral wrist, thumb side)

Ulnar (medial wrist, 5th finger side)

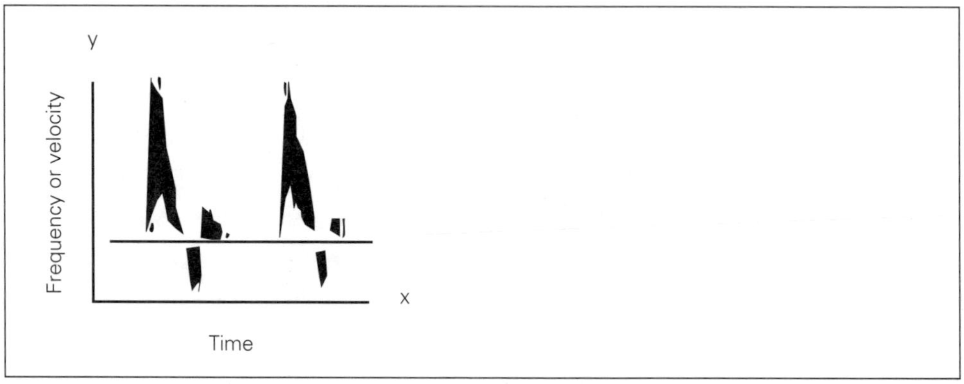

Figure 4-3. Spectral analysis of the same triphasic signal displayed in figure 4-2.

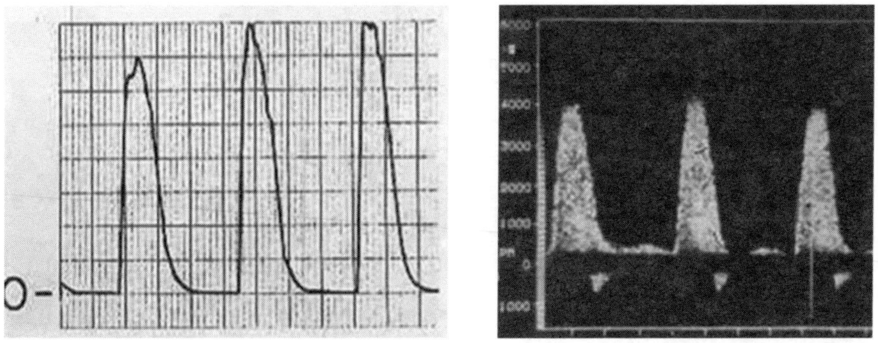

A B

Figure 4-4. Analog and spectral Doppler waveforms contrasted. **A** Analog waveform from a common femoral artery with an absence of reversal. **B** The Doppler spectral waveform has much more frequency content; because of this increased sensitivity, it documents the reversal of flow. This particular Doppler signal is triphasic in quality.

- For the lower extremities, Doppler velocity waveforms are recorded from the following arteries bilaterally:

  Common femoral (CFA) (groin level)

  Superficial femoral (SFA) (mid thigh, approximately)

  Popliteal (figures 4-5 and 4-6)

  Posterior tibial (PTA) (medial malleolus)

  Dorsalis pedis (DPA) (top of foot)

  Peroneal (if necessary) (lateral malleolus)

- Auditory signals are obtained. If the examiner is using a headset, the right earphone provides forward (antegrade) flow signals, while the left earphone provides reverse (retrograde) flow signals.

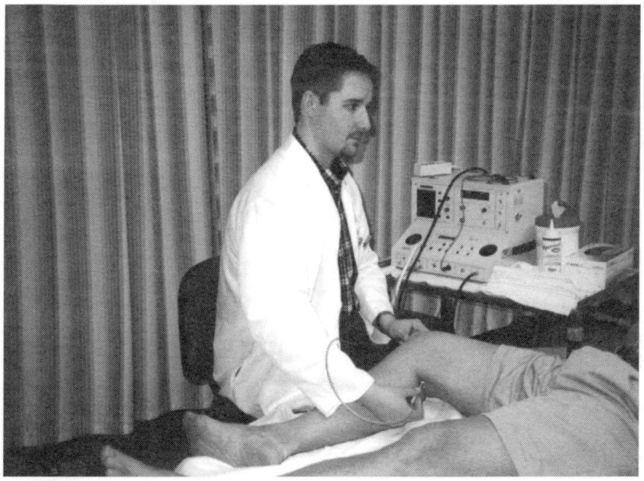

Figure 4-5. Obtaining Doppler waveforms requires optimal positioning of both the extremity and the probe to ensure a good signal. In this example the popliteal artery is being evaluated. The extremity is externally rotated with the knee slightly flexed to maximize patient comfort and to optimize the approach for the technologist. The technologist is seated and positioned close to the site under evaluation.

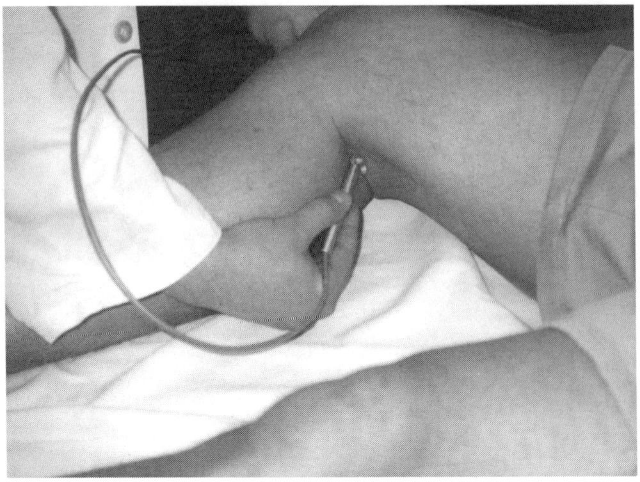

Figure 4-6.
When accessing the popliteal space, the technologist does not position the Doppler probe as far cephalad as usual because of the anatomic location of the popliteal artery. With the posterior approach, the artery actually curves toward the skin line, creating a ready-made angle for the Doppler probe, which is positioned at an angle of nearly 80–90 degrees to the skin.

- The qualities of the auditory signals and waveforms are observed and documented. These waveforms are usually combined with Doppler segmental pressures (see chapter 5).

- Potential sources of technical error when performing Doppler arterial survey include the following:

  Improper probe position

  Inadvertent probe motion

  Incorrect incident angle

  Inadequate amount of gel

  Excessive pressure on probe tip

  Insufficient period of rest before testing

6 Interpretation—Qualitative:

- Normal signals are triphasic or multiphasic with a rapid upstroke, sharp peak, rapid downstroke, a short peak below the baseline representing reverse flow, and resumption of forward flow (above baseline). (See figures 4-7A and 4-8.) The upper and lower extremity arteries have higher-resistance flow patterns.

- Abnormal signals are monophasic (figure 4-7C), nonpulsatile, or absent. In addition, biphasic signals can also be considered abnormal (figure 4-7B).

It is essential, however, to observe for deterioration of the Doppler signal quality from one level to the next (e.g., triphasic to biphasic or triphasic to monophasic).

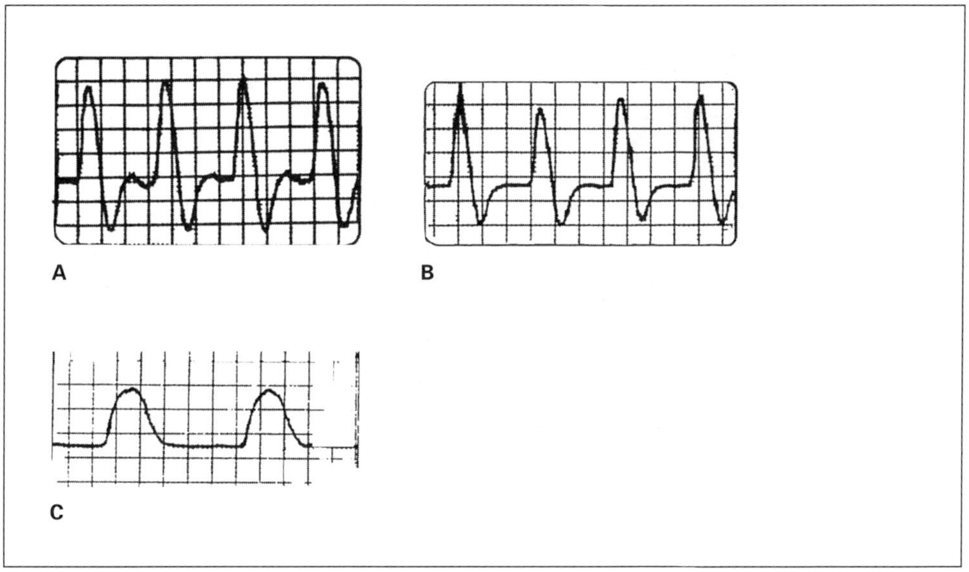

Figure 4-7. Peripheral arterial Doppler signals. **A** Triphasic (forward, reverse, forward). **B** Biphasic (forward, reverse). **C** Monophasic (forward only).

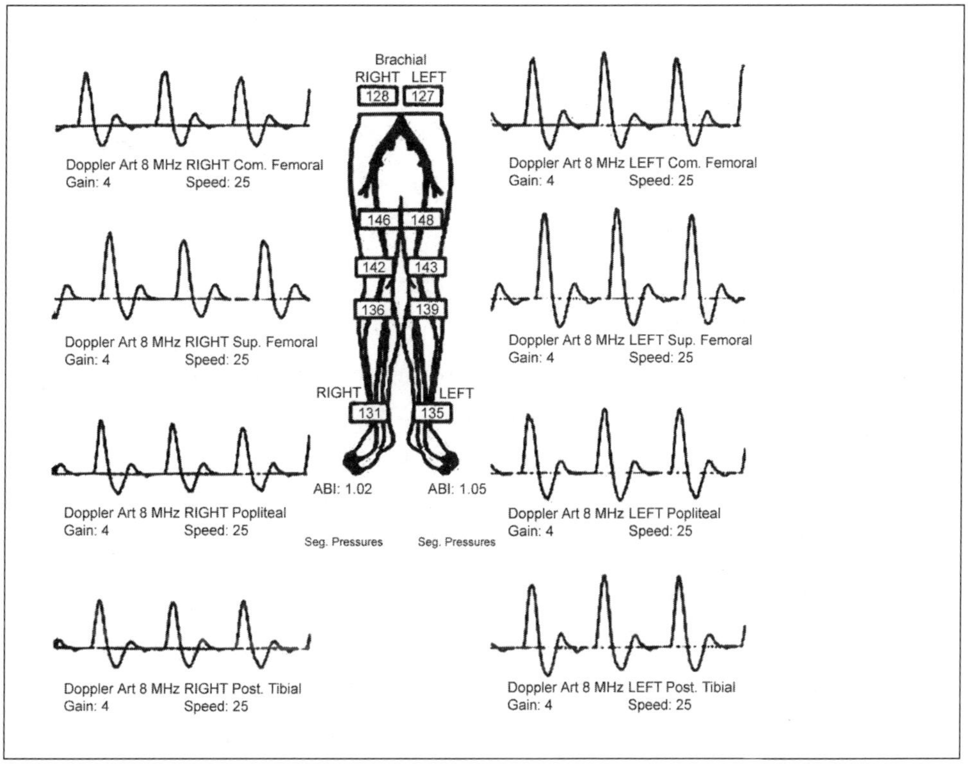

Figure 4-8. Normal peripheral arterial Doppler triphasic signals obtained from the common femoral, superficial femoral, popliteal and posterior tibial arteries.

- It is unwise to determine the degree of obstruction (stenosis versus occlusion) on the basis of waveforms alone because flow distal to well-collateralized occlusions can appear similar to flow distal to a stenosis.

- A monophasic and dampened signal can be obtained proximal to an obstruction as well as distal to it. (See figures 4-7C, 4-9.) In the absence of additional obstructions, the distal signal may normalize somewhat.

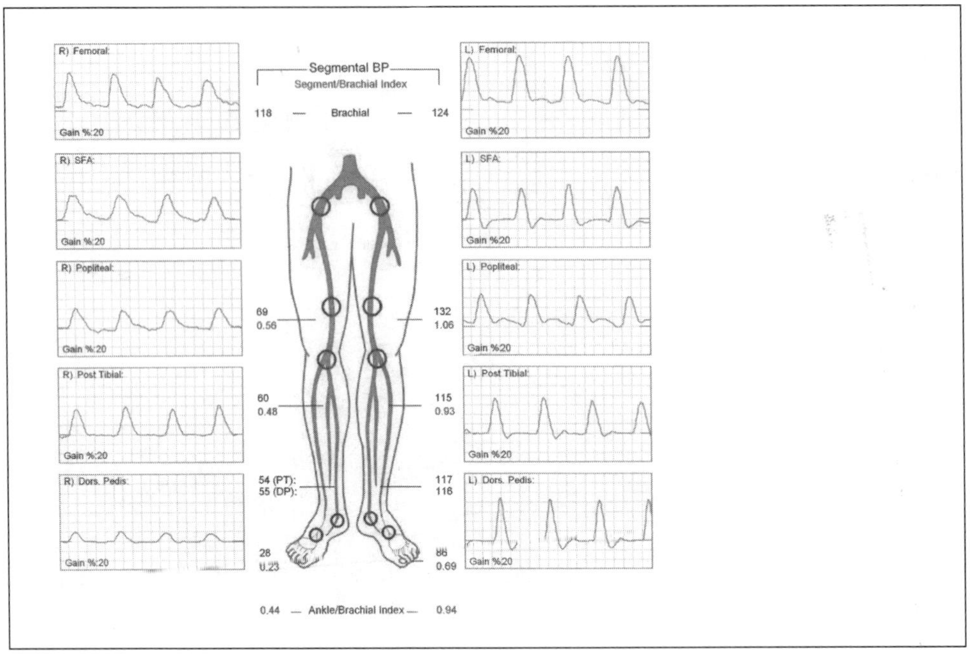

Figure 4-9.
On the right, abnormal peripheral arterial Doppler monophasic signals are obtained from the common femoral, superficial femoral (SFA), popliteal, dorsalis pedis, and posterior tibial arteries. Note the rounded appearance of the waveform peaks distal to the common femoral artery signals coupled with an abnormal ankle-brachial index of 0.44. On the left, biphasic Doppler signals are consistently present throughout, and there is a minimally abnormal (0.94) ankle-brachial index.

- Remember that vasodilatation of the distal vessels can reduce the pulsatility of the arterial signal, causing it to assume the quality of low-resistance vessels. The examples in figure 4-10 compare a normal high-resistance signal (figure 4-10A) to a signal from a vessel taking blood to a low-resistance, vasodilated bed (figure 4-10B.)
- Other abnormal examples include:

  Doppler signals obtained from a common femoral artery with a proximal iliac artery occlusion (figure 4-11A).

  Doppler signals obtained from a posterior tibial artery distal to a popliteal artery occlusion (figure 4-11B).
- Postexercise waveforms normally maintain or augment preexercise waveforms with all of the waveform components usually depicted above

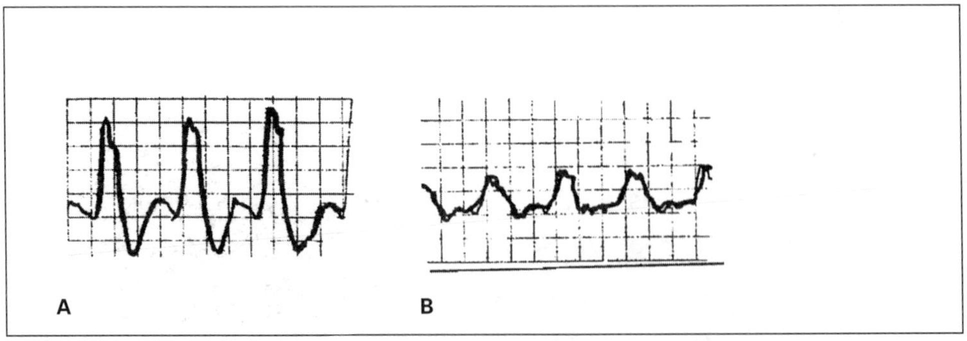

Figure 4-10.
Analog peripheral Doppler arterial signals. **A** High-resistance, normal, triphasic signal. **B** Lower-resistance Doppler signal. Note the absence of the reversed signal and the continuous diastolic quality.

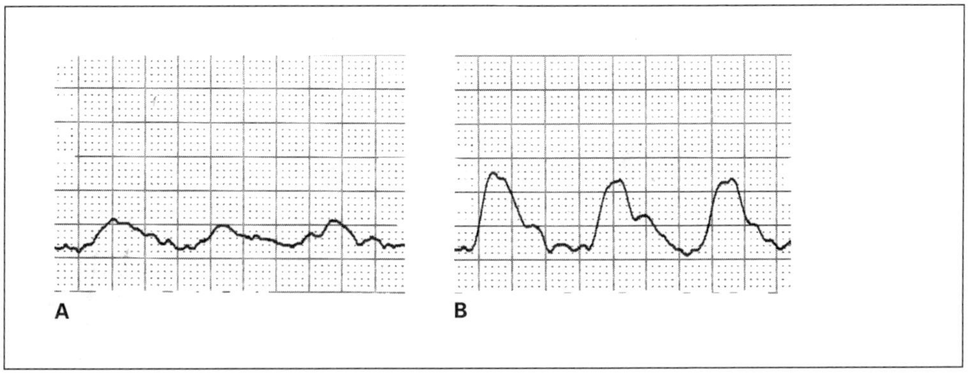

Figure 4-11. Abnormal peripheral arterial Doppler signals. **A** Note the poor quality of the signal (poor upslope and downslope, with a rounded peak). **B** In this example the signal strength is slightly better than the previous figure, but it is representative of arterial flow feeding a vasodilated arterial bed in the presence of proximal arterial obstruction.

the baseline as demonstrated in figure 4-12A. Abnormally, the following changes can occur (see figure 4-12B):

Slow upstroke with more rounded peak

Slow downstroke

No reverse component

As previously discussed, exercise produces a demand for blood flow to the muscles, which does cause an element of arterial vasodilation and—even in a normal postexercise state—low-resistance flow is observed (figure 4-12A). When this demand is not met, the distal peripheral vascular beds dilate in response to inadequate blood flow, thereby changing the quality of the arterial Doppler signals (figure 4-12B). **Note:** Usually just Doppler pressures are obtained postexercise.

● The inability to elicit Doppler signals may suggest occlusion, but the sensitivity to slow flow by the CW Doppler—which cannot detect velocities of less than 6 cm/sec—may not be adequate to detect flow in a preocclusive vessel. This preocclusive lesion may still have very slow velocities moving through it. This type of condition is known as a *string sign.* Some of the more sophisticated duplex/color flow systems have the capability to display these slow flows.

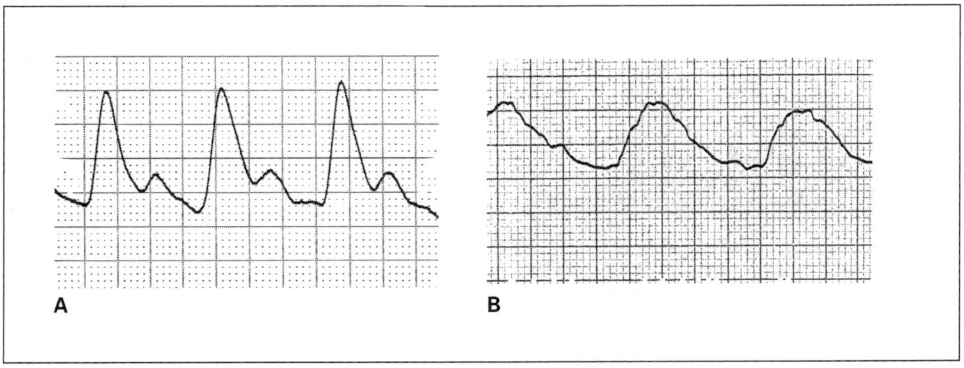

Figure 4-12. Peripheral arterial analog Doppler signals. **A** Postexercise waveforms maintain the same quality with all waveform components depicted above the baseline when compared to pre-exercise quality. **B** Abnormally, postexercise waveforms lose the phasic quality, becoming more monophasic and less resistant.

- Regarding the upper extremities:

  The subclavian artery is normally characterized by multiphasic, high-resistance flow. The qualities of its abnormal Doppler waveform are quite similar to those of lower extremity arteries with proximal or distal arterial obstructions.

  According to the work by Strandness and Sumner entitled *Hemodynamics for Surgeons*, the arteriovenous shunts in the skin of the fingertips cause the flow patterns in the hand to vary tremendously. It is not uncommon to see continuous low-resistance Doppler signals in the brachial, radial, and ulnar arteries of a patient who is relaxed and warm. If the patient's hand is cool, blood flow in the upper extremity arteries will assume qualities of high-resistance (pulsatile) flow.

- Troubleshooting:

  Problem: The recorder stylus is not recording any waveforms.
  Action: Check to make certain the proper test and probe (i.e., 8 MHz Doppler) have been selected.

  Problem: Bands of noise are displayed on the tracing. See figure 4-13.
  Action: This problem could be a 60 cycle noise and definitely a challenge to get rid of. This kind of noise occurs in spectral tracings more often than in analog tracings. Initially, increasing the filter on the Doppler system and/or reducing the gain (if a control is available) may help. Sometimes turning the system off and then on again helps, as does plugging the system into another outlet.

  Problem: No auditory signal is obtained, but there are recordable signals on the tracing.
  Action: Adjust the volume control and/or the headset connection.

  Problem: There is an auditory Doppler signal but no tracing.
  Action: Ascertain that the recorder is on and not in pause or freeze mode.

Figure 4-13.
Example of 60 cycle noise at baseline.

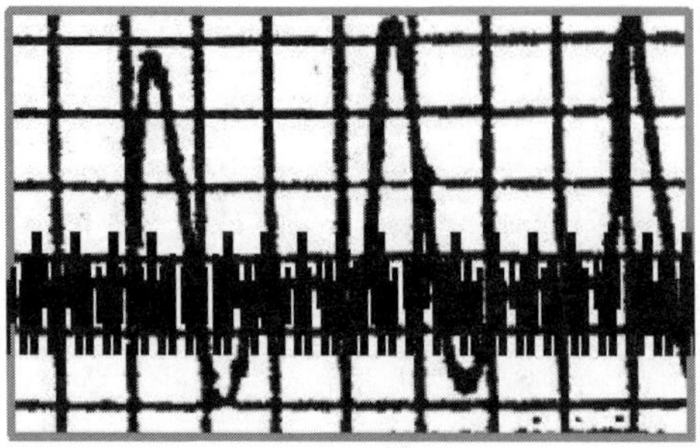

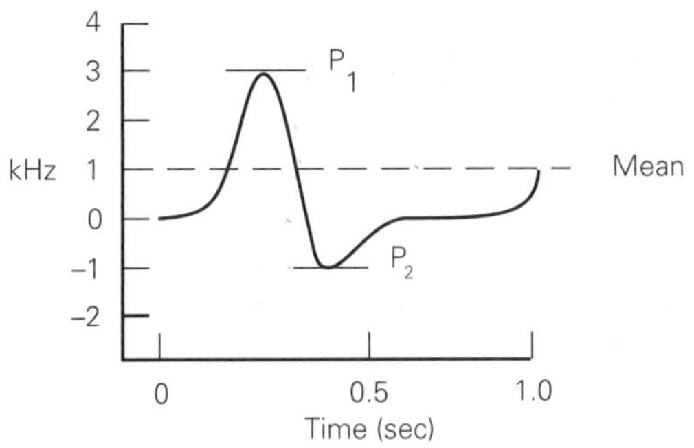

Figure 4-14. Elements of the pulsatility index.

7 Interpretation—Quantitative:

- *Pulsatility index* (PI) is calculated by dividing the peak-to-peak frequency (i.e., from P_1 to P_2 in figure 4-14) by the mean (average) frequency as depicted in the following equation and in figure 4-14. This ratio provides quantitative data, although it is independent of the beam-to-vessel angle when using handheld Doppler equipment. Spectral analysis is more sensitive than analog waveforms. See figure 4-15.

$$\frac{\text{Peak-to-Peak Frequency}}{\text{Mean Frequency}}$$

- This equation is applied to high-resistance beds. Normally the values of the index increase from the central to the peripheral arteries. *Examples:* A PI of greater than 5.5 is normal for the common femoral artery, while a normal PI for the popliteal artery is approximately 8.0. These values decrease in the presence of proximal occlusive disease. *Example:* A PI of less than 5.0 (some even say less than 4.0) in the common femoral artery with a patent superficial femoral artery, has a pretty good predictive value and indicates proximal aortoiliac occlusive disease. If the superficial femoral artery is occluded, however, the same reduced PI finding is not diagnostic.

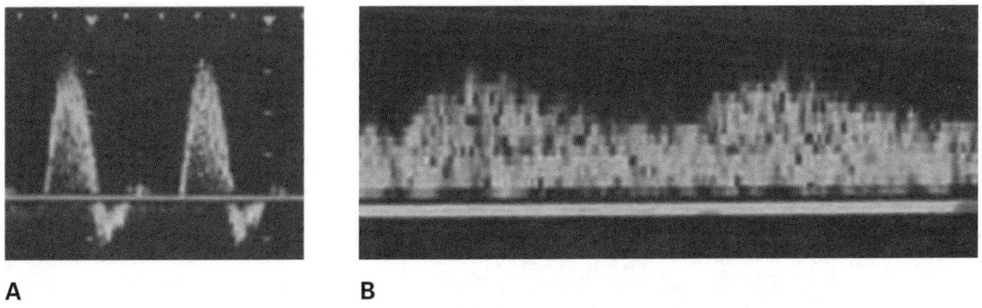

Figure 4-15. **A** Normal triphasic Doppler signal that would provide a normal PI as evidenced by figure 4-14. **B** Abnormal Doppler waveform that would provide an abnormal PI value based on a rounded peak and the lack of flow reversal.

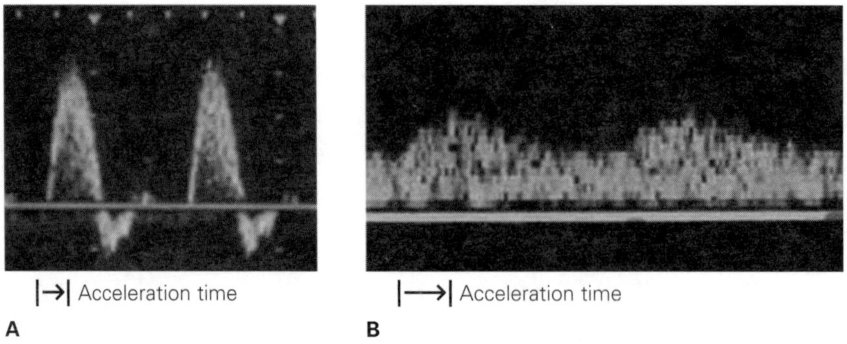

Figure 4-16. Acceleration time represents the time period between the onset of systole and the point of maximum peak velocity. **A** The quick systolic upslope of this waveform from a normal common femoral artery represents normal acceleration time. **B** In this common femoral artery the slower upslope from onset of systole to maximal peak is abnormal.

- *Inverse damping factor* is the ratio of the distal pulsatility index to the proximal pulsatility index of an arterial segment. It indicates the degree to which the wave is dampened as it moves through an arterial segment. *Example:* Superficial femoral artery occlusion or severe stenosis is usually present when the inverse femoral popliteal damping factor is less than 0.9. Normally it would be 0.9–1.1.

- *Acceleration time* helps to differentiate inflow from outflow disease. It is based on the principle that arterial obstruction proximal to the site of the Doppler probe lengthens the time between the onset of systolic flow to the point of maximum peak in waveforms at the probe site. Acceleration time is not prolonged when there is disease distal to the probe. The acceleration time is applied to those signals evaluated by spectral analysis because it is necessary to maximize sensitivity and minimize artifacts.

  False positives can occur with technical errors (e.g., large Doppler angle such as 80 degrees dampen the Doppler signal qualities) and in the presence of poor cardiac output since the Doppler flow signals will already be attenuated with the waveform depicting slow upstroke, rounded peak, and slow downslope.

  In general, an acceleration time of $\geq$ 133 msec suggests the presence of significant iliac disease. It is essential, however, to use this criterion with waveform evaluation. To compare Doppler waveforms associated with normal and abnormal acceleration times, refer to figure 4-16. Normal acceleration time is characterized by a rapid upslope; abnormal acceleration time is characterized by a slowed acceleration upslope.

- *Transit time*: Systole should be simultaneously evident at a specific site bilaterally. A delay on one side may indicate a more proximal occlusive process. Always compare signals bilaterally at the same site.

Doppler Segmental Pressures—Lower Extremities

CHAPTER 5

1 Capabilities:

- Identifies the presence and severity of arterial occlusive disease.
- Provides an objective baseline to follow the progression of the disease process and/or the postoperative course.
- Objectively evaluates the treatment plan.
- Results are usually combined with Doppler velocity waveform analysis or volume pulse waveforms.

2 Limitations:

- Cannot discriminate between stenosis and occlusion and cannot precisely localize the area of obstruction, although it can identify a general location. Difficult to discriminate between common femoral and external iliac disease.
- May show falsely elevated Doppler pressures in those patients with calcified vessels—diabetics and patients with end stage renal disease, for example.
- May show decreased ankle/brachial indices after exercise in patients with uncompensated congestive heart failure.
- May show artifactually elevated high-thigh pressures when the thigh cuff is too narrow.
- Multilevel disease makes it difficult to interpret segmental pressures.
- See also previous limitations for the Doppler velocity waveforms evaluation on page 51.

3 Patient preparation and positioning:

- It is essential that the patient has rested at least 20 minutes prior to the test, especially if he or she has vascular disease. Just walking in for the test can reduce the amount of blood flow to the extremities if in fact the patient does have vascular claudication. Testing the patient without this resting period would lead to initially poorer results at rest and then, after the patient is walked during the test process, an apparently less significant pressure decrease than one would expect in the patient with disease.

- The patient is positioned supine with the extremities at the same level as the heart so that hydrostatic pressure (which would affect a sitting patient) cannot falsely elevate the blood pressure measurement. The head of the bed can be elevated slightly, and the patient's head can rest on a pillow.
- The patient's hip is externally rotated with the knee slightly bent to help facilitate the lower extremity evaluation.
- Alternative positions for lower extremity examinations include right or left lateral decubitus (patient on his or her side) or prone for access to the popliteal space.

4 Physical principles:

- The *Doppler effect:* When a wave is reflected from a moving target, the frequency of the wave received differs from that of the transmitted wave. This difference is known as the *Doppler shift.* There is a Doppler effect whenever there is relative motion between the source and the receiver of the sound.
- A continuous wave (CW) Doppler is used for this study. Blood is the moving target and the transducer the stationary source. Depending on direction of flow relative to the Doppler beam, the reflected frequency is higher or lower than the transmitted frequency (Doppler shift). There are two crystals in a CW Doppler, one that transmits the signal and one that receives the reflected frequencies.
- The Doppler probe (transducer) must be positioned on the long axis of the vessel. An angle of insonation of 45–60 degrees to the skin surface is usually appropriate for this study. This angle tends to work fairly well since most peripheral major arteries are located parallel to the skin.
- Cuff artifact resulting in inaccurate pressure measurement can occur if the cuff size is inappropriate for the size of the extremity. If the cuff is too large, blood pressure is artifactually lower; if the cuff is too narrow, the blood pressure is artifactually higher.

5 Technique:

- Bilateral brachial blood pressures are obtained using blood pressure cuffs 12 × 40 cm bladders. An 8–10 MHz Doppler is utilized.

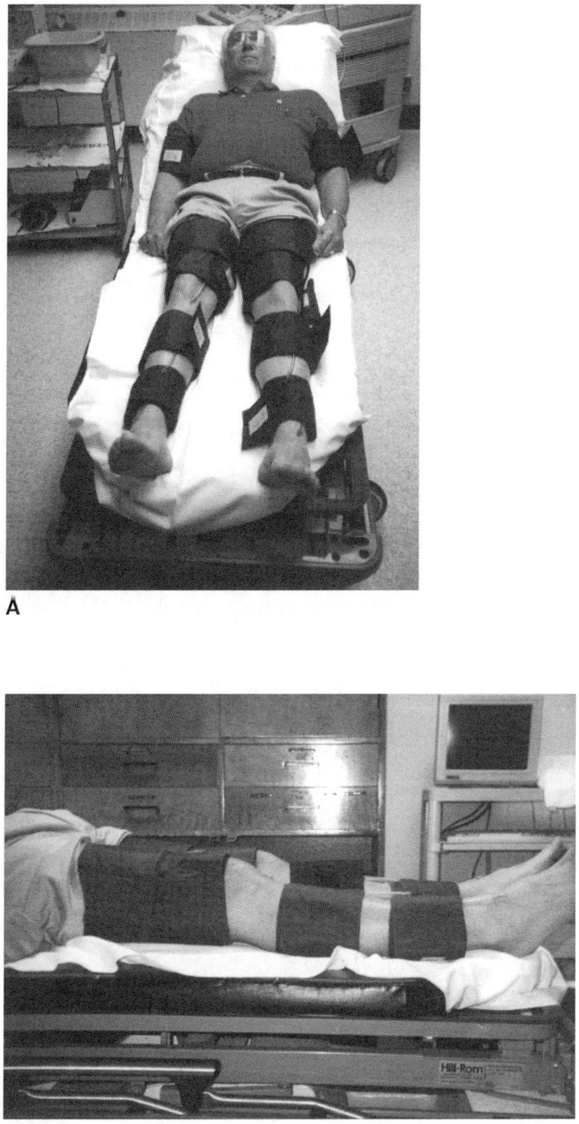

Figure 5-1. Cuff placement for the four-cuff method of obtaining segmental Doppler pressures. **A** Two narrow thigh cuffs are used on the thigh, one at the high thigh and one at the low thigh. The low-thigh position is also referred to as *above knee*. **B** It is essential that the cuffs not be placed around a bony prominence. Here the above-knee cuff is somewhat wrapped around the proximal knee. The ankle cuff must also be placed so that it is above the medial malleolus.

- Appropriately sized blood pressure cuffs (12 × 40 cm bladders) are utilized as follows (four-cuff technique): high thigh, low thigh (above knee proximal to patella), below knee (just distal to the tibial tubercle), and at the ankle as depicted in figure 5-1. Cuff bladder sizes vary somewhat; these are average bladder dimensions.

  All cuffs should be placed "straight" on the extremity site and fit snugly so that bladder inflation quickly transmits the pressure into the tissue. A loose cuff creates additional space between the skin and the bladder, necessitating additional inflation and possibly leading to falsely elevated pressures.

The width of the pneumatic cuff should be at least 20% (or 1.2 times) greater than the diameter of the limb so that the artery under evaluation can be compressed when the bladder is inflated. Tissue, not bony structures, must be compressed; otherwise the pressures obtained may be falsely elevated.

If the cuff bladder is too wide for the extremity segment being evaluated, the blood pressure measurement will be falsely low. Conversely, if the cuff bladder is too narrow for the extremity segment, the blood pressure will be falsely elevated.

The relationship between the size of the extremity and the size of the cuff bladder encircling the extremity is critical for accurate blood pressure measurements. A falsely elevated blood pressure is expected when the narrow cuffs are purposely used on the thigh in order to obtain two separate blood pressure readings. However, a narrow cuff secondary to body habitus will also produce a falsely elevated blood pressure reading.

The use of two thigh cuffs is advantageous in that it provides proximal and distal thigh pressure measurements. The disadvantage is that artifactually elevated pressures are obtained (i.e., the pressure at the high-thigh level is normally 30 mmHg or higher than the highest brachial pressure). This occurs because the thigh cuff width is much too narrow for the usually large girth of the thigh.

The bladder of the cuff should be placed over the artery. This is especially important when the bladder does not encircle the limb. Ideally, the bladder of the cuff should encircle the limb's circumference.

Another technique—the three-cuff technique—involves the use of one large cuff placed as high as possible on the thigh. Although one cuff (19 × 40 cm) satisfies the recommended width-girth relationship, it is so wide that only one can fit on the thigh. The advantage is that a more accurate thigh pressure is obtained (normally this thigh pressure is very similar to the higher brachial pressure). A contoured thigh cuff is optimal, based on the shape of the thigh. The remaining two cuffs in the three-cuff technique are applied to the calf and ankle, as depicted in figure 5-2.

- If not already evaluated, Doppler signals are obtained and evaluated for the posterior tibial artery, dorsalis pedis artery, and peroneal artery (if needed). A 45–60 degree angle of insonation to the vessel is optimal and necessary to obtain the clearest and best possible signal. Behind the knee (popliteal artery) the angle of insonation in relation to the skin surface may well be 90 degrees because of vessel angle. Probe angulation may be required.

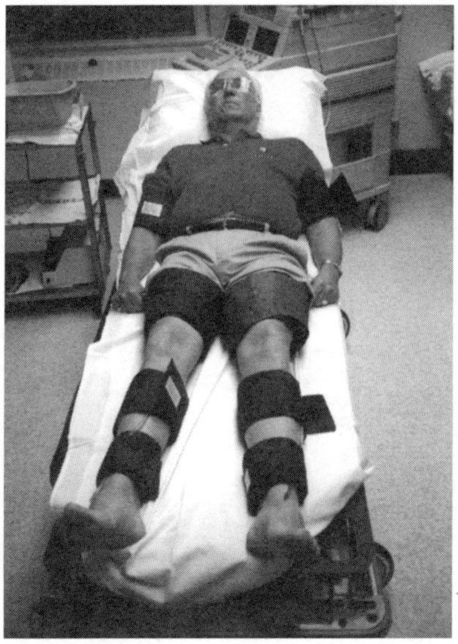

Figure 5-2. Cuff placement for the three-cuff method of obtaining segmental Doppler pressures. In this example one large thigh cuff is placed high on the thigh.

● Segmental Doppler pressures are obtained bilaterally (one leg at a time) at the sites illustrated in figure 5-3 and in the order indicated below using a handheld sphygmomanometer with manual inflation or a computerized system with automatic inflation and digital display.

1. Ankle (use the posterior tibial artery and dorsalis pedis artery). See figures 5-4 and 5-5. **Note:** Usually the higher of the two pedal Doppler pressures is used to obtain the remainder of the segmental pressures in the leg beginning with the calf level.

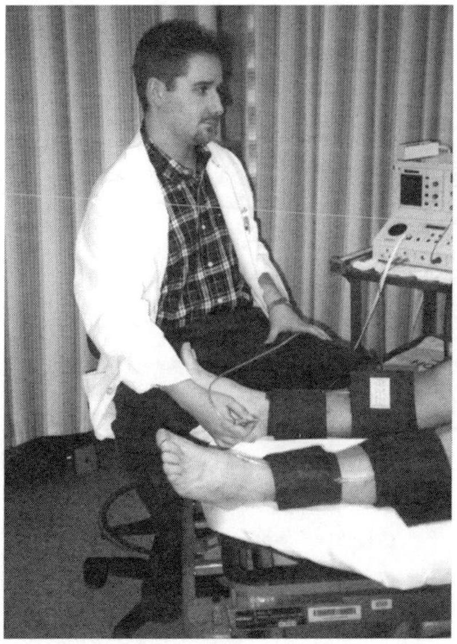

Figure 5-3. Position of technologist during a segmental Doppler pressures examination. The technologist should be as comfortable as possible and maintain a secure hold on the probe to obtain clear Doppler signals. Normally, all of the segmental pressures in the extremity are obtained from the posterior tibial artery (PTA, as shown here) or dorsalis pedis artery (DPA), whichever has the highest ankle pressure.

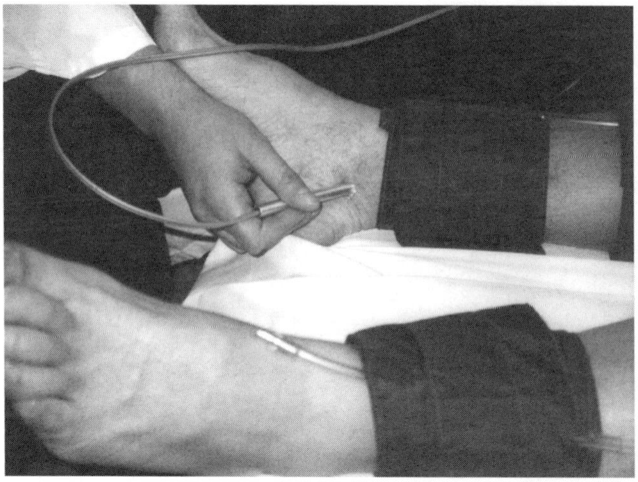

Figure 5-4. Optimal placement of the Doppler probe is essential for obtaining the strongest and clearest Doppler signal, in this case from the PTA. The probe is held at an angle of approximately 45 degrees to the skin, adequate acoustic coupling gel is applied to the skin, and the probe is positioned so that arterial blood flow is antegrade (toward the probe).

2. Calf/below knee (BK) (use the higher of the posterior tibial artery and dorsalis pedis artery pressures; some examiners use both sites to help differentiate popliteal from proximal tibial disease).
3. Above knee (AK) (same as above, although the popliteal artery can be used if these Doppler signals are difficult to obtain).
4. High thigh (HT) (same as above knee).

Notes:

- It is important to start at the ankle level and then to move more proximally one level at a time to eliminate the possibility of underestimating the systolic pressure measurement. For example, the high-thigh cuff is

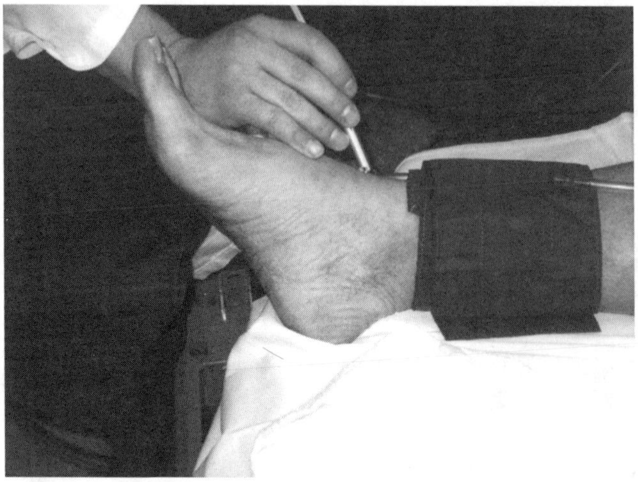

Figure 5-5. The dorsalis pedis artery (DPA) is difficult to assess because of its location on the foot. There must be enough coupling gel on the skin so that the probe tip can simply lie on top of the gel at an angle of 45 degrees to the skin and with the probe angled cephalad so that blood flow is antegrade (toward the probe). Lack of adequate gel creates an uncoupled space between the skin and the tip of the probe, reducing the ability to obtain an optimal Doppler signal.

inflated to above systole and then slowly deflated to determine the blood pressure. If the cuff is quickly deflated, immediately followed by inflation of the next (above-knee) cuff, there is not enough time for the arterial blood to completely normalize in the leg. So the blood pressure obtained in this way would likely be falsely lower than it should be.

- When a full extremity study is not required, the brachial and ankle pressures are obtained bilaterally. The posterior tibial artery and dorsalis pedis artery are used for the ankle pressures. With this information, the ankle/brachial index (ABI) is calculated. The peroneal artery may be used if one or both of the previous sites are unavailable. (Digits are discussed below in this section.)

- Complete cessation of blood flow should be accomplished by inflating the cuffs 20–30 mmHg beyond the last audible Doppler arterial signal.

- In order to determine how high to take the inflation pressure, note the higher brachial systolic pressure and then increase inflation pressure 20–30 mmHg higher than it is. (The higher brachial pressure is used because if there were a subclavian stenosis present unilaterally, there would be a systolic pressure difference of 15–20 mmHg or greater, with the affected side being lower.)

- If pressure measurements need to be repeated, the cuff should be fully deflated for about a minute prior to repeat inflation so arterial flow stabilizes; otherwise, the segmental pressures can be underestimated.

- The systolic pressure is recorded as the pressure at which the first audible Doppler arterial signal returns. This is similar to listening for the first audible (low-pitched) sound when taking a brachial blood pressure with a stethoscope.

- Blood pressure and comments regarding difficulty with the study should be documented (e.g., inability to clearly discriminate a poor arterial signal from a venous signal at the PTA or DPA site, which could produce unreliable results).

6 Interpretation:

- The *ankle/brachial index* (ABI) is calculated by dividing the ankle pressure by the higher of the two brachial pressures. Additional terms for this index include *ankle/arm pressure index* (API) and *ankle/arm index* (AAI), both of which utilize the same formula. The interpretation criteria listed in table 5-1 do vary somewhat in the literature.

Note: Incompressible vessels have falsely elevated and therefore inaccurate pressures. An ABI of > 1.3–1.5 is considered to be the result of

Table 5-1. Normal and abnormal ankle/brachial (ABI) indices.

Ankle/brachial index	***Findings***
> 1.0	Normal.
> 0.9–1.0	Asymptomatic obstructive disease. Also referred to as minimal arterial disease.
0.5–0.9	Claudication.
< 0.5	Rest pain; severe arterial disease.

incompressible vessels, as are pressures that are not reproducible (e.g., 158 mmHg one time, but 112 mmHg when repeated) and/or a large difference in the pressure observed when the signal disappears (suprasystolic) and when pulsations are evident with cuff deflation.

- Normally, the ankle systolic pressure is the same as or greater than the highest brachial blood pressure.
- Remember the limitation of Doppler segmental pressures in patients with incompressible vessels (e.g., diabetics): That is, pressures are falsely elevated and therefore inaccurate.
- The Doppler segmental pressure study is usually combined with Doppler velocity waveforms or plethysmographic waveforms (e.g., pulse volume recordings or volume pulse waveforms).
- Additional observations about ABIs:

  Patients with rest pain usually have ABIs between 0.25 and 0.5.

  There are those who feel that an absolute ankle pressure (less than 50 mmHg approximately) rather than the ABI of 0.5 is better at predicting symptoms at rest.

  Strandness' work suggests that an ABI ≥ 0.5 often represents single-segment involvement and that lesser values are more indicative of multiple lesions.

- A decrease in pressure of > 30 mmHg between two consecutive levels is considered significant and would suggest significant obstruction. **Note:** Other sources indicate that a decrease in pressure of ≥ 20 mmHg suggests significant obstruction.
- A horizontal difference of 20–30 mmHg or more suggests obstructive disease at or above the level in the leg with the lower pressure. Although useful, horizontal pressure differences are less important when compared to vertical segmental pressure differences. Also, horizontal pressure differences are no longer valid once proximal disease has been detected.

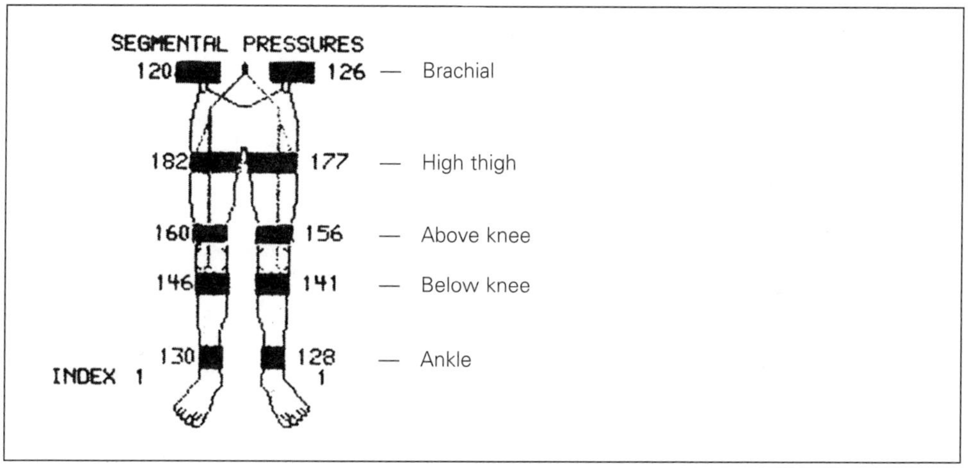

Figure 5-6.
No significant pressure gradients are observed in this normal patient.

● Normal segmental pressure levels are presented in figure 5-6 and examples of abnormality in figures 5-7 and 5-8.

● As limb girth increases from the ankle to the thigh, pressure measurements also increase. In the four-cuff method, the high-thigh pressure in the average-size limb is normally at least 30 mmHg greater than the highest brachial pressure. The above-knee and below-knee systolic pressures should be the same as or greater than the brachial systolic pressure.

● Thigh pressure indexes (thigh pressure/higher brachial pressure) are normally greater than 1.2, while 0.8–1.2 suggests aortoiliac disease and < 0.8 indicates that proximal occlusion is likely.

● With the three-cuff technique the large, single thigh cuff segmental pressure is normally similar to the highest brachial pressure. Because the single thigh cuff has a longer bladder, the transmission of pressure to the underlying limb segment is more uniform and accurate. Although this wide-girth cuff usually yields a more accurate pressure measurement, it does not allow differentiation of proximal and distal thigh pressures.

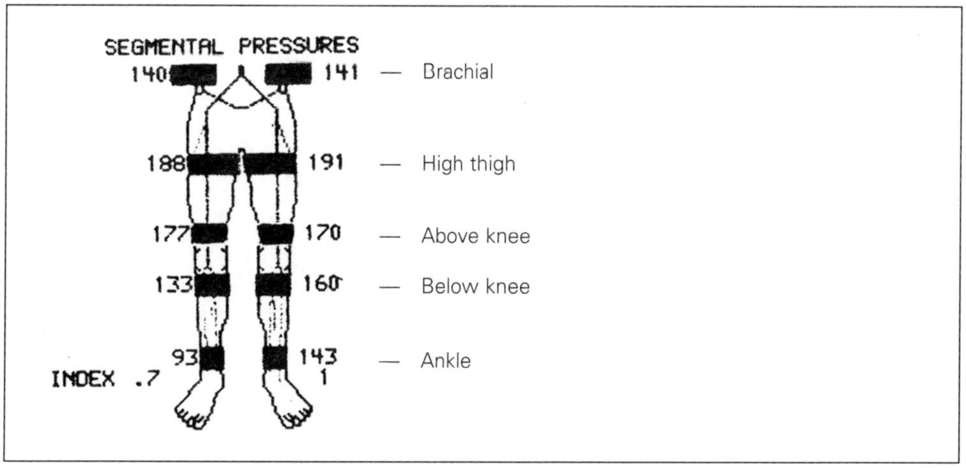

Figure 5-7.
There is a > 30 mmHg pressure difference between the right above-knee (AK) level to the calf and calf to ankle. This suggests right femoral-popliteal and tibial vessel obstructive disease.

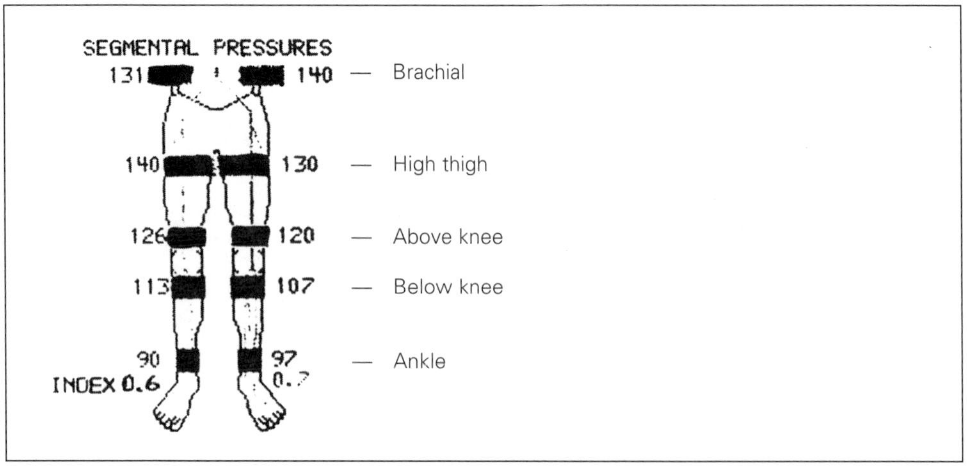

Figure 5-8. There are reduced pressures at both high-thigh levels without further pressure gradients bilaterally. The high-thigh pressure levels should be approximately 30 mmHg higher than the highest brachial pressure. This suggests an aortoiliac obstruction.

- Pressure levels for foot ulcer healing:

  It has been shown that toe pressures of 30 mmHg or less are evident in cases of foot and toe ulcers that fail to heal.

  Ankle pressures are not always reliable (in diabetic patients, for example). Toe pressures are more reliable and less likely to be misleading.

Doppler Segmental Pressures—Lower Extremity Exercise and Reactive Hyperemia

1 Capabilities:

Resting values are compared to those obtained after exercise (if no limitations or contraindications exist) or after reactive hyperemia (an alternate means of stressing the peripheral circulatory system) for the following reasons:

- To help differentiate between true claudication and pseudoclaudication.
- To determine the presence or absence of collaterals.

Treadmill testing is the preferable test because it produces a physiologic stress that reproduces a patient's ischemic symptoms.

2 Contraindications and limitations:

- Shortness of breath
- Hypertension
- Cardiac problems
- Stroke
- Walking problems*

* Some labs do not perform the exercise component if resting ABI < 0.5, but this criterion can vary.

3 Technique for exercise testing:

- After resting values are obtained, the patient walks on a constant-load treadmill at ≤12% elevation and 1.5 MPH for a maximum of 5 minutes or until his or her symptoms increase to such severity that the patient must stop. A standardized exercise such as treadmill can determine maximum walking time and distance walked while documenting symptoms, if any. However, some patients who complain of claudication have to stop walking because of conditions other than leg pain (e.g., shortness of breath).

Because exercise usually increases the patient's blood pressure, ankle/brachial index calculations—rather than absolute ankle blood pressure measurements—are obtained following exercise.*

- Duration of walking, miles/hour, and onset, location, and progression of symptoms are documented.

- Postexercise Doppler pressures are obtained from the higher brachial blood pressure (taken at rest) and both ankles using the same technique as previously described for patients at rest. To capture the maximum pressure drop, pressure is first measured in the more symptomatic lower extremity ankle, then the contralateral ankle, and finally the arm with the higher brachial blood pressure.

- Postexercise ABIs are obtained immediately and then every two minutes until preexercise pressures are once again attained, a process that can take as long as 20 minutes. Although many diagnostic facilities are interested in only the immediate pressure change after exercise and stop there, other protocols require monitoring pressures for up to 10–12 minutes to determine recovery time. Normally, pressures increase after exercise. Abnormally, they may decrease minimally or severely.

4 Interpretation for exercise testing:

- The length of time it takes to recover, combined with the symptoms, duration of exercise, and pressure changes (if any) from pre- to post-exercise status, form the basis of the interpretation.

- Ankle pressures that drop to low or unrecordable levels immediately after exercise and then increase to resting levels in 2 to 6 minutes suggest obstruction at a single level.

- When such ankle pressures remain reduced or unrecordable for up to 12 minutes or more, multilevel obstructions are usually present.

* Other forms of exercise include walking a predetermined distance in the hallway or toe raises. Toe raises have limitations such as inability to determine maximum distance walked over a period of time and inability to elicit an overall cardiovascular response to determine whether the patient stops walking because of claudication or for other reasons.

Patients with significant arterial obstructive disease have a drop in pressure proportional to the severity of disease.

5 Technique for reactive hyperemia testing:

- This alternate method for stressing the peripheral circulation may be indicated when the patient cannot walk long enough, uses a cane or walker, has pulmonary problems, has poor cardiac status, or other situations. *Example:* One situation in which reactive hyperemia could be used is in the patient who has known severe arterial disease in one extremity and moderate arterial disease in the other. With walking, the extremity with the severe disease will force the patient to stop sooner, making it impossible to determine what the actual response would be in the other extremity. The application of reactive hyperemia in this situation would essentially test both extremities simultaneously for a truer comparison.
- Patients with an extremity bypass graft or stent placement in a thigh artery would most likely not be candidates for this procedure because of the high pressure required in the thigh cuff.
- Thigh cuffs (19 × 40 cm) are inflated to suprasystolic pressure levels (usually 20–30 mmHg above the higher brachial blood pressure) with the pressure maintained for 3 to 5 minutes. This is a difficult process for the patient because of the discomfort.
- This produces ischemia and ultimately vasodilation distal to the cuff.
- Upon release of cuff occlusion, the changes in ankle pressures are similar to those observed after exercise. There is a difference, however: After treadmill exercise, ankle systolic pressures in normal limbs do not decrease, whereas a transient pressure decrease in the range of 17% to 34% does occur at the ankles of normal limbs after reactive hyperemia.

6 Interpretation for reactive hyperemia testing:

- Patients who have single-level disease experience a less than 50% drop in ankle pressure.
- Patients with multilevel arterial disease experience a pressure drop greater than 50%.
- Treadmill testing is considered by many to be the preferable test because it produces a "physiologic stress" that reproduces a patient's ischemic symptoms.

Doppler Segmental Pressures—Upper Extremities

CHAPTER 6

1 Capabilities:

The capabilities are the same as those for Doppler segmental pressures in the lower extremities:

- Identifies the presence and severity of arterial occlusive disease.
- Provides an objective baseline to follow the progression of the disease process and/or the postoperative course.
- Objectively evaluates the treatment plan.
- Results are usually combined with Doppler velocity waveform analysis or volume pulse plethysmographic waveforms.

2 Limitations:

- Cannot discriminate between stenosis and occlusion and cannot precisely localize the area of obstruction, although it can identify a general location.
- May show falsely elevated Doppler pressures in patients with calcified vessels—diabetics and patients with end-stage renal disease, for example, although these findings would be unusual in the upper extremities of such patients.
- As noted above under *Doppler Waveform Analysis in the Upper and Lower Extremities* (see page 52 for additional information about the Doppler effect, probe position, etc.), limitations also may include the following:

  Patients with casts or extensive bandages that cannot be removed.

  Waveforms may be affected by ambient temperature.

This test cannot discriminate stenosis from occlusion or precisely localize the obstruction.

The test is technologist-dependent. It is essential that the technologist understands the importance of maintaining an adequate Doppler angle.

3 Patient positioning:

- The arms should be at the patient's sides in a relaxed state.
- The patient's head can be slightly elevated.
- The cuffs should be applied snugly over the muscular portion of the arm (upper and forearm).

4 Physical principles:

- The *Doppler effect*: When a wave is reflected from a moving target, the frequency of the wave received differs from that of the transmitted wave. This difference is known as the *Doppler shift*. There is a Doppler effect whenever there is relative motion between the source and the receiver of the sound.
- A continuous wave (CW) Doppler is used for this study. Blood is the moving target and the transducer is the stationary source. Depending on the direction of flow relative to the Doppler beam, the reflected frequency is higher or lower than the transmitted frequency (Doppler shift). There are two crystals in a CW Doppler, one transmitting the signal and the other receiving the reflected frequencies.
- The Doppler probe (transducer) must be positioned on the long axis of the vessel. An angle of insonation of 45–60 degrees to the skin surface is usually appropriate for this study since most major peripheral arteries are parallel to the skin.

5 Technique:

Approximately sized blood pressure cuffs are utilized as follows:

- A cuff with a 12 × 40 cm bladder is placed snugly on the upper arm and a cuff with a 10 × 40 cm bladder on the forearm bilaterally. The cuff size may vary according to the size of the extremity.
- All cuffs should be placed "straight" on the extremity site and fit snugly so that bladder inflation quickly transmits the pressure into the tissue. A loose cuff creates space between the skin and the bladder, necessitating additional inflation and, possibly, falsely elevated pressures. See figure 6-1.

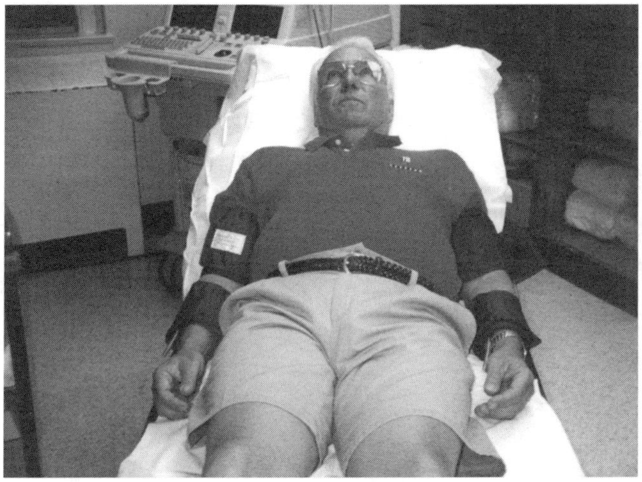

Figure 6-1.
Cuff set-up for the upper extremity peripheral arterial segmental Doppler pressure exam. In this example, the usual 12 x 40 cm cuff accommodates the patient's forearm size. When the patient's forearm is small, a smaller cuff size (10 x 40 cm) is used to avoid a falsely lower blood pressure.

- The width of the pneumatic cuffs should be at least 20% greater than the diameter of the limb so that the artery under evaluation can be compressed when the bladder is inflated. Tissue, not bony structures, must be compressed; otherwise the pressures obtained may be falsely elevated.

- The bladder of the cuff should be placed over the artery. This is especially important when the bladder does not encircle the limb. Ideally the bladder of the cuff should encircle the limb's circumference.

- Upper arm pressure is obtained from the brachial artery (figure 6-2); forearm pressure is obtained from the radial and ulnar arteries (figure 6-3).

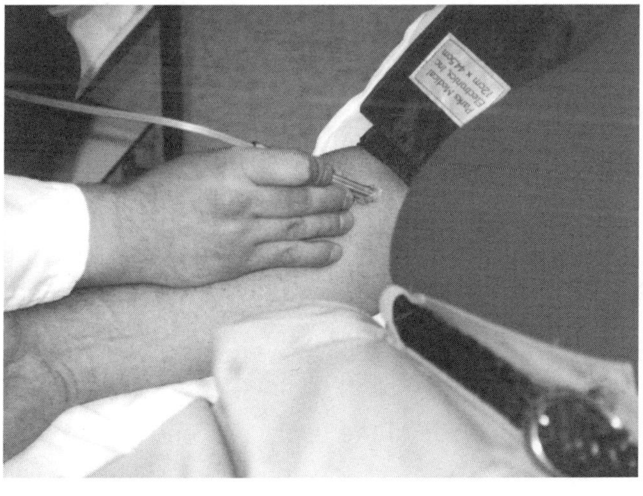

Figure 6-2.
The brachial artery is used to obtain the upper arm blood pressure (brachial pressure). Note the position of the Doppler probe—approximately 45 degrees to the skin, angled cephalad so that arterial blood flow is antegrade. An adequate amount of acoustic coupling gel has been applied to the skin.

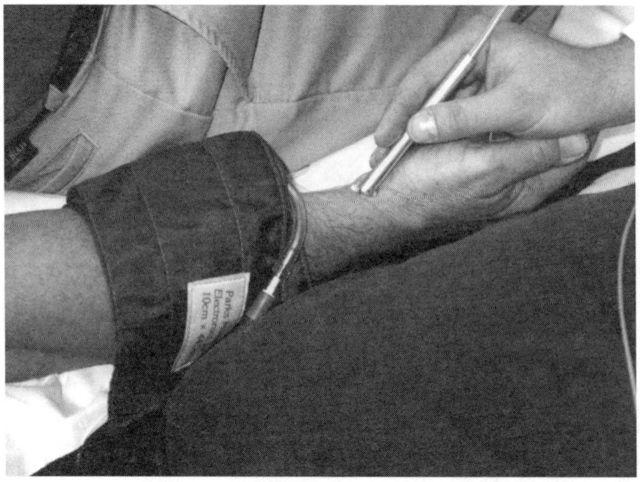

Figure 6-3. Both the radial and ulnar arteries are used to obtain the forearm pressures during an upper extremity peripheral arterial examination. Note that Doppler probe position and angulation are similar to those previously described.

- The pressures are combined with the Doppler velocity waveforms from the sites described in *Doppler Waveform Analysis in the Upper and Lower Extremities* beginning above on page 51. For the upper extremities, Doppler velocity waveforms are recorded from the following arteries bilaterally:

  Subclavian

  Axillary (axilla)

  Brachial (at elbow—antecubital fossa)

  Radial (thumb side, at wrist)

  Ulnar (wrist level, 5th finger side)

- See *Digital Pressures and Plethysmography* on page 103 for evaluation of digits.

6 Interpretation:

- If there is a 15–20 mmHg difference from one brachial pressure to the other, then the lower side would suggest a greater than 50% diameter reduction of the subclavian artery and/or the vessel under the cuff. (For additional information regarding subclavian stenosis and subclavian steal, see page 247.)

- Normally, decreases in blood pressure between the upper arm and the forearm should not exceed 15–20 mmHg. When the decrease is greater than this, it may suggest one of the following:

  A brachial artery obstruction distal to the upper cuff and/or one of the following:

Obstruction in both the radial and ulnar arteries.

Obstruction in the single forearm artery which has a decreased pressure.

- Radial and ulnar artery pressures should be within 5–10 mmHg of each other. If there is a difference of $\geq$ 20 mmHg, the vessel with the lower pressure may be obstructed.
- Arterial occlusive disease occurs in the upper extremity much less frequently than in the lower extremity.

Allen Test

Although optional, the standard or modified Allen test can be performed in conjunction with the upper extremity arterial Doppler examination (i.e., segmental pressures and waveforms).

1 Capabilities:

- Evaluates patency of the wrist arteries and the palmar arch.
- The modified Allen test is useful in determining hand viability if the radial artery is to be removed for use as a coronary artery bypass graft.

2 Limitations:

- Excessive dorsiflexion of the wrist may compress the radial or ulnar arteries, but more often the ulnar artery, as they cross the wrist, leading to a false-positive result.
- If the hand is opened and fingers forcibly extended, the skin over the palm can be stretched, causing compression of the small vessels. Relative pallor can occur because of this interference with circulation.

3 Patient positioning:

- The arm to be evaluated is in a relaxed position at the patient's side.
- The patient's head can be slightly elevated.

4 Technique and documentation:

Standard Allen Test

- The technologist palpates the radial artery at the lateral wrist and then applies manual compression (figure 6-4A).
- While the technologist continues to manually compress the artery, the patient is asked to clench the hand into a tight fist (figure 6-4B) for about a minute, inducing pallor.
- As the radial artery remains compressed, the patient relaxes the hand (figure 6-4C). See Limitations above.

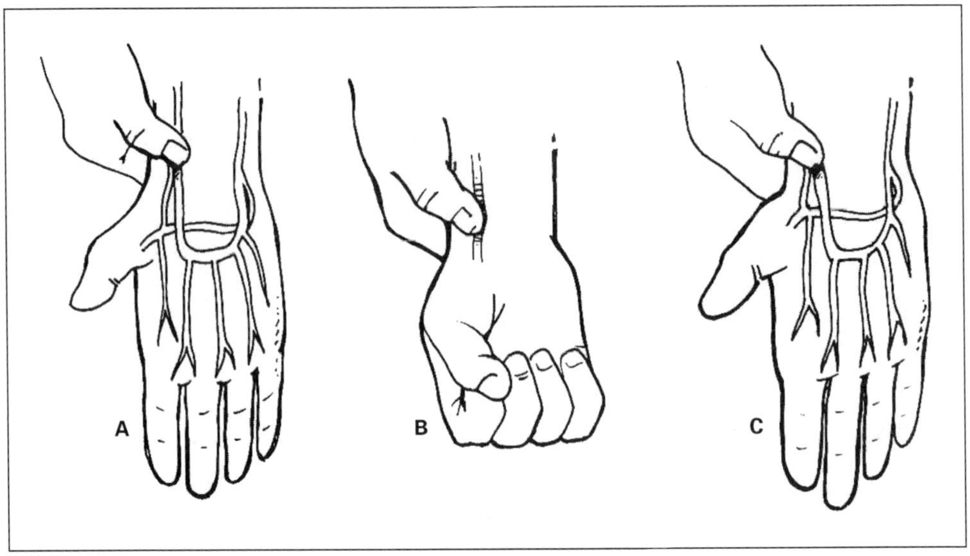

Figure 6-4.
Allen test technique. **A** Compress the radial artery. **B** Patient makes a fist. **C** Patient opens hand while the radial artery is compressed.

- For documentation once the hand is relaxed, see the *Modified Allen Test* below.

Modified Allen Test

The modified version of the Allen test is often used to document the presence or absence of arterial flow in the digits during manual compression of the radial artery, as follows:

- A photoplethysmographic (PPG) sensor is affixed to 2–3 fingers, one at a time, so that waveforms can be obtained both prior to and during the radial artery compression maneuver.
- PPG waveforms are obtained as described in chapter 10, Digital Pressures and Plethysmography.
- The PPG sensor is affixed to the digit using double-stick tape.
- The equipment is set to PPG function (AC mode).
- The strip-chart analog recorder is set at 5 mm/sec speed if only monitoring is required. The speed is set for 25 mm/sec for waveform quality assessment.
- A baseline tracing of each digit is obtained before compression of the radial artery.
- A tracing of flow patterns of each digit is obtained during the radial artery compression maneuver.
- To simplify the process, obtain the waveforms both before and during compression of the radial artery while the PPG sensor is affixed to each digit.

5 Intrepretation:

Standard Allen Test

- When the hand is clenched into a tight fist, it will lose its normal coloration and turn somewhat pale.

- Normal: (Radial artery continues to be manually compressed.) Reappearance of the normal skin color when the hand is relaxed, indicating that the palmar arch is patent with the ulnar artery providing inflow.

- Abnormal: (Radial artery continues to be manually compressed.) If the hand does not return to its normal color during relaxation, it would indicate an obstruction of either the distal ulnar artery or palmar arch. This finding is described as being "radial artery dependent" since without inflow from the radial artery (due to the manual compression) the palmar arch does not receive adequate blood supply, making the hand ischemic.

- See Limitations above.

Modified Allen Test

- The plethysmographic waveforms obtained during radial artery compression should be within normal limits, as defined in the Interpretation section of chapter 10, *Digital Plethysmography and Pressures.* This suggests that there is adequate (normal) arterial flow into the palmar arches and digits via the ulnar artery.

- If the PPG waveform quality decreases appreciably or disappears completely during the radial artery compression maneuver, this suggests that without the radial artery, the hand would not be adequately perfused. In other words, the radial artery could not be removed for use as a coronary artery bypass graft.

6 Miscellaneous:

- It is possible to use a continuous-wave Doppler probe instead of a PPG sensor to obtain waveforms, but doing so is more difficult because of the inability to hold the probe steady on the very small digital artery.

- This test can also be performed while compressing the ulnar artery in the same manner to assess radial arterial inflow to the palmar arch.

CHAPTER 7

Laser Doppler

1 Capabilities:

- Laser Doppler helps to determine the healing potential of a wound or ulcer.
- Laser Doppler helps to determine the healing potential by amputation level.

2 Limitations:

- Ulcerated skin surface contraindicates this test; skin must be intact for placement of probe.
- The inability of the patient to rest quietly makes testing difficult or impossible.

3 Patient positioning:

- Patient is supine with head on a pillow.
- The extremity is positioned comfortably. Positioning is extremely important because patient movement can alter exam results dramatically.

4 Physical principles:

- The Doppler effect is described in chapter 4.
- The Doppler principle is routinely used for the transcutaneous measurement of the movement (flow) of blood. The laser Doppler is used to assess the characteristics of the microvascular blood volume in the capillary beds of the skin. Optical systems are limited to measurements of superficial layers of tissue. In contrast, ultrasound-based systems can measure blood flow velocities in deep arteries and veins because the longer wavelengths permit the ultrasound wave to remain coherent for longer distances through tissue.
- Laser Doppler uses optical (light) waves, and because of their relatively short wavelength they do not remain coherent for distances longer than a fraction of a millimeter in tissue. A low-power beam of light is delivered to the skin through a fiber optic cable. A volume of tissue, which includes both moving red cells and stationary tissue cells, is illuminated.

- Photons (light energy) are immediately scattered in random directions by both cell types. The photons scattered by moving red blood cells are Doppler-shifted in frequency. The photons scattered by stationary cells are not.
- A portion of the scattered laser light is collected by a receiving optical fiber and returned to a photodetector. The photodetector converts this optical signal into an electronic signal that is processed to determine microvascular blood volume.
- The pulse volume recording (air plethysmography) principle is the same as described in chapter 9 (*Plethysmography—Upper and Lower Extremities*).

5 Technique:

- Obtain brachial blood pressure bilaterally.
- Determine the test site(s).
- Prior to placement of the laser Doppler (embedded in the cuff), prepare the skin surface using an alcohol wipe.
- The appropriate laser Doppler cuff is selected, and then:
  1. A pulse volume (PV) waveform is obtained. (Toe cuff inflated to 40 mmHg; all other cuffs inflated to 65 mmHg.) If the result is abnormal, it may not be possible to perform the remainder of the exam.
  2. Skin perfusion pressure (SPP) mode is activated. Cuff is inflated to 100 mmHg.
  3. The pressure is held for 10 seconds, watching for the volume indicator to drop below 0.1%. If it doesn't, the transducer is inflated another 20 mmHg. Wait 10 seconds.
  4. Repeat the process if the volume indicator has not dropped to 0.1%. If necessary, the next step is performed.
  5. The patient's leg is elevated to approximately 60 degrees above horizontal, holding it there until the volume drops to below 0.1%.
  6. Once the volume is < 0.1%, obtain the patient's SPP.
- Obtain the patient's SPP:
  1. Slowly deflate the cuff, using the bleed valve (bleeding 10 mmHg for every 5 seconds).

2. Optimally, the examiner sees a steady baseline, which is followed by a doubling or more of the graph's pressure scale. This indicates the patient's actual pressure. One must be careful that artifact (e.g., patient movement) does not cause the SPP level to be read too soon. Confirmation of the actual SPP is the increasing size change of the bars on the scale during the deflation process.

6 Interpretation:

- *Normal:* The PV waveform is graded similarly to the photoplethysmographic (PPG) waveforms for the digital exam (chapter 10). SPP of > 30 mmHg indicates likely to heal.
- *Abnormal:* The PV waveform may be dampened or diminished (chapter 10). SPP of < 30 mmHg indicates less likely to heal.

Penile Pressures and Waveforms, Duplex/Color Flow Imaging Evaluation

CHAPTER 8

1 Capabilities:

- Helps to determine whether an individual's failure to attain or maintain a sufficient erection is related to peripheral vascular insufficiency.
- Combined with ergometry (a form of exercise), this test may indicate pelvic steal, arterial vasospasm, or ipsilateral arterial occlusive disease.

2 Limitations:

- Highly anxious or antagonistic patients are difficult or impossible to study.
- The exercise portion of the study should be terminated if the patient develops shortness of breath, hypertension, or cardiac problems.
- Only duplex and color flow imaging provide data about arterial or venous flow rates and describe some anatomic pathologic conditions.
- If a patient is sensitive to injectable medication or is receiving anticoagulation therapy, he may be unable to undergo the injection component.

3 Patient positioning:

Patient is supine with head on a pillow and appropriate draping to maintain as much privacy as possible during the testing.

4 Physical principles:

Each of the following techniques can be utilized to obtain pressures and waveforms, but only one would be selected for the nonimaging protocol. The physical principles of the plethysmographic techniques are discussed in the next chapter.

- Continuous wave Doppler ultrasound. An 8–10 MHz Doppler probe can be utilized. Doppler waveforms and pressures obtained.
- Volume plethysmography
- Photoplethysmography (PPG) (appears to be used most often)
- Strain gauge plethysmography (SPG)
- Duplex scanning, which includes B-mode imaging and Doppler spectral analysis with or without color flow imaging, also may be used before and after the injection of a medication such as papaverine and/or prostaglandin.

5 Nonimaging technique and interpretation:

Technique

- Appropriately sized blood pressure cuffs are placed on the upper arms, ankles, and proximal shaft of the penis. Penile cuff sizes are 2.5 × 12.5 cm or 2.5 × 9 cm.
- Calculate the ankle/brachial index (ABI) as previously described to be certain that resting ankle pressures are within normal limits. (If a patient has poor arterial flow into the legs because of a proximal obstruction, poor arterial flow into the penis can also exist, depending on where the disease is located. For example, aortoiliac occlusive disease and/or internal iliac artery occlusive disease can affect arterial inflow to the penis.)
- Penile pressures are obtained with any of the following end-point detectors: Photoplethysmography (PPG), strain gauge, or pulse volume plethysmography; continuous wave (CW) Doppler can also be used (figure 8-1). Because these are nonimaging and therefore blind techniques, it is important to obtain pressures from two or three sites (e.g., dorsal, ventral, and medial aspects).

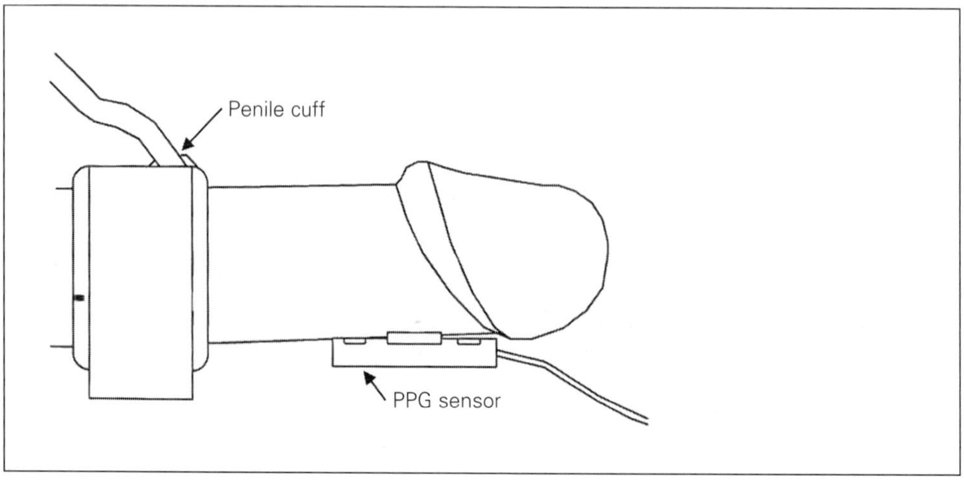

Figure 8-1. Penile pressure obtained by using a small blood pressure cuff placed proximally, while the PPG sensor is placed distally to act as the end-point detector to obtain the blood pressure. Courtesy of Imex Medical Systems.

● Plethysmographic waveforms are obtained as necessary utilizing one of the above techniques. Doppler waveforms are rarely obtained.

Interpretation

● Penile/brachial index (PBI) is calculated using the higher brachial pressure.

● Normal PBI = 0.75 or greater

Marginal = 0.65–0.74

Abnormal = less than 0.65, which is consistent with vasculogenic impotence

● The penile/brachial index *least* expected in a young adult male with normal sexual response would be anything less than 0.65.

● Plethysmographic waveforms are evaluated. Examples of these waveforms can be found in figure 9-6. Description of these waveform qualities follows:

Normal: Sharp, systolic peak with prominent dicrotic wave. (Other terms used to describe this are *reflected diastolic wave, reflected wave*, and *dicrotic notch*.) The plethysmographic waveform contour is qualitatively evaluated. Emphasis on the dicrotic wave is important since its presence usually excludes significant proximal arterial disease.

Mildly abnormal: Sharp peak, absent dicrotic wave, downslope is bowed away from baseline.

Moderately abnormal: Flattened systolic peak, upslope and downslope time decreased and nearly equal, dicrotic wave absent.

Severely abnormal: The pulse wave has low amplitude (which has equal upslope or downslope time) or may be absent.

6 Imaging technique and interpretation:

Technique

● In most cases informed consent is obtained from the patient. A 7 or 10 MHz transducer is selected and light probe pressure employed.

● Before injection, utilizing duplex and/or color flow imaging, the cavernous arteries are measured bilaterally in an A/P transverse orientation. In addition, using a longitudinal approach, the examiner measures peak systolic velocities in the cavernous arteries bilaterally (and possibly the dorsal arteries as well). The literature varies regarding the specific protocol.

- Utilizing the lateral aspect of the proximal shaft of the penis, the urologist injects a special medication (e.g., papaverine and/or prostaglandin) to induce erection. It is important to note, however, that clinical trials continue to be performed and that some medications are being used investigationally because they have not yet been specifically approved for this use.
- Repeat measurements are obtained postinjection. Protocol varies in the literature regarding how soon the postinjection velocities and cavernous artery dimensions are measured. For example, the data may be obtained at 1 or 2 minutes after injection; multiple measurements may be obtained at various increments for up to 6 minutes after the injection.

  Peak systolic and end diastolic measurements are obtained from the proximal cavernous arteries (as possible) before full erection is achieved. This may require taking several measurements to obtain the highest velocity recordings.

  The deep dorsal vein flow velocity is measured from a dorsal approach, with light probe pressure.

  Both sagittal and transverse approaches are used to evaluate the vasculature.

  The dimensions of the cavernous arteries are measured in an A/P transverse view during systole.

  The examiner monitors the time elapsed since injection and documents when velocities are recorded.
- It is important to instruct the patient about priapism. If a rigid erection is maintained for a period of 3 hours (approximately) after injection (the criterion varies from lab to lab), the patient is instructed to contact his urologist immediately for further instruction and/or medical intervention to reverse the priapism.

Interpretation

- Preinjection (pre-erection), blood flow in the penile arteries has a high-resistance quality as seen in figure 8-2. Postinjection, blood flow assumes a lower-resistance quality as seen in figure 8-3.
- There are various schools of thought concerning the importance of peak systolic and end diastolic velocities. It has been noted that peak systolic velocities increase after injection (the following data may vary among investigators and published references):

  Normal ≥ 30 cm/sec

  Marginal = 25–29 cm/sec

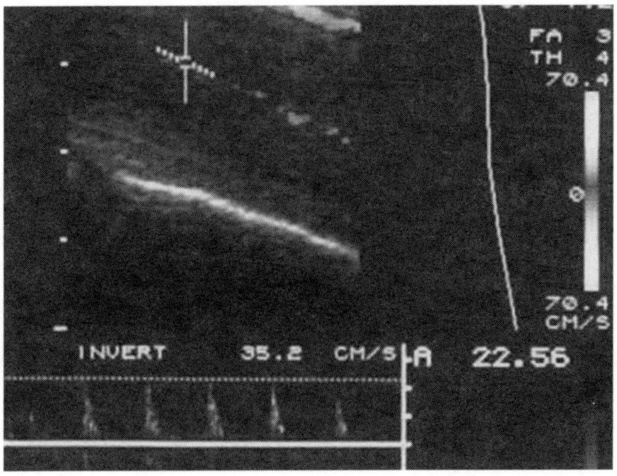

Figure 8-2. High-resistance flow quality (note increased pulsatility) in the preinjection blood flow Doppler signals.

Reduced < 25 cm/sec

End diastolic velocities increase as well. For some, this data is also used for interpretation.

- Postinjection, the deep dorsal venous flow velocities should not increase:

Normal < 3 cm/sec

Moderately increased = 10–20 cm/sec

Markedly increased > 20 cm/sec

It has been suggested that an increase to greater than 4 cm/sec may indicate a venous leak, which could contribute to the erectile dysfunction. These data may vary depending on the published reference.

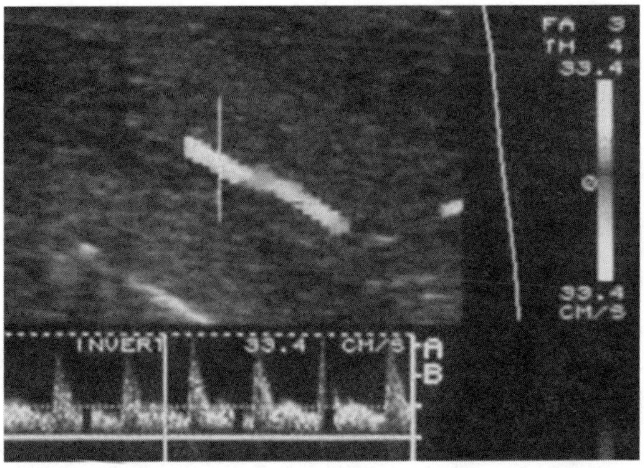

Figure 8-3. Low-resistance flow quality (note the peak and end diastolic velocities) in the postinjection blood flow Doppler signals.

- The diameter of the cavernous arteries normally increases (dilates) after injection, changing previously high-resistance flow to continuous, steady, low-resistance flow.

- Patient history and observations during the pre- and postpapaverine injection study are valuable elements of this procedure.

7 Varicocele:

- A varicocele is an enlargement of the veins of the spermatic cord, a condition which occurs most often in young men and adolescents. It is most common on the left side.

- B-mode imaging, utilizing a 7 or 10 MHz transducer, has been used to determine the presence of a varicocele. Color flow imaging has been of benefit in imaging these varicosed, enlarged veins. Whichever imaging technique is used, it is also important to have the patient perform the Valsalva maneuver in order to document the direction of venous flow and the pattern of filling in this bundle of veins. Reverse flow, noted during the Valsalva maneuver, is diagnostic of a varicocele.

Plethysmography—Upper and Lower Extremities

CHAPTER 9

1 Capabilities:

- In combination with Doppler segmental pressures, helps to differentiate true arterial claudication from nonvascular sources.
- Helps to document the functional aspects of vascular disease.
- Detects presence/absence of arterial disease.
- Helps to localize the level of obstruction.
- Assesses the results of medical and/or exercise therapy.
- Overall accuracy enhanced when used in combination with Doppler segmental pressures.
- Photoplethysmography and strain gauge plethysmography most often used to evaluate the digits.
- Photoplethysmography often used to obtain penile pressures.

2 Limitations:

- Cannot be specific to a single vessel because it measures volume changes in a large segment of a limb.
- May be difficult to discriminate between major arteries and collateral branches.
- Obesity makes it difficult to perform volume and strain gauge plethysmography.

3 Patient positioning:

- Most forms of plethysmography can be adequately performed with the patient supine and the limbs in a resting position, although the patient can be sitting during evaluations of upper extremity digits.

- Special care must be taken not to twist or improperly position strain gauge tubing, as this can cause inaccurate information.
- When using PPG, it is necessary to cover the skin completely with the PPG device and to ensure its proper adherence with double-stick tape or a small Velcro strap.

4 Physical principles:

Volume Plethysmography

- Other terms used to describe this type of plethysmography are *pneumo- (air)*, *volume*, and *true plethysmography*.
- Terms to describe the volume plethysmographic waveforms include *pulse volume recordings* (PVR), *volume pulse recordings* (VPR), and *pulse contour recordings* (PCR).
- Pneumatic cuffs are placed around specific levels of the extremities or digits. A measured amount of air is sequentially introduced into a cuff up to a pressure ranging from 10 to 65 mmHg, depending on the site (digit or extremity) and level of the cuff. See figure 9-1.
- As the arterial flow moves underneath the cuff through arteries, branches, small vessels, and any collateral branches, momentary increases in the limb segment volume occur during systole, when arterial flow peaks. These systolic increases in limb volume put pressure against the air-filled cuff bladder, as represented diagrammatically in figures 9-2A and B.

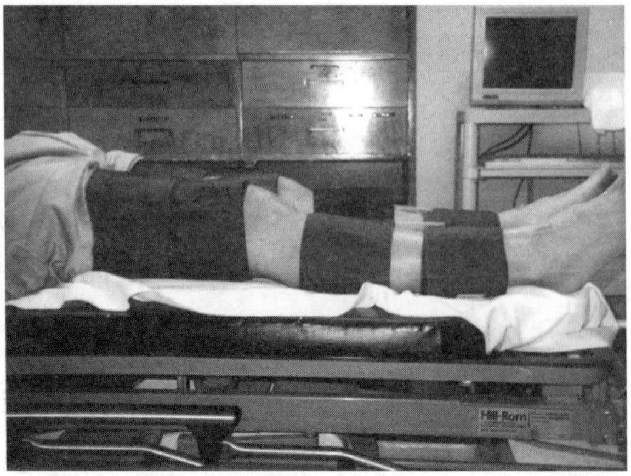

Figure 9-1.
Air-filled cuffs are used to obtain plethysmographic waveforms. These same cuffs may also be used to obtain subsequent segmental Doppler pressures.

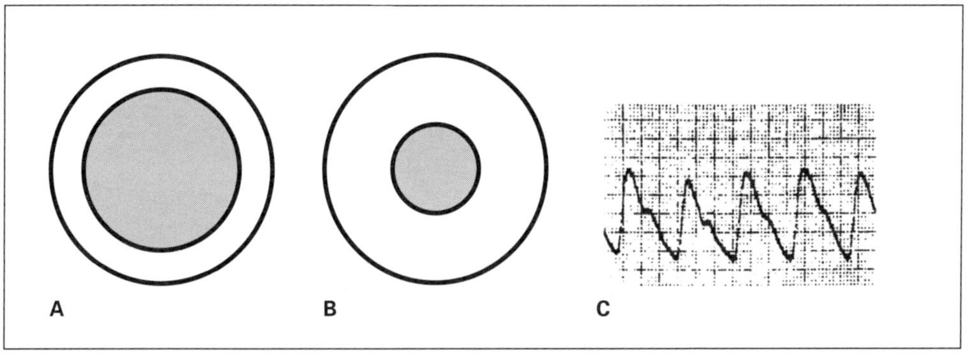

Figure 9-2. Pneumatic cuff bladder (white) and arterial system (gray) during systole (**A**), when arterial pressure is high, and diastole (**B**), when pressure is low. **C** Normal volume plethysmographic waveform. Note how similar the qualities are in all three types of plethysmographic waveforms; compare with figures 9-3 and 9-5.

- The pulsatile pressure changes occurring inside the air-filled bladder are converted by a pressure transducer into an analog waveform display such as that in figure 9-2C.

- The more arterial blood flow present moving underneath the air-filled cuff bladder, the greater the analog waveform excursion. When there is a decrease in arterial flow underneath the cuff (due to obstructive disease), the waveform changes in appearance. Other characteristics of the volume plethysmographic waveform are described below in point #6—*Interpretation.*

Photoplethysmography (PPG)

- Consists of a transducer, amplifier, and strip-chart recorder.

- Detects cutaneous blood flow and records pulsations (see figure 9-3 below) rather than recording volume changes. This is not true plethysmography (i.e., volumetric), although it is still considered a plethysmographic technique. In AC mode, PPG records pulse waveforms. The rapid changes in blood content of the skin are recorded as pulsatile waveforms with each heartbeat.

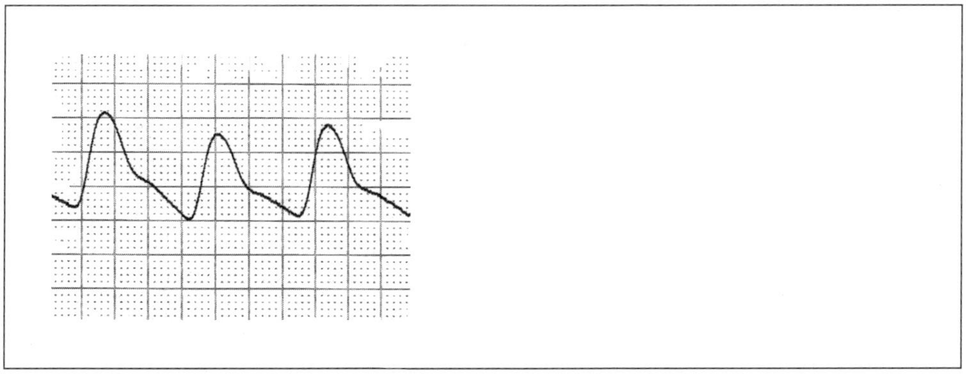

Figure 9-3. Normal photoplethysmographic waveform.

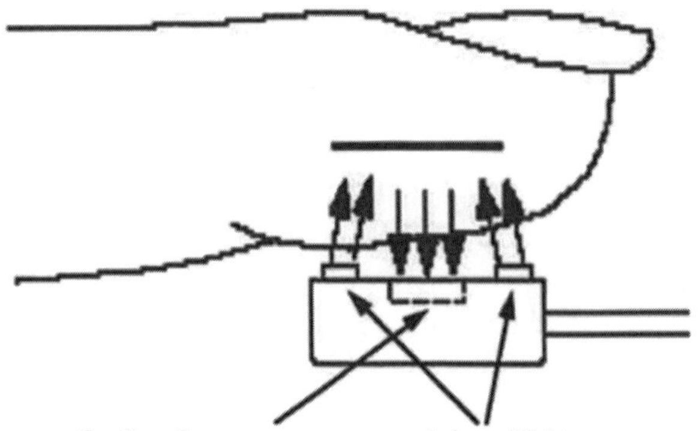

Figure 9-4. Drawing shows the PPG positioned on the underside of the finger, depicting the signal transmission/reflection. Courtesy of Imex Medical Systems.

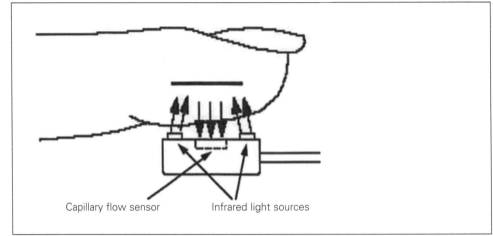

- Sends infrared light into the underlying tissue with a light-emitting diode, and the adjacent photodetector (photocell) receives the backscattered infrared light and measures this reflection of light. See figure 9-4.
- As light is directed into the skin, the tissue and the blood in the cutaneous vessels attenuates a portion of it. PPG uses the infrared light as the sensor because measurements made in the infrared range have a constant but minimal attenuation.
- The cutaneous blood flow determines the reflection. Although PPG cannot quantitate flow, its waveform display provides qualitative information.

Strain Gauge Plethysmography (SPG)

- Utilizes a mercury-filled silicone-like tube that makes contact with copper electrodes at either end. The strain gauge is wrapped around the limb part being evaluated with just enough stretch to ensure good contact. The length of the extremity strain gauge is approximately 1–3 cm shorter than the circumference of the extremity. With digits, the gauge is approximately 0.5 cm shorter.
- As the limb contracts and expands, the length of the tube changes as well. Because the resistance of the gauge varies with its length, the variations in the voltage change across the gauge reflect limb circumference changes.
- Normal limb expansion produces a relatively tall waveform amplitude and good quality (see figure 9-5), whereas abnormally reduced limb

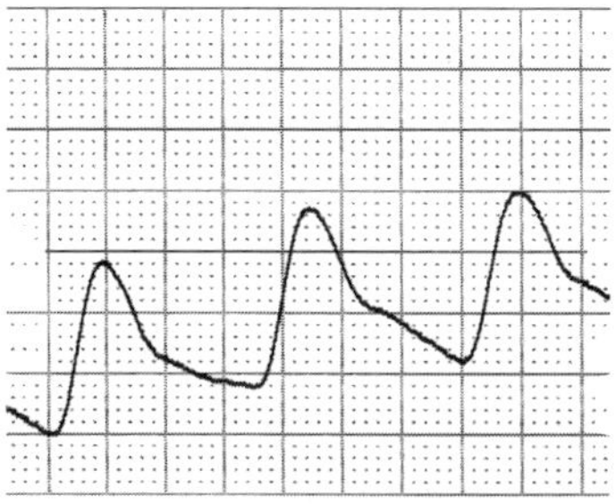

Figure 9-5. Normal strain gauge plethysmographic waveform.

expansion caused by poor arterial inflow produces a waveform of diminished amplitude and quality (see figures 9-6C and 9-6D on page 98). All pulse contours and sizes are evaluated.

5 Technique:

Volume Plethysmography

- Patient is supine, with heels slightly elevated on a hard cushion.

- Appropriately sized pneumatic cuffs are applied snugly to the thigh, calf, and ankle bilaterally. Cuff sizes are the same as for pressures (3 cuff versus 4 cuff).

- When the machine is activated it self-calibrates. **Note:** If during the recording of the waveforms the stylus pen is not centered but rather located at either the top or the bottom of the chart paper, activate the RE-SET control to center the stylus pen.

- Bilateral brachial pressures are obtained as previously described.

- An appropriate amount of air is used to inflate cuff pressure to predetermined levels. The examiner begins with the upper part of the extremity and moves distally. At least three pulse cycles are recorded at each level. The cuff must be snugly wrapped, but not too tight. Artifacts caused by improper cuff application are not uncommon. For example, if the cuff is applied too loosely, it will take much more air to inflate the cuff bladder to the predetermined amount. This can attenuate the waveform.

- A similar GAIN setting is maintained throughout the study for each level where pulse waveforms are recorded. If a different GAIN setting must be used, that setting is indicated on the recording paper.

- Doppler segmental pressures are obtained as previously described. The volume pulse plethysmography and pressures are complementary procedures.

Photoplethysmography (PPG)

Because photoplethysmography is most often applied to the evaluation of digital arterial disease, it is discussed in chapter 10, *Digital Pressures and Plethysmography*.

Strain Gauge Plethysmography (SPG)

- Appropriately sized strain gauges are applied to various levels of the extremity (e.g., high and low thigh, below knee [calf], and ankle) with careful attention to extremity positioning and gauge application.
- At least three pulse cycles are recorded at each level.
- As in volume plethysmography, a similar GAIN setting is maintained during the procedure. If the setting is changed, that setting is documented as well.
- Doppler segmental pressures are obtained as previously described. More often than not, these are complementary procedures.
- This modality is also used to evaluate digital arteries; see chapter 10, *Digital Pressures and Plethysmography*.

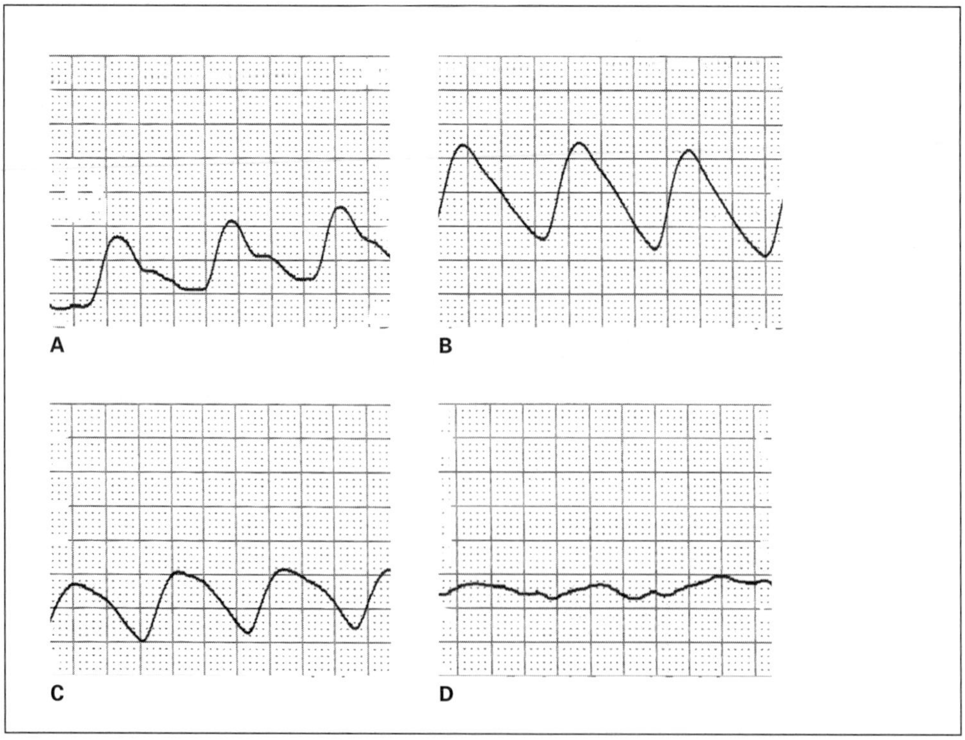

Figure 9-6. Volume plethysmographic waveforms. **A** Normal. **B** Mildly abnormal. **C** Moderately abnormal. **D** Severely abnormal.

6 Interpretation:

Volume Plethysmography

● There are different schools of thought concerning the interpretation of plethysmographic waveforms. Some have described qualitative criteria for the waveforms; others have described criteria of a semiquantitative nature, which considers both the contour and amplitude of the pulse waveforms. Generally accepted criteria follow:

Normal: Sharp systolic peak with prominent dicrotic wave (figure 9-6A). See also figure 9-7.

Mildly abnormal: Sharp peak, absent dicrotic reflective wave; downslope is bowed away from baseline (figure 9-6B).

Moderately abnormal: Flattened systolic peak, upslope and downslope time decreased and nearly equal, absent dicrotic wave (figure 9-6C).

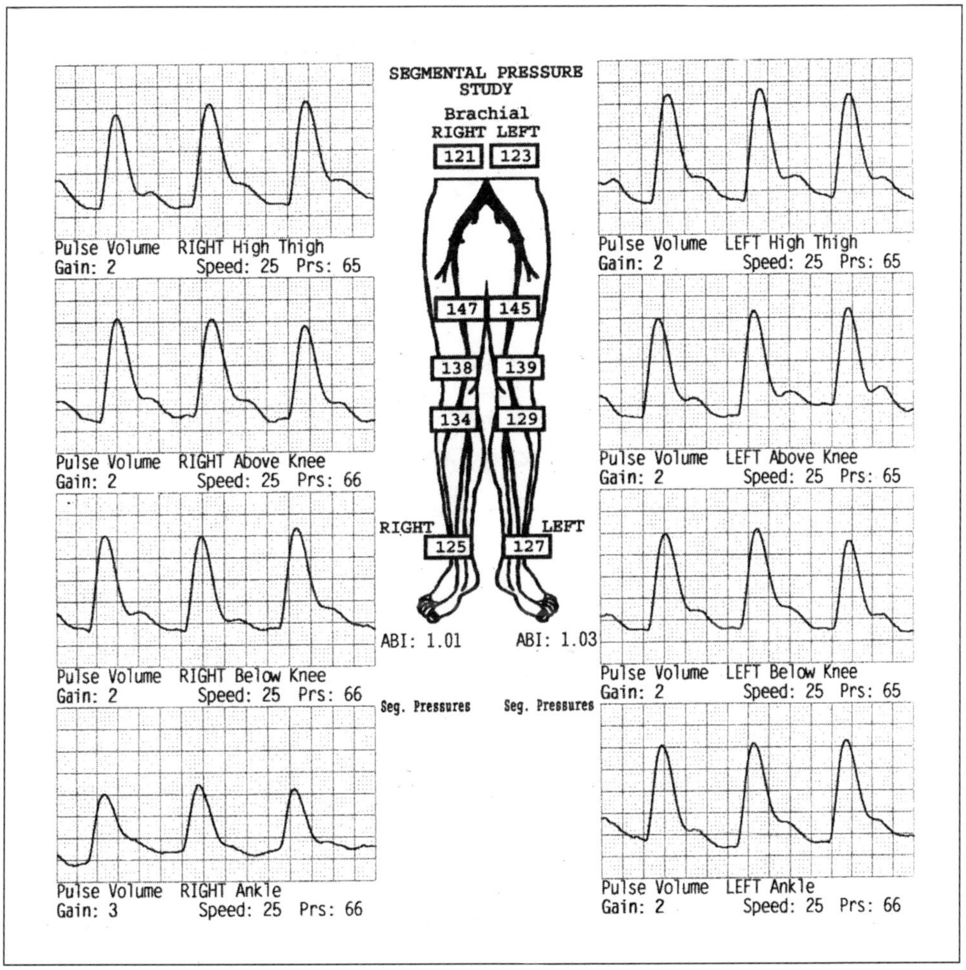

Figure 9-7. Normal volume plethysmographic waveforms as well as Doppler systolic blood pressures at all levels—bilaterally.

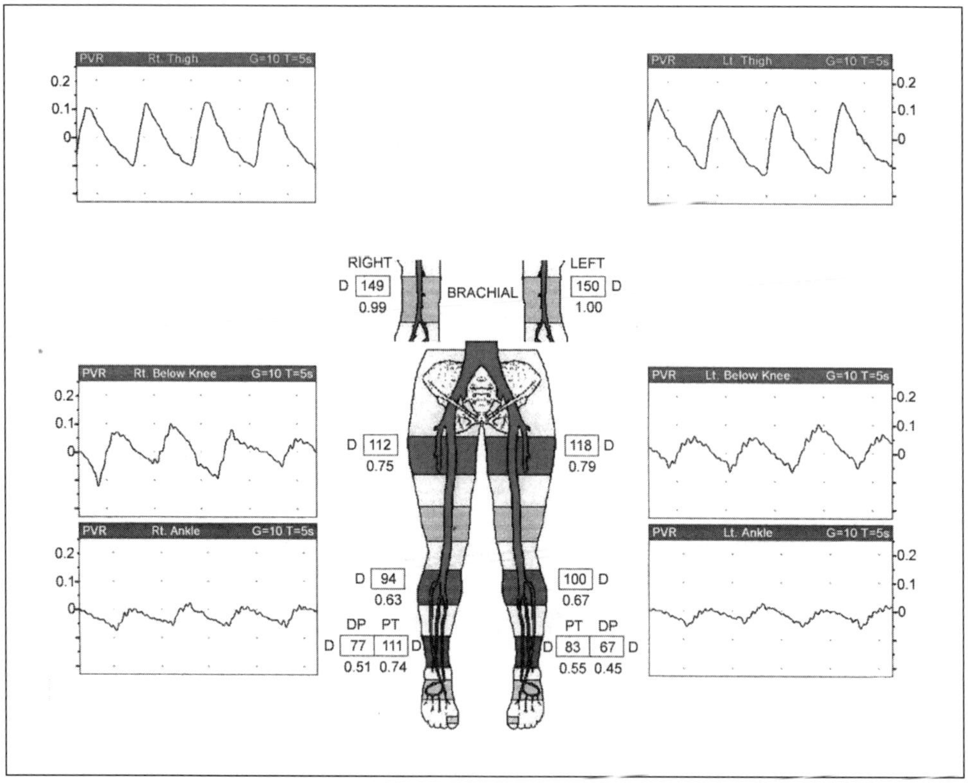

Figure 9-8. Moderately abnormal volume pulse recordings beginning at the thigh level, which become more abnormal distally at the calf and ankle levels bilaterally. Based on the waveform quality throughout the extremities, there is likely inflow (aortoiliac) disease as well as outflow arterial disease.

Severely abnormal: Absent or low-amplitude pulse wave with equal upslope and downslope time (figure 9-6D). See also figure 9-8.

- Moderately or severely abnormal volume plethysmographic waveforms always reflect hemodynamically significant disease proximal to the level of the tracing. Severity of disease is generally underestimated.
- If amplitude is reduced but there are no changes in the contour of the waveform, the finding is likely to be not significant unless it is unilateral.
- It is not uncommon to obtain a waveform of fair quality but an abnormal Doppler segmental pressure at the same level. In the presence of collaterals, plethysmography alone can underestimate the significance of an obstruction. This fact is a key reason for combining Doppler pressures with this technique.
- The comparative plethysmographic waveforms in figure 9-9 reveal the difference between a normal waveform (figure 9-9A) and a hyperemic waveform that is affected by collateralization (figure 9-9B). The amplitudes are similar, but the characteristics of the waveform differ.

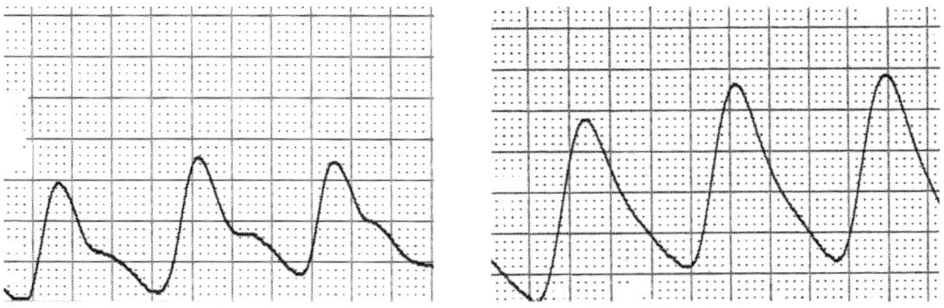

Figure 9-9. Volume plethysmographic waveforms. **A** Normal. Note the "reflection" in the downslope (catacrotic limb) of the waveform. **B** Abnormal. Although the waveform maintains its amplitude, the loss of reflection is highly suggestive that this waveform is produced by the presence of collateral vessels. Although the main artery is obstructed, collateral vessels continue to provide arterial perfusion.

- Troubleshooting

  Problem: Recorder stylus cannot be centered.
  Action: Confirm that the machine is in the right mode (i.e., arterial studies require the AC mode; venous studies, the DC mode).

  Problem: Recorder stylus is stuck at the bottom or top of the paper (the machine is in the correct mode).
  Action: Activate the RE-SET control to center the stylus.

  Problem: An acceptable waveform cannot be recorded (i.e., too much stylus movement).
  Action: Reattach the PPG or strain gauge if necessary; the patient may have tremors and be unable to hold still.

  Problem: There is no tracing at all.
  Action: Ascertain that the correct mode (AC, DC) has been selected, that the correct test name has been entered, and that the proper paper (i.e., heat sensitive) has been correctly installed.

Photoplethysmography (PPG)

Qualitative criteria are similar to those previously listed above under *Volume Plethysmography.*

Strain Gauge Plethysmography (SPG)

Qualitative criteria are similar to those previously listed above under *Volume Plethysmography.*

Displacement Plethysmography

- Consists of a water-filled, watertight container in which the body part is immersed. The water temperature must remain constant.

Figure 9-10.
Water-filled plethysmograph. Reprinted with permission from Bernstein EF: *Noninvasive Diagnostic Techniques in Vascular Disease*, 3rd edition. St. Louis, Mosby, 1985.

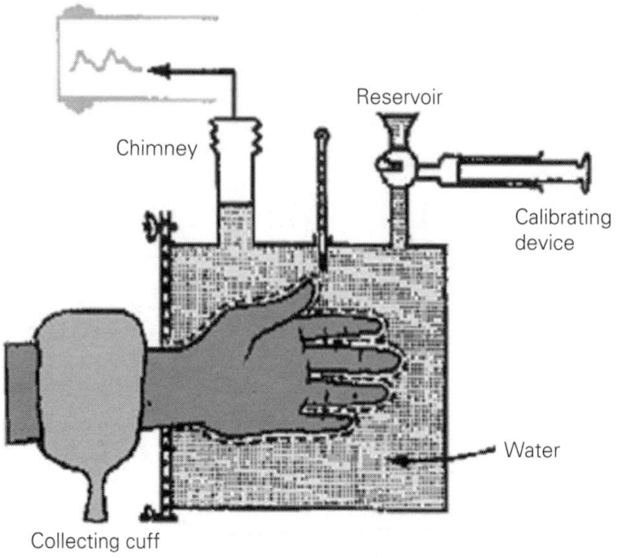

- When the volume of the immersed body part increases, the water is displaced by an equal amount.
- Pulse plethysmography refers to transient changes in limb volume related to the pulse-by-pulse activity of the left ventricle, the body part expanding when arterial inflow exceeds venous outflow.
- As depicted in figure 9-10, when the immersed part expands, the water is displaced into the glass chimney (upper left hand side of drawing) where a column of air is compressed, which in turn activates a spirometer. A recording is made of the amount of displacement into the chimney.
- In the example shown, the hand is inside a loose-fitting surgical rubber glove (dotted lines). Hydrostatic pressure exerted by the surrounding fluid keeps the glove in close contact with the skin.

CHAPTER 10

Digital Pressures and Plethysmography

1 Capabilities:

- Helps to detect the presence of arterial disease.
- Differentiates fixed arterial obstruction from vasospasm.
- Assesses effects of treatment.
- Helps to determine the presence/absence of sympathetic activity.

2 Limitations:

- Vasoconstriction greatly affects the quality of the results. Before testing begins, the examiner should ask such important questions as these: Has the patient just come in from cold weather? Has the patient been smoking? Is the patient nervous?
- With volume plethysmography, cuffs that are applied too tightly can obliterate or diminish the pulse waveforms.
- With photoplethysmography, a photocell that is incorrectly applied to the skin will cause both artifact and poor results.
- With strain gauge plethysmography, kinked or cracked tubing and poor positioning of the gauge will yield poor results.
- The presence of ulcerations and/or gangrene may prevent placement of the cuff, photocell, or strain gauge.
- Extensive bandages that cannot be removed make the placement of cuffs, photocells, or strain gauges impossible.

3 Patient positioning:

- For the evaluation of toes, the patient should be supine. Patient's head can be elevated 10–20 degrees.

- For the evaluation of fingers, the patient can be sitting. A pillow is placed on the patient's lap, with his or her arms resting on it with palms up.
- Keep the patient comfortable and warm.

4 Physical principles:

The principles for volume, photo-, and strain gauge plethysmography have been discussed previously on pages 94–96.

5 Technique for toes:

- This digital study is often done in combination with a physiologic lower extremity arterial test. If not, bilateral brachial and ankle Doppler pressures are obtained and the ABI determined in order to rule out any proximal disease in the extremity.
- An appropriately sized cuff, the width of which should be at least 1.2 times that of the toe, is applied to the base of the toe(s). A 2.0–2.5 cm cuff is usually used for fingers and a 2.5–3.0 cm cuff for the great toe.
- There is no specific manual "calibration" done for volume or photoplethysmography prior to usage, but strain gauge equipment requires manual calibration by the technologist.
- Plethysmographic waveforms are recorded as previously described for the technique being used (volume, photo, strain gauge) for the great toe bilaterally and additional toes as necessary. See figures 10-1 to 10-3.

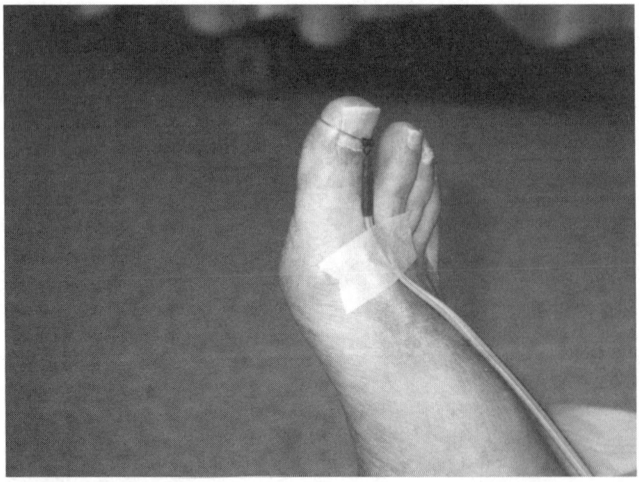

Figure 10-1. A strain gauge applied to the distal end of the great toe to record arterial pulsations.

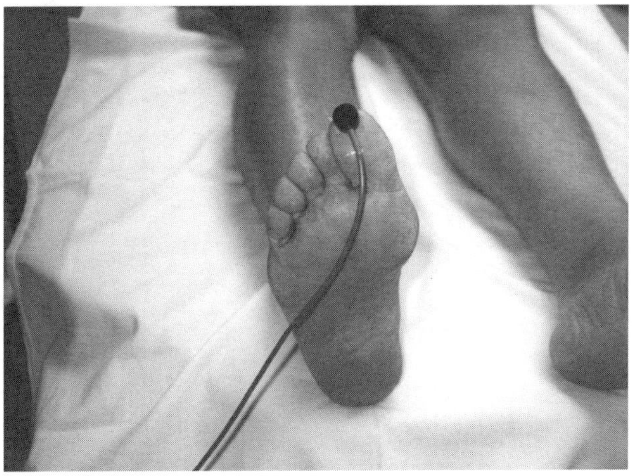

Figure 10-2. PPG sensor applied to the great toe with double-stick tape, the most common method of securing this type of PPG sensor.

- Great toe pressures are obtained using "end-point detectors" such as the photocell, strain gauge, or a digit pneumatic cuff.
- The following procedure describes the use of PPG for obtaining a blood pressure. A continuous wave Doppler can be used to obtain pressures, but it is a much more difficult technique, because it requires holding the probe steady on a very small digital artery. This procedure applies to strain gauge and volume plethysmography as well as to photoplethysmography:

  Digit cuff is at the base of the toe.

  Photocell is securely attached with double-stick tape to the underside of the distal portion of the great toe. See figures 10-2 and 10-3.

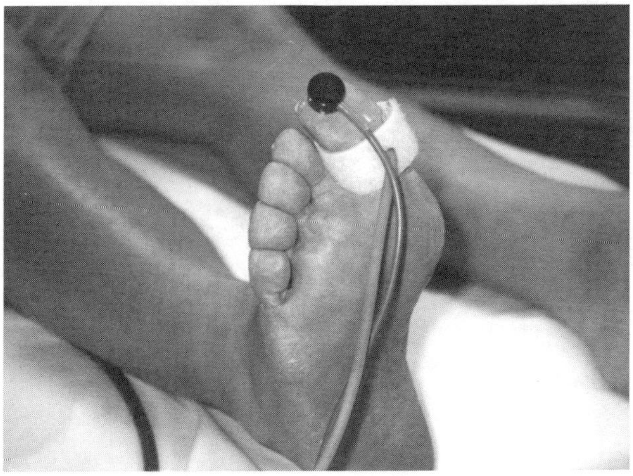

Figure 10-3. The great toe has a small digital cuff applied to the base of the toe and a PPG sensor taped to the distal end of the toe (plantar side). In this case the PPG sensor acts as the end-point detector and monitors the arterial pulsations at rest on the analog recording paper. Then, with inflation, no tracings are seen. Finally, as the cuff deflates, the point at which blood flow returns is documented on the recording paper.

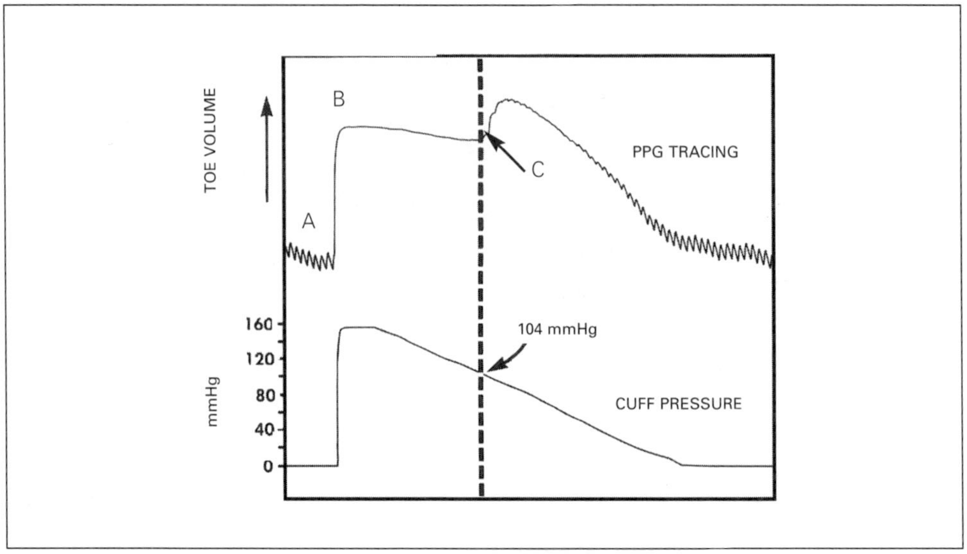

Figure 10-4. Simultaneous pneumatic cuff pressure and PPG tracing in a normal toe. Note the sharp increase in toe volume and the reappearance of digital pulses when the cuff pressure reaches the systolic pressure in the toe. From Hershey FB, Barnes RW, Sumner DS (eds): *Noninvasive Diagnosis of Vascular Disease.* Pasadena, Appleton Davies, 1984.

The pulse is recorded with slow paper speed (e.g., 5 mm/sec) (figure 10-4, point A).

While the pulsations are being recorded, the examiner inflates the cuff to suprasystolic pressure (about 20–30 mmHg higher than the ankle pressure), at which point there should be no pulsations (figure 10-4, point B). The increase in toe volume (rise from A to B) is secondary to venous outflow obstruction. Plethysmography measures all volume changes.

The examiner slowly deflates the cuff, watching for return of the first pulsation. Once this pulsation is observed, the pressure level (figure 10-4, point C) at which it occurred is recorded as the toe pressure.

Elevation of the recording pen (heat-sensitive stylus) on the paper after point C is likely related to a sudden slight movement of the toe and is therefore considered artifactual. The key to determining the pressure with this method is to look for the first small pulsation that is followed by pulsations of greater amplitude as the cuff continues to deflate. This discriminates the actual real first pulsation from artifact (such as the one just described) on the recording paper. See figure 10-5.

- *Analogy:* When a brachial blood pressure is taken, the arm cuff is normally inflated to about 200 mmHg while a stethoscope is positioned over the brachial artery. If no pulsations are heard, the cuff is slowly deflated.

When the *first* signal is heard during the deflation process, that is called the systolic pressure. This is exactly what happens when the photocell is

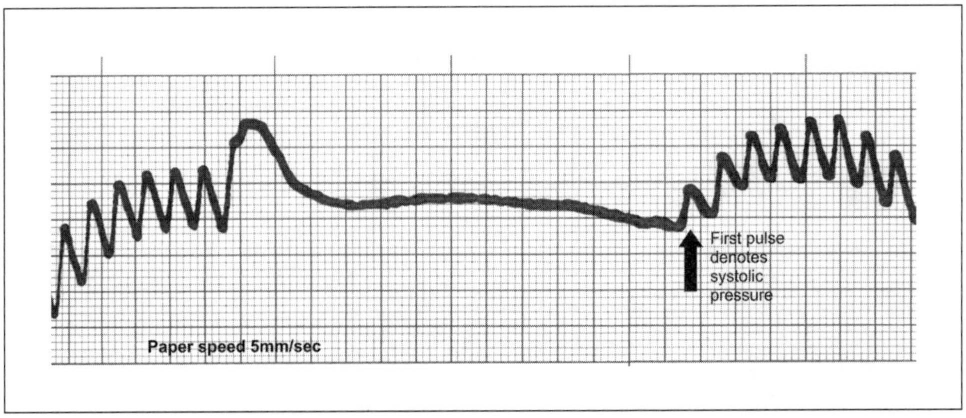

Figure 10-5. Tracing showing an actual recording while obtaining a systolic pressure of the great toe. Paper speed is 5 mm/sec, making it easier to define the first pulse, which is recorded as the great toe pressure.

used. It is essentially acting as the stethoscope does; instead of auditory signals, however, the pulsations are depicted on recording paper.

6 Technique for fingers (without cold stress):

- Unless this study is performed as part of the upper extremity arterial study (in which case the following pressures would be obtained routinely), both brachial and forearm pressures are measured bilaterally to rule out any proximal disease:

  Evaluate the Doppler signal (triphasic, biphasic, monophasic).

  Obtain Doppler signals of the palmar arch to verify patency. The palmar arch is normally supplied by the radial and ulnar arteries.

- A 2.0–2.5 cm cuff is usually used for fingers. As with toes, the width of the cuff should be at least 1.2 times that of the digit.
- With PPG, a waveform and pressure are obtained from the distal ends of selected fingers.
- Because fingers are long enough to allow it, two cuffs can be applied to obtain two levels of volume plethysmographic waveforms, if desired.
- Pressures are obtained using either method (i.e., volume, PPG, or strain gauge) described above for the toes. (See figure 10-6, which shows a different method for placement of the PPG sensor.)

7 Technique for fingers (with cold stress):

When a patient's symptoms suggest cold sensitivity, the following procedure is used to document the presence or absence of intermittent digital ischemia in response to cold exposure:

- After the resting study is performed, the feet or hands (depending on which digits are tested) are immersed in ice-cold water for 3 minutes if the patient can tolerate it.

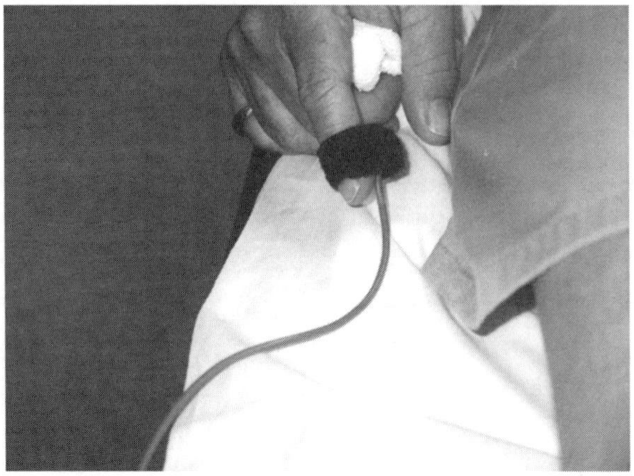

Figure 10-6. Use of the PPG sensor to obtain a plethysmographic waveform of the index finger. In this example a small Velcro strap—rather than double-stick tape—gently but securely keeps the PPG in place. Oftentimes a patient's skin is quite moist, making it difficult to keep the PPG sensor on the skin with tape.

- The skin is towel dried and waveforms and pressures obtained as described previously for the resting study (i.e., item #6, technique for fingers without cold stress) at the following times:

  Immediately after cold stress.

  Five minutes after cold stress.

  If the resting study indicates that one side has more significant disease, obtain the post–cold stress information from that side first.

- Document patient symptoms, skin color observations, and other pertinent findings on the report form.

8 Interpretation:

- Normal waveform qualities (figure 10-7):

  Sharp upstroke during peak systole.

  Prolonged downstroke with dicrotic notch (reflected wave) approximately half-way down.

  Amplitude is usually greater in the fingers than in the toe tracings.

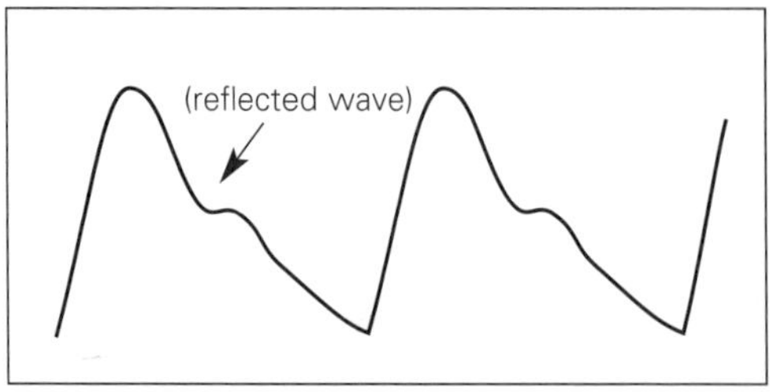

Figure 10-7. Normal plethysmographic waveform as noted by the presence of a reflection (may also be referred to by some authors as a dicrotic notch) in the downslope of the waveform.

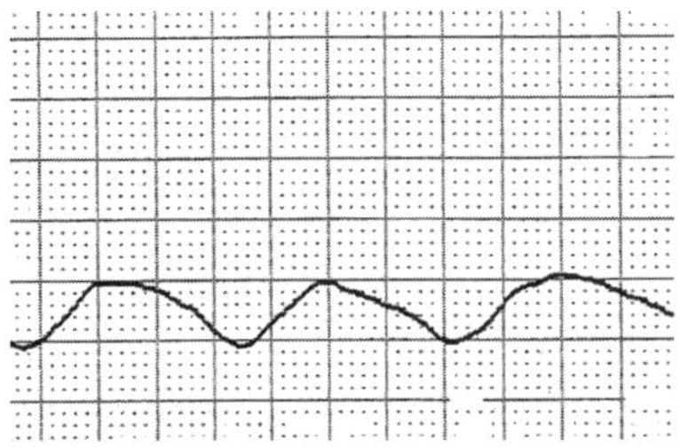

Figure 10-8. Abnormal obstructive plethysmographic waveform.

- Abnormal obstructive waveform quality (figure 10-8):

  This waveform is present with functional (intermittent) obstructive disease.

  Occlusion located anywhere proximally to the end of the digit causes the pulse to assume an "obstructive" pattern, i.e., slow upslope to a rounded peak and a downslope that bows away from the baseline.

- Abnormal peaked waveform quality (figure 10-9):

  Upslope is slower than normal (A).

  Sharp, anacrotic notch is present (B).

  Dicrotic notch located high on the downslope (C).

  Has characteristics of both the normal and the obstructive waveform.

- Other observations:

  An obstructive toe pulse in an extremity with a normal ankle pressure localizes the disease to the pedal or digital arteries.

  When arterial calcification makes it impossible to obtain a reliable ankle pressure, an abnormal toe pulse may provide the only objective physiologic indication for the presence of clinically suspected arterial disease.

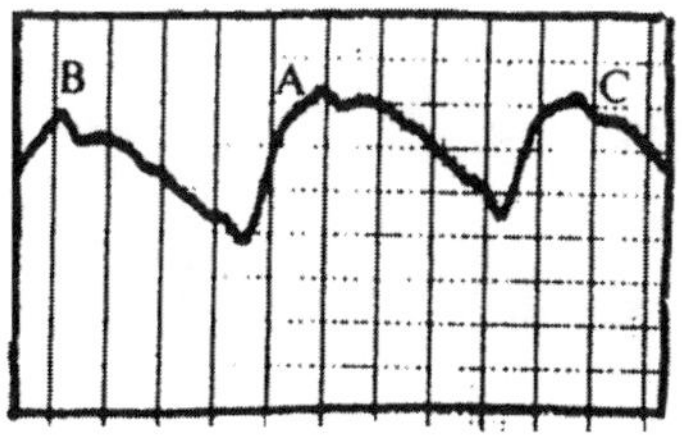

Figure 10-9. Abnormal peaked plethysmographic waveform.

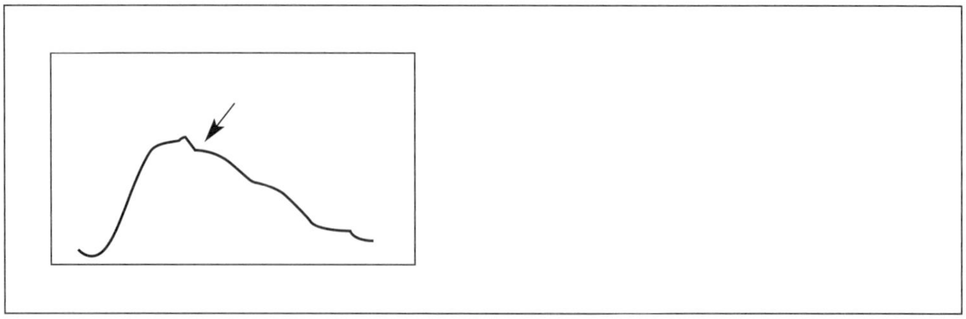

Figure 10-10. A second rendition of the peaked pulse characteristically seen in the digital pulse contours of patients with Raynaud's phenomenon. Reflection (may also be referred to by some authors as a dicrotic notch) is located high on the downslope (arrow).

Organic (fixed) occlusive disease produces abnormal Doppler arterial signals, systolic pressure measurements, and PPG tracings.

Functional (intermittent) obstructive disease is characterized by normal Doppler arterial signals, systolic pressure measurements, and/or PPG tracings, but abnormal findings after cold stimulation.

The "peaked" pulse described by Sumner and Strandness* and demonstrated in figure 10-10 is characteristically seen in the digital volume pulse contours of patients with Raynaud's phenomenon. This waveform contrasts with those in patients with true vasospasm, where the contour is normal in quality but of decreased amplitude.

The patients with true vasospasm are ideal cases for cold stress testing. At room temperature, patients with early vasospastic disease can have normal digital perfusion. After immersion in ice water for 3 minutes, serial plethysmographic waveforms are obtained as rewarming occurs. If the amplitude fails to return to baseline levels within 5 minutes, abnormal cold sensitivity is likely. Figures 10-11A–C provide examples of what occurs after cold stress in this type of patient.

- Systolic pressure measurements:

  Finger/brachial indices of about 0.8–0.9 characterize normal upper extremity digits. Approximately 90% of normal values exceed an index of 0.79.

  According to the literature, normal toe pressures vary from 60% to 80% of the brachial pressure. Values significantly less than this signify digital arterial occlusive disease. The presence of artifactually high ankle pressures from arterial calcinosis usually negates a toe/ankle pressure index.

* Sumner DS, Strandness DE Jr: An abnormal finger pulse associated with cold sensitivity. Ann Surg 175:294–298, 1972.

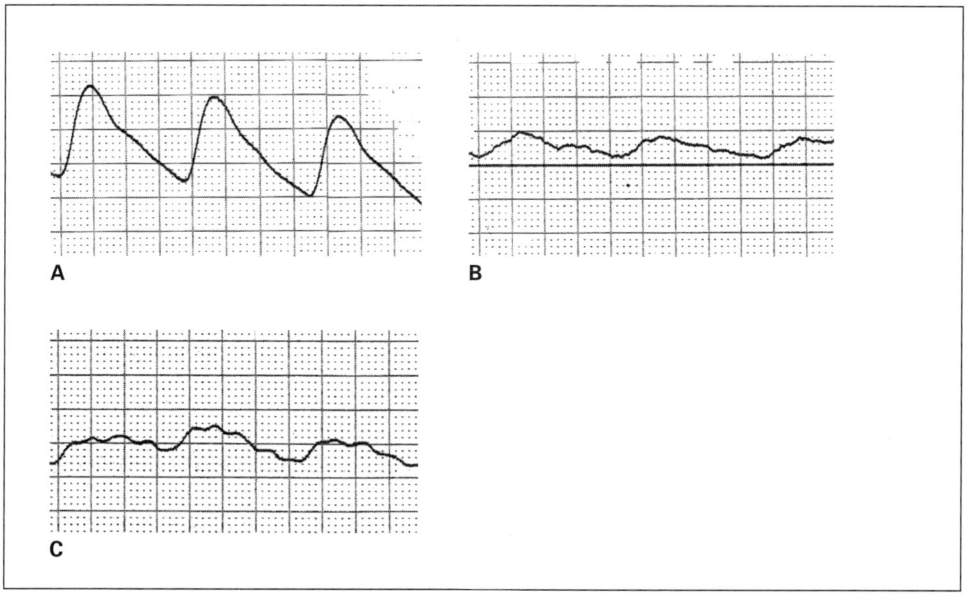

Figure 10-11. Plethysmographic waveforms. **A** Normal resting waveform in a digit prior to any cold stress. **B** Abnormal waveforms seen immediately after 3 minutes of digits' immersion in ice-cold water. Note the attenuation and rounded peak contour, which suggests vasospasm. **C** Abnormal waveforms continue to be seen 5 minutes after the cold stress. Under normal circumstances, the waveform would not be greatly affected by the cold. In this example vasospasm is seen (B), but it does not return to pre-cold stress levels, which tends to suggest more severe disease.

There seems to be little difference between toe pressures in diabetics and nondiabetics, although as previously discussed there is substantial difference in their ankle pressures.

Notes:

- Although some advocate use of the PPG as a monitor (used on the great toe) to obtain ankle and other segmental lower extremity systolic blood pressures, both before and after exercise, discrepancies may occur. There are several sources of error, including peripheral vasoconstriction and room temperature (room too cold). Both play a very important role in how sensitive the exam can be.
- In addition, patients with chronic moderate to severe arterial insufficiency often have very dry skin and/or hypertrophic skin, making good contact between the PPG probe and the skin difficult.

CHAPTER 11

Transcutaneous Oximetry ($tcPO_2$)

1 Capabilities:

- Helps determine wound healing and amputation level.
- Reflects the tissue oxygen tension, which depends on the balance between oxygen supply and consumption.

2 Limitations:

- Inability to keep the electrode fairly flat on the skin surface.
- Electrode cannot be placed on skin that is not intact or is edematous.
- Inability of a patient to lie quietly for as long as 20 minutes (usually the maximum amount of time required to obtain a reading).

3 Patient positioning:

- Patient can be supine, with head of bed slightly elevated.
- Patient should be warm and quiet.
- Ambient room temperature should exceed 20 degrees C.

4 Physical principles:

- Transcutaneous oximetry ($tcPO_2$) reflects tissue PO_2, which depends on a balance between oxygen consumption and oxygen supply. On the surface of the skin, PO_2 can be measured to within 1–2% of its true value if a correctly calibrated electrode is used.
- The electrode houses a heating element that heats the skin to a temperature of 44–45 degrees C, increasing blood flow and melting a lipid layer in the fatty tissue. Since the PO_2 on the surface of the skin is near zero, the vasodilation increases this value by increasing blood flow to the capillaries, which raises their oxygen content and provides for the diffusion of oxygen to the surface where it can be measured.

- A sensor in the electrode measures how much oxygen comes through the skin.

  Located between a membrane (attached to the electrode) and the skin surface is an electrolyte solution. As the arterioles and capillaries dilate, the blood gives up its oxygen more quickly, the oxygen diffusing through the skin, entering the electrolyte solution where a chemical reaction develops, and finally reaching the PO_2 electrode.

  The electrode converts the chemical reaction to a "current" reading that in turn is converted to a measurement of PO_2 in mmHg.

5 Technique:

- The skin site is cleansed with an alcohol wipe and air dried.
- A self-adhesive molded plastic fixation ring is applied to the intact skin. See figure 11-1.
- A few drops of electrolyte solution are put inside the plastic ring.
- The electrode/sensor is gently placed on the skin and then turned securely into the fixation ring. It is important that the electrode be as flat as possible against the skin so that the electrolyte solution covers the skin inside the fixation ring. See figure 11-2.

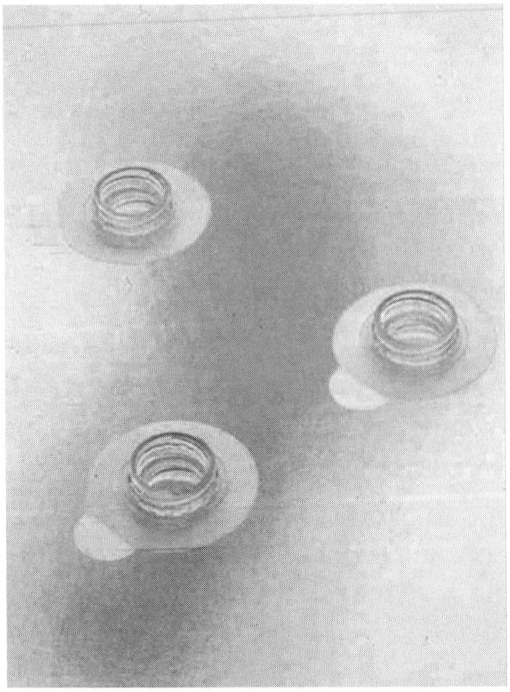

Figure 11-1. The fixation ring is applied to intact skin. It adheres to the skin at its outer circular tape edge only.

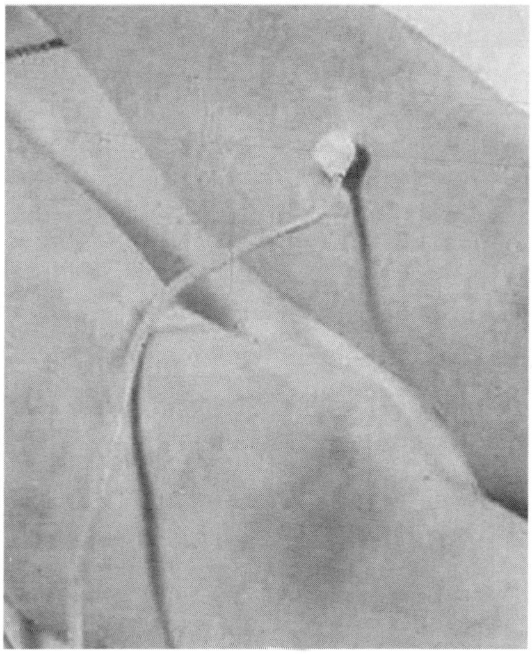

Figure 11-2. The electrode twists onto the fixation ring after the electrolyte solution is put inside the ring.

- After the required manual calibration, PO_2 readings are noted within 15–20 minutes after stabilization, depending on the particular equipment used.
- The reference reading (usually upper left chest) is obtained first, followed by other specific sites.
- The electode is applied near a wound (to determine the probability of healing) or at the anticipated level of amputation (such as 5 cm from the toes or approximately 10 cm below and/or above the patella) to determine healing potential.
- The sensor should not be placed on edematous skin, ulcers, areas of cellulitis, or skin that is close to the bone.
- If the PO_2 level is less than normal or poor, transcutaneous oximetry with oxygen challenge can be performed.

6 Technique with oxygen challenge:

- Electrodes are applied to the chest (reference site) and to the skin site of interest.
- Oxygen per face mask is administered to the patient.
- PO_2 values are obtained.

7 Interpretation:

- Normal PO_2: 60–80 mmHg. (Some references indicate that 30–40 mmHg is normal.) Oxygen challenge should increase this measurement by 10–20 mmHg.
- Poor PO_2: 10–15 mmHg. Oxygen challenge would not increase this value to normal; there may be a slight increase or none at all.
- When determining amputation level, a poor reading would call for the electrode to be moved a bit more proximally until a better reading is attained. A poor PO_2 reading below the knee but a better one above the knee suggests that above-knee amputation is more likely to heal. A poor PO_2 reading at the site of a wound or arterial ulceration suggests that the wound will not heal.
- If a patient's PO_2 falls in between normal and poor, oxygen challenge is the appropriate next step. *Example:* If the initial reading is 20 mmHg, it is possible that it could increase to 30 mmHg following oxygen challenge. The chest electrode (reference) should always increase after oxygen challenge.
- Factors that affect the measured $tcPO_2$ value include arterial PO_2, skin blood flow, skin composition (e.g., thickness), and capillary temperature under the sensor. Arterial PO_2 and skin blood flow are the two most important factors. A low $tcPO_2$, for instance, can be interpreted as reduced arterial PO_2, as in cases of cardiopulmonary disease.
- The values at which healing occurs do vary, depending on the type of study performed and on the level at which PO_2 is being evaluated (e.g., toe, foot, below knee, above knee). In some cases healing takes place at levels less than 30 mmHg.

Duplex Scanning and Color Flow Imaging of the Upper Extremities

CHAPTER 12

1 Capabilities:

- Localize arterial stenosis or occlusion and evaluate degree of stenosis.
- Determine the presence or absence of aneurysm.
- Evaluate hemodialysis access graft or arterial bypass graft postoperatively.
- Detect arteriovenous fistulas or other unusual abnormality.
- Evaluate medical treatment or surgery on follow-up basis.

2 Limitations:

- Presence of dressings, skin staples, sutures, or open wounds.
- Presence of IV site, around which imaging is difficult.
- Hemodialysis access grafts:

  Difficult to assess the anastomotic sites because of graft angulation.

  Difficult to adequately evaluate the outflow vein in an obese patient.
- Diagnosis of Raynaud's syndrome. Previously described physiologic studies combined with the clinical presentation constitute the most useful diagnostic approach.

3 Patient positioning:

- Patient is supine with small pillow under the head.
- The extremity is positioned close to the examiner.
- The arm to be evaluated is externally rotated and positioned at approximately a 45 degree angle from the body in what has been called the pledge position.

4 Physical principles:

- Duplex scanning combines real-time B-mode imaging (gray-scale evaluation) and Doppler spectral analysis (which is an analysis and display of the Doppler-shifted frequencies).
- Color flow imaging continues to provide the duplex information described above and in addition evaluates the Doppler flow information for its *phase* (direction toward or away from the transducer, on which basis color is assigned) and its *frequency content* (which determines the hue or shade of the assigned color).
- More in-depth coverage of this topic should be gleaned from additional sources focusing on physical principles.

5 Technique:

For the patient with a dialysis graft, auscultate for a bruit and/or palpate for a "thrill" (vibration), bearing in mind that both graft stenosis and the high volume of blood flow through a patent dialysis graft will produce this effect.

- A 7 or 5 MHz linear array transducer is used.
- The neck vessels are identified, with attention given to the innominate artery on the right. The left common carotid artery arises from the aortic arch.
- Duplex scanning (with or without color flow imaging) is performed at the following anatomic sites in this order:
  1. Subclavian artery
  2. Axillary artery
  3. Brachial artery
  4. Radial artery
  5. Ulnar artery
  6. Palmar arch if necessary

Gray-scale imaging is used to observe vessel walls and to identify the presence of plaque and its morphology. Spectral analysis of these vessels, including evaluation of waveforms and measurement of peak systolic velocities, is used to assess blood flow. (See figure 12-1.) Color flow imaging is useful to discriminate flow deviations, flow channel narrowing within the vessel, and absence of flow where it should be, as well as assessing whether there is flow where there should not be (e.g., pseudo- or true aneurysm or inflamed lymph node).

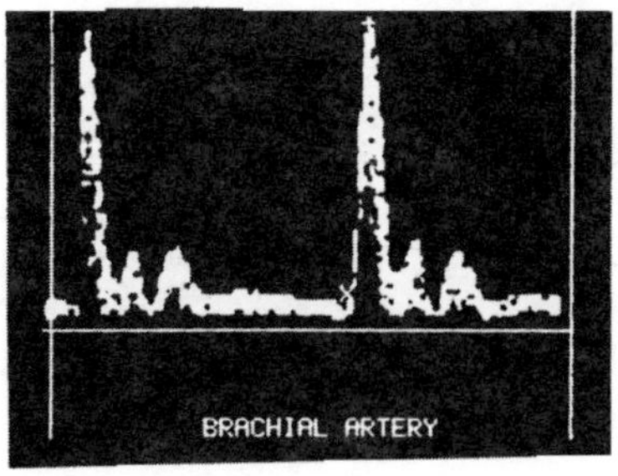

Figure 12-1.
Brachial artery spectral analytic waveform.

- It is rather uncommon for arteries in the upper extremities to become stenotic. The main use for duplex and color flow imaging in the upper extremity is to evaluate hemodialysis access grafts, although other applications exist as well, including vein mapping and evaluation for thrombosis.
- Hemodialysis grafts are evaluated as follows:
  1. Identify and evaluate inflow artery.
  2. Identify and evaluate arterial anastomosis.
  3. Identify and evaluate body of the graft.
  4. Observe for aneurysm, puncture site leaks, perigraft fluid collection, and the like.
  5. If color flow imaging is available, observe the image for frequency increases, turbulence, and flow channel changes, to name a few.
  6. Identify and evaluate venous anastomosis.
  7. Identify and evaluate outflow vein.
- Types of dialysis access include Brescia-Cimino fistula (e.g., radial artery and cephalic vein) (figure 12-2), straight (see figure 12-3 on the following page), and looped synthetic grafts.

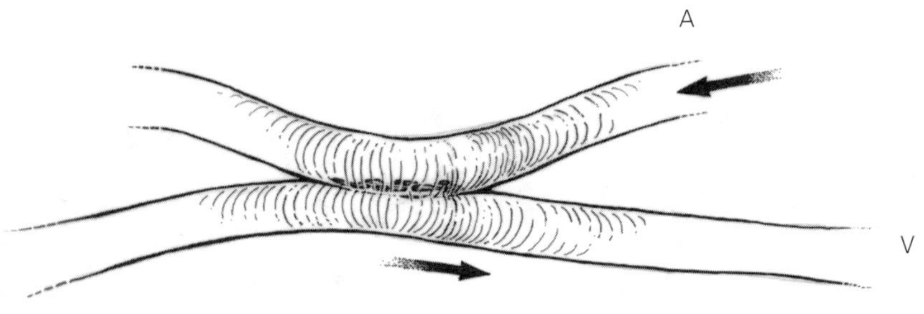

Figure 12-2.
Brescia-Cimino fistula.

Figure 12-3. Straight arteriovenous fistula (dialysis graft).

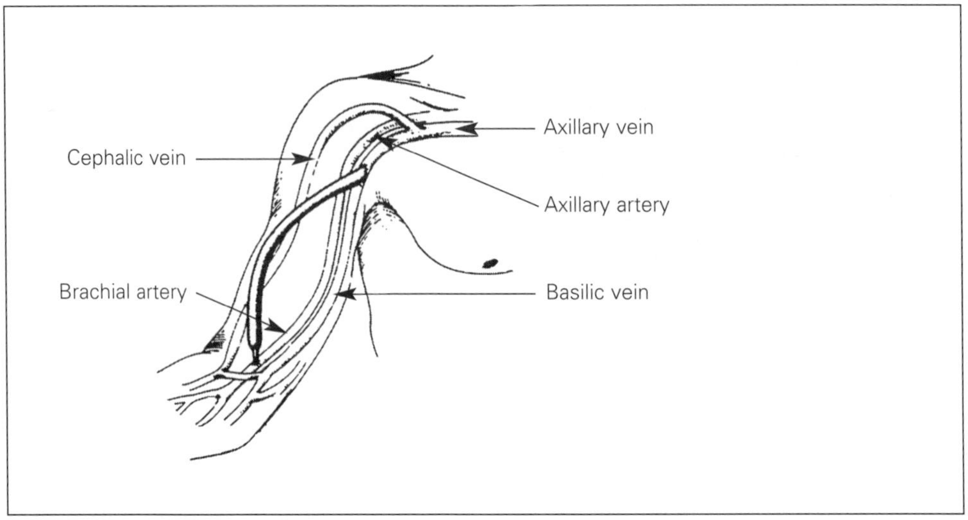

- Longitudinal and transverse approaches are used to evaluate the gray-scale image for thrombus, wall thickening, and other changes.

- Doppler peak systolic velocities (PSVs) are measured at the aforementioned sites (depending on which study is being performed) and also at other locations of interest (e.g., stenosis, preocclusion).

- Documentation consists of videotaping the duplex and color flow images and/or making hardcopy prints/film.

6 Interpretation:

Stenosis

- Currently, there are no criteria for classifying upper extremity disease as there are for the lower extremities.

- Peak systolic velocities vary widely with changes in skin temperature. The Doppler signal quality is usually triphasic (figure 12-4A), but when the hand is quite warm or the arm has been exercised the Doppler signal quality becomes lower-resistant (continuous through diastole). (See figure 12-4B.) The effects of cooling were previously discussed on page 110.

Figure 12-4. Upper extremity Doppler signal. **A** Normal triphasic quality. **B** The same signal in the presence of a vasodilated distal vascular arterial bed due to hyperemia.

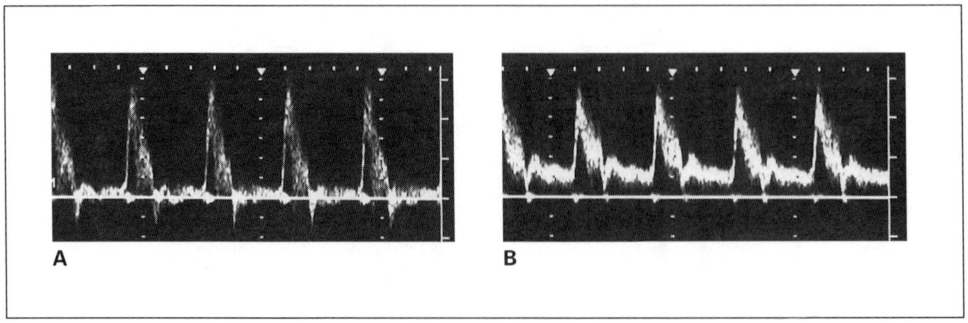

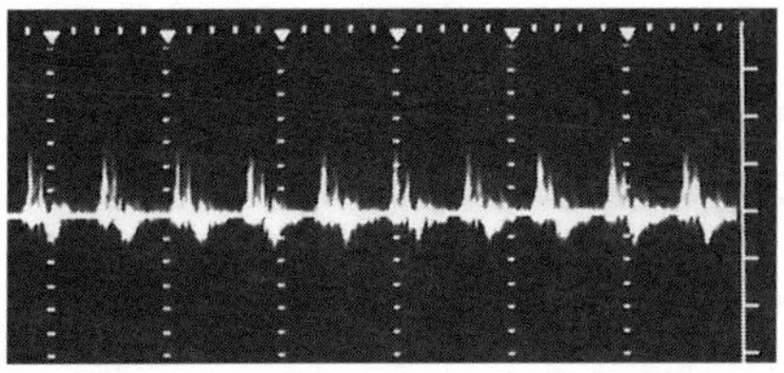

Figure 12-5. High-resistance Doppler arterial signal with bidirectional quality found just proximal to an occlusion.

● Interpretation of duplex and color flow findings is similar to that for other arterial systems. If there is a hemodynamically significant stenosis present, the stenosis profile described on pages 36–37 should exist.

Occlusion

● The examiner watches for the absence of Doppler signals (image and/or waveform) and for the proverbial "thump" that is obtained proximal to the occlusion, the waveform appearing like the example in figure 12-5.

Note: It is not uncommon for nerves, tendons, and veins to be mistaken for an occluded artery. Warming the extremity can decrease the possibility of false-positive information. Color flow imaging is useful in searching for collateral vessels. Gray scale is used to evaluate the walls and/or thrombus, if present.

Aneurysm

● If present, the aneurysm appears as a dilation of the vessel, which can result from degeneration and/or weakening of the wall.

● Ulnar arterial aneurysms can form in response to using the palm as a hammer.

● Subclavian aneurysms often are associated with embolization to the digits.

Hemodialysis Access Graft Findings

● As previously described, the following should be identified and documented as to location, extent, and type:

Aneurysmal changes, including pseudoaneurysm

Puncture sites

Perigraft fluid

Thrombus

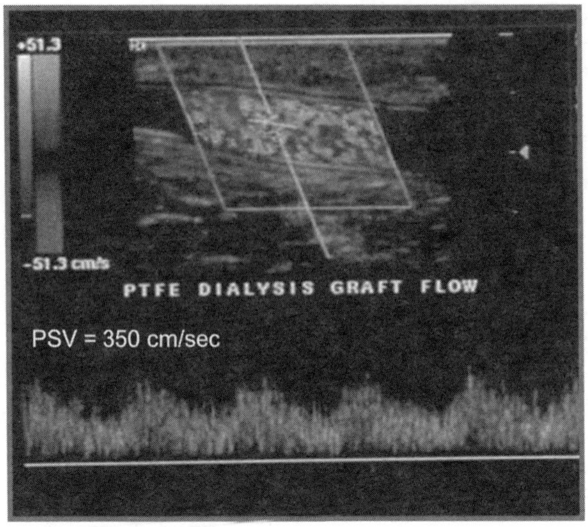

Figure 12-6. Longitudinal view of a synthetic dialysis access graft located in the forearm. The Doppler spectral waveforms depict severe spectral broadening and increased peak systolic velocities (PSVs) of 350 cm/sec. These higher velocities are normally seen throughout the access graft. Color flow imaging would reveal a mixture of color hues.

- Peak systolic velocities vary according to the graft type and normally can be quite elevated. See example in figure 12-6. There are no standardized velocity criteria at this time. Follow-up studies provide specific comparison to previous studies on the same patient.
- A low peak systolic velocity associated with poor-quality Doppler signals obtained throughout the access graft could suggest an arterial inflow problem.
- Occlusion of a fistula or graft obviously prevents successful dialysis. Oftentimes, the dialysis nurse will have trouble accessing the fistula or graft. The patient is then referred for duplex evaluation to determine if there is an occlusion as seen in figure 12-7.
- Volume flow measurement is another method to evaluate dialysis access function. However, volume flow can be quite variable and depends on several factors.
- Regarding dialysis access grafts, the venous anastomosis and outflow vein are the most common sites for stenosis, probably caused by increased arterial pressure introduced to the vein and/or intimal hyperplasia.
- Other hemodynamic complications:

  Because large blood volumes are shunted from the artery to the low-resistance venous circulation, the increased venous return can cause congestive heart failure.

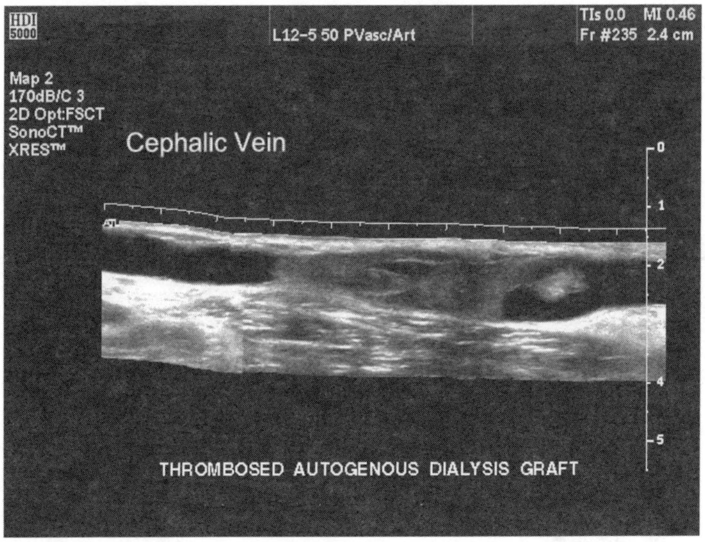

Figure 12-7. Longitudinal view of an occluded cephalic vein that had served as a dialysis access graft. When dialysis access sites are evaluated, complete observation is necessary to determine the presence of abnormalities such as aneurysms, puncture site leaks, and, as here, thrombosis. It is likely that this patient had a Brescia-Cimino fistula. See figure 12-2.

Occasionally there can be a "steal syndrome" whereby the distal arterial (high-pressure) blood flow is reversed into the low-resistance (low-pressure) venous circulation.

Symptoms include pain on exertion of the affected extremity as well as pallor and coolness of the skin (e.g., hand) distal to the shunt.

7 Brachial artery reactivity testing*:

- Assesses the ability of the endothelium to regulate vasodilation by measuring the degree to which the brachial artery dilates in response to reactive hyperemia. Endothelial dysfunction (as documented by a positive test) correlates with a higher-than-normal incidence of long-term coronary events. Although brachial artery reactivity testing is often used to assess endothelial function, the clinical utility of this test remains unclear.

- The patient should be fasting, relaxed, and supine. Reactive hyperemia is induced by inflating a cuff on the forearm (or the upper arm) to 250 mmHg for 4.5–5 minutes.

- The diameter of a brachial artery that has healthy endothelium should increase by $\geq$ 10% when the cuff is on the upper arm. When the cuff is on the forearm, the increase should be at least 6%.

*Source: Vogel RA: Measurement of endothelial function by brachial artery flow-mediated vasodilatation. AM J Cardiology 88:31–34, 2001.

CHAPTER 13

Duplex Scanning and Color Flow Imaging of the Lower Extremities

1 Capabilities:

- Determine presence/absence of greater than 50% diameter reduction stenosis or occlusions.
- Determine presence/absence of aneurysms.
- Evaluate bypass grafts to establish baseline studies for follow-up examinations and to identify grafts with problems.
- To localize the stenotic lesion prior to endovascular procedures.

2 Limitations:

- Presence of dressings, skin staples, sutures, or open wounds.
- Incisional tenderness, hematomas.
- Obesity—difficult to image calf vessels.
- Diabetic patients with calcified arteries as well as atherosclerotic arterial walls, which can cause artifactual shadowing.

3 Patient positioning:

- The patient is supine with pillow under the head.
- The extremity to be examined is positioned as close as possible to the examiner.
- The patient's hip is minimally rotated externally, with his or her knee slightly flexed.

- A prone position may be necessary to assess the popliteal artery, although positioning the patient on his or her side (right or left lateral decubitus) also provides access to the popliteal space.

4 Physical principles:

- Duplex/color flow imaging was briefly described in the previous chapter. The simplified explanation of the Doppler equation that follows serves as a reminder of the importance of understanding each component as well as the significance of utilizing an appropriate Doppler angle. This is applicable specifically to any peripheral or abdominal vessel, arteries in particular. The reader is encouraged to study other references as needed.

- Depending on the reference, the Doppler equation may be written as follows:

$$\Delta f = \frac{2Fo\ V\ Cos\ \Theta}{c} \quad \text{or} \quad \Delta f = \frac{2Fo\ V}{c}\ Cos\ \Theta$$

where:

Δf = Doppler frequency shift. This is a known value measured by the duplex system; it is proportional to the velocity of the source.

Fo = carrier frequency.

c = speed of ultrasound in tissue (1540 m/sec); this is a constant, not a variable.

V = velocity. This is a *calculated* value which represents the velocity of the flow. To calculate velocity, the frequency (Δf) must be known.

Cos Θ = the cosine of the angle between the ultrasound beam and the blood flow. Angle (Θ) is determined from the image; then cosine Θ is calculated (e.g., cosine Θ of 90° = 0).

Note: The number 2 in the Doppler equation represents the "round trip" of the ultrasound. Initially, the red blood cells act as a target for the ultrasound coming from the transducer. (Other terms for the transducer are: stationary source or carrier frequency.) The red blood cells then reflect the ultrasound back to the transducer/receiver.

- The Doppler frequency shift (Δf) is directly proportional to variables in the numerator of the Doppler equation.

- The following variation of the Doppler equation shows how to solve for (i.e., calculate) velocity. To calculate velocity, the duplex system first measures the frequency.

$$V = \frac{c\ \Delta f}{2\ Fo\ Cos\ \Theta}$$

Notes:

- The source of error in the above equation (solving for velocity) is the Doppler angle or cosine Θ, which increases its nonlinear influence as the angle becomes closer to 90 degrees. There is more error production in velocity calculations with large Doppler angles.

- Velocity is directly proportional to c and Δf, and inversely proportional to Fo and Cos Θ.

- The question that comes to mind is, *What should the Doppler angle be when acquiring velocity information in vascular ultrasound?* Since the late 1970s, the ideal Doppler angle for acquiring consistent Doppler flow data in duplex exams is 60 degrees (obtained center stream and parallel to the vessel walls).

- The well-established criteria for interpreting Doppler peak systolic and end diastolic velocity findings, e.g., for carotid, lower extremity, and aortoiliac duplex, are based on using a 60 degree angle of insonation. In order to correctly utilize that criteria, you must duplicate the conditions under which it was developed.

- It is well known that vessel tortuosity or curvature will often prevent usage of the 60 degree angle and, in fact, will likely produce angles of much less than 60 degrees. Nevertheless, as the Doppler angle decreases toward 0 degrees, error production is miniscule. A Doppler angle of greater than 60 degrees is not reliable.

- Regardless of the criteria being used, it is essential to validate that criteria by correlating the results of noninvasive vascular exams with those of angiography to ensure the criteria provide accurate conclusions. Previously used or even old, established criteria cannot be used and relied on in new and different ways with the assumption that they will produce accurate results.

5 Technique for native arteries:

- The examiner scans the following vessels with a 7 or 5 MHz linear array transducer using duplex ultrasonography with or without color imaging:

  Distal external iliac artery

  Common femoral artery (CFA)

  Bifurcation of the common femoral artery

    Superficial femoral artery (SFA)

    Deep femoral artery (profunda femoris)

  Superficial femoral artery (proximal–distal)

  Popliteal artery (proximal–distal)

  Trifurcation

    Anterior tibial artery (ATA)

    Posterior tibial artery (PTA)

    Peroneal artery

- The artery is assessed in gray scale for plaque or other abnormalities. If available, color flow imaging is used to evaluate flow patterns.

- Peak systolic velocities are measured in each major vessel (proximal, mid, and distal if the vessel has great length) and more often if disease is suspected.

  Characterize the Doppler signals as triphasic (normal), biphasic (can be normal or abnormal), or monophasic (abnormal).

  Evaluate the waveform for spectral broadening (minimal, moderate, or severe).

- When a greater than 50% diameter reduction is suspected on the basis of peak systolic velocity and the color flow and/or gray-scale image, obtain the following:

  Prestenotic peak systolic velocity (i.e., approaching the stenosis)

  Peak systolic velocity in the stenotic segment (highest PSV can be at the entrance to or exit from the stenosis). See figure 13-1.

  Poststenotic signals (turbulence)

- Documentation, which includes the duplex color flow images with spectral waveforms, consists of storage on a PACS system, hardcopy prints, or videotape.

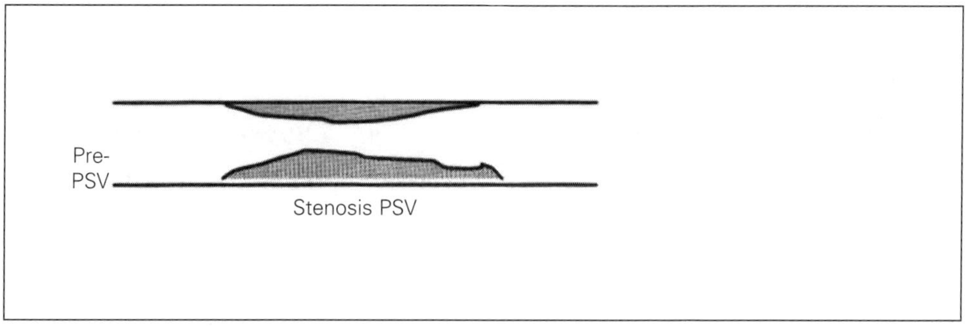

Figure 13-1. Pre-PSV represents the site at which Doppler signals approaching the stenosis should be obtained. Stenosis PSV is where the highest peak systolic velocity is found.

6 Technique for bypass grafts:

- Types of grafts:

  Synthetic graft (e.g., Gore-Tex)

  Reversed saphenous vein graft: After surgical removal and ligation of branches, the saphenous vein is reversed prior to anastomosis to the proximal and distal portions of the artery (i.e., small end of vein to proximal segment of artery, larger end of vein to distal segment of artery). The venous valves are not disrupted because the arterial flow forces them to stay in the open position.

  In situ vein graft: The great saphenous vein (GSV) is left in place with the proximal and distal ends anastomosed to the required artery segments. Prior to anastomosis, the valves are broken up with a special instrument and the vein branches are ligated.

- This protocol is usually combined with ABIs (with or without exercise).

- For synthetic bypass grafts, the following sites are imaged in gray scale and color (if available), and peak systolic velocities are obtained:

  Proximal native (inflow) artery

  Proximal anastomosis

  Mid graft

  Distal anastomosis

  Distal native (outflow) artery

- The evaluation of anastomotic sites is crucial; this is where pseudoaneurysms (possibly due to blood flow leaking through the suture lines) as well as stenoses can occur. Figure 13-2 on the following page presents an example of a bypass graft anastomosis. Figure 13-3 is an example of a pseudoaneurysm of a common femoral artery.

Figure 13-2.
Bypass graft with end-to-side anastomosis. From Ridgway DP: *Introduction to Vascular Scanning: A Guide for the Complete Beginner,* 3rd edition. Pasadena, Davies Publishing, 2004.

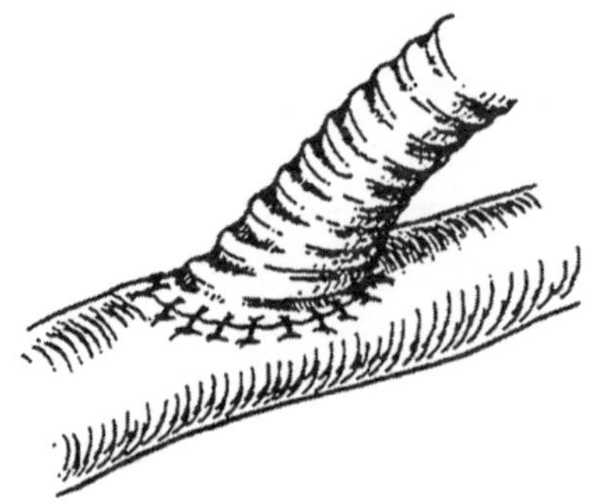

- For vein grafts, all of the aforementioned sites are examined. In addition, the length of the graft is carefully assessed and checked for the presence of patent branches that could possibly form AV fistulas, stenosis, and valve cusp sites.
- To optimize the study while decreasing the frustration level of the technologist/sonographer, it is helpful to know what type of bypass graft was implanted (i.e., in situ, reversed saphenous vein, or synthetic) and where the bypass graft was anastomosed proximally and distally. In addition, it is also helpful to know if there were previous bypass grafts and/or a harvesting of the great saphenous vein. Careful documentation of previous operations is useful for future follow-up studies.
- In general, no matter which type of vein graft has been placed, the entire length of the graft must be assessed. In addition, when evaluating

Figure 13-3.
Synthetic "ringed" Gore-Tex femoral-popliteal bypass graft with a pseudoaneurysm (large arrow) located at the proximal anastomosis, where the pseudoaneurysm originates (small arrow).

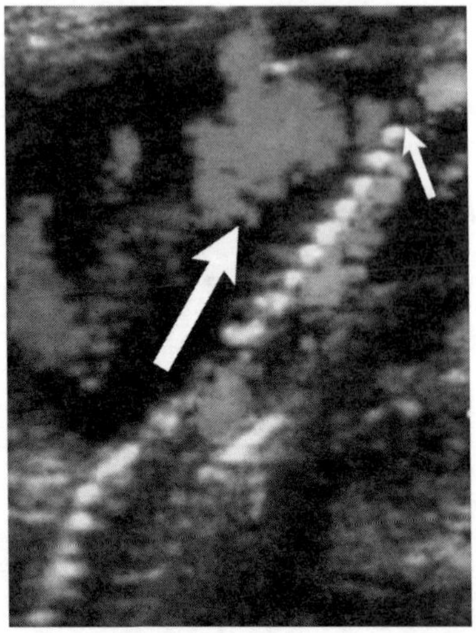

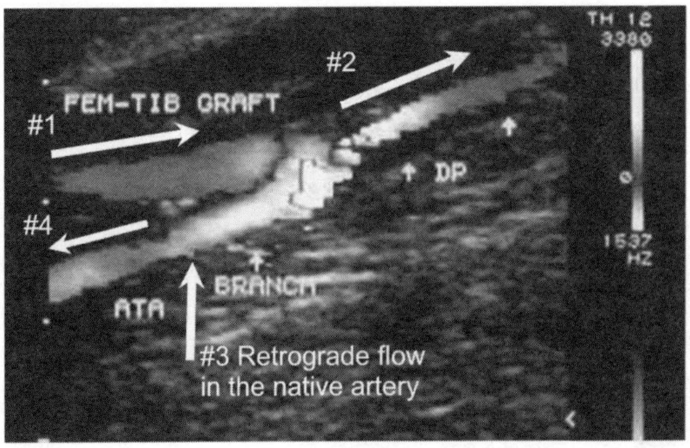

Figure 13-4. **Arrow #1** Flow moving in the vein bypass graft (from left to right on the image) toward the patient's foot. **Arrow #2** Flow moving in the same direction as #1, but in the native dorsalis pedis artery. **Arrow #3** The native anterior tibial artery. **Arrow #4** Flow moving from right to left on the image as it moves proximally in the native anterior tibial artery to fill any branches that may be providing collateral flow around the proximal occlusion or high-grade stenosis.

a reversed saphenous vein bypass graft, the anastomotic sites must be carefully evaluated. For the in situ bypass graft, it is important to observe for branches that may not have been ligated and include observation of the valve cusp sites.

7 Interpretation:

Doppler Signals

- Normal Doppler signals are triphasic. Biphasic signals can be significant, although some patients will normally have biphasic flow signals without any evidence of significant arterial disease.

- Some retrograde Doppler flow in the native vessel at the distal anastomosis may be evident. Because blood moves only from high pressure to low pressure, the normally higher pressure within the graft promotes the flow of blood into the lower-pressure state of the native artery. This retrograde flow provides additional vascular benefit to the patient. See figure 13-4. See also color plate 7.

Stenosis of Native Arteries

- The prestenotic peak systolic velocity is compared to the stenotic peak systolic velocity and a ratio calculated:

  An increase in velocity greater than 100% (i.e., a 2:1 ratio) suggests a greater than 50% diameter reduction.* See figure 13-5.

  A prestenotic to stenotic PSV ratio greater than 4:1 suggests a greater than 75% diameter reduction.** See figure 13-6.

* Moneta GL, Strandness DE: Peripheral arterial duplex scanning. J Clin Ultrasound 15:645–651, 1987.

** Cossman D, Ellison J, et al: Comparison of contrast arteriography to arterial mapping with color-flow duplex imaging in the lower extremities. J Vasc Surgery 10:522–529, 1989.

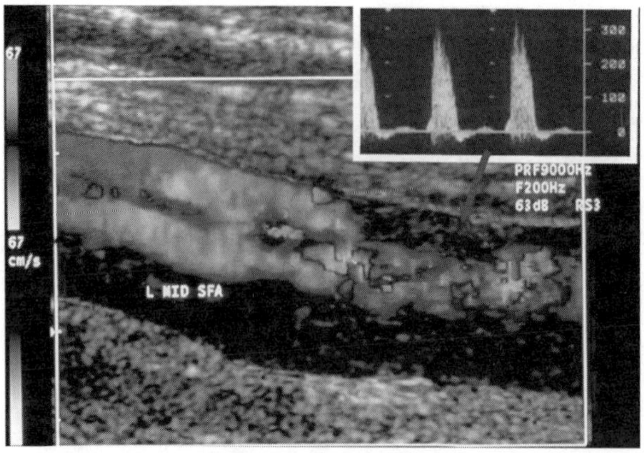

Figure 13-5. Longitudinal image of the superficial femoral artery. The flow channel change depicted by the lighter shade of gray scale is easily noted in this scan of the native artery. This stenosis appears to be longer than the image of the in situ vein bypass stenosis (figure 13-9), which may explain the waveform's more severe spectral broadening and elevated peak systolic velocities. Although not presented here, the prestenotic Doppler peak systolic velocities were 100 cm/sec, which makes the ratio greater than 2:1 since peak systolic velocities at the stenosis are close to 300 cm/sec.

A stenotic PSV greater than 400 cm/sec also suggests a greater than 75% diameter reduction.* See figure 13-6.

- In combination with the aforementioned criteria, there should be post-stenotic turbulence. One should never rely on the "numbers" alone, but rather combine the Doppler spectral data with the image information. (See figure 13-7.)
- Significant stenoses and occlusions commonly occur at the level of the adductor canal (or *Hunter's canal*) in the distal superficial femoral artery and proximal popliteal artery.

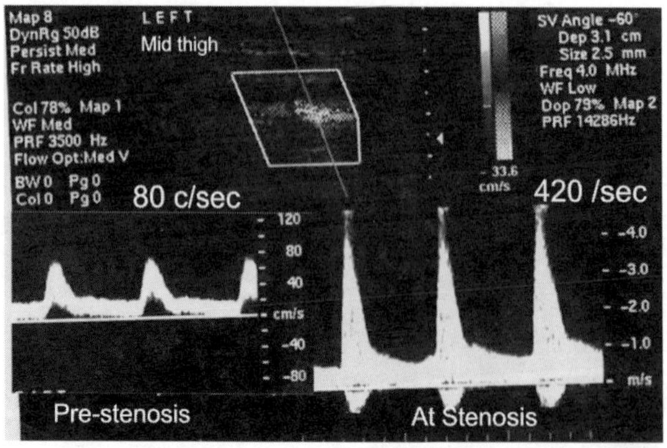

Figure 13-6. Longitudinal image of a bypass graft superficial femoral artery at mid thigh level. Comparison of prestenotic Doppler peak systolic velocities (80 cm/sec) to stenotic peak systolic velocities (420 cm/sec) suggests a greater than 75% diameter reduction stenosis.

* Cossman D, Ellison J, et al: Comparison of contrast arteriography to arterial mapping with color-flow duplex imaging in the lower extremities. J Vasc Surgery 10:522–529, 1989.

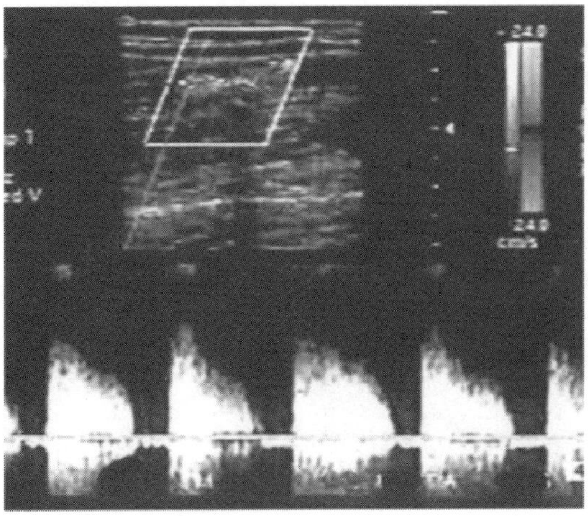

Figure 13-7. Doppler waveforms obtained distal to a significant stenosis (≥50%) of the superficial femoral artery. This pattern of poststenotic turbulence is characteristic of hemodynamically significant stenosis. Flow moves away from the stenosis in random bidirectional patterns of disturbance.

Stenosis and Other Observations of Vein Bypass Grafts

- The most important element in the serial assessment of bypass grafts is the comparison of the current study to previous studies. Observe for significant changes, such as:

  A decrease of 30 cm/sec PSV in any graft segment

  A change in quality (e.g., from triphasic to biphasic waveforms, from biphasic to monophasic, or from triphasic to monophasic). See figure 13-8.

  A decrease in ABI of greater than 0.15*

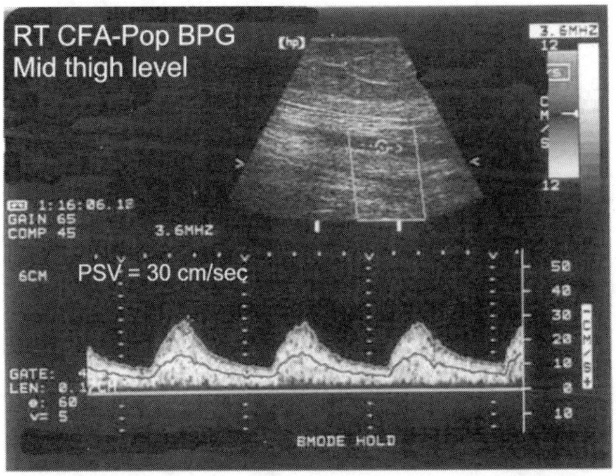

Figure 13-8. Longitudinal image of a femoral-popliteal bypass graft showing poor-quality Doppler signals with PSVs of 30 cm/sec. While it is important to compare velocities to those from previous exams, the shape of the Doppler waveform is equally important. In this case the rounded peak and continuous (low-resistance) flow in diastole suggest a significant proximal problem.

* Cato RF: Bypass graft duplex evaluation with or without color flow imaging. In Rumwell CB, McPharlin MM, Strandness DE, et al (eds): *Vascular Laboratory Policies & Procedures Manual.* Pasadena, Davies Publishing, 2000.

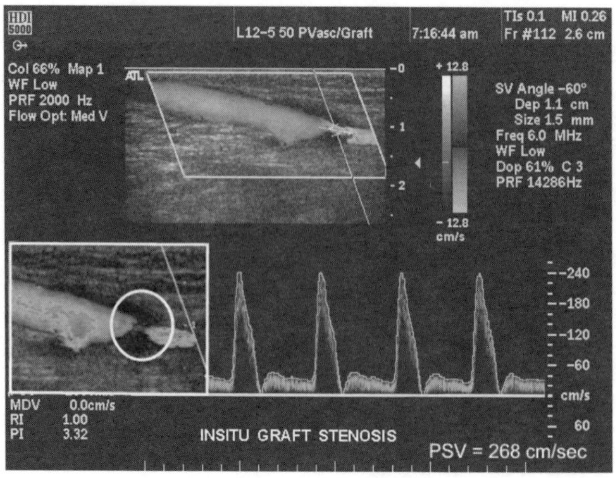

Figure 13-9. Longitudinal image of a highly stenotic in situ vein bypass graft. The significant focal reduction in blood flow filling within the bypass graft immediately suggests there will be Doppler velocity abnormality. The Doppler peak systolic velocities are 268 cm/sec, associated with a monophasic waveform. A smooth, localized narrowing can cause a lack of spectral broadening in the Doppler waveforms, as depicted here.

Note: Bear in mind the fact that discrepancies in size (vessel diameter) between the graft and the native artery—as well as variations within the vein graft itself—will affect the velocity of blood flow.

- The previous criteria and observations for determining whether there is a greater than 50% diameter reduction in native arteries can be applied to the evaluation of grafts (see *Stenosis of Native Arteries* above). See figure 13-9.

- Postoperative complications of a vein bypass graft include arteriovenous fistula (which can siphon off graft blood volume) and an intact valve cusp (which can eventually produce stenosis or occlusion).

- There are currently no standardized interpretation criteria for synthetic grafts; however, the interpretation criteria applied to native arteries (see *Stenosis of Native Arteries* above) and to vein bypass grafts have been applied in a nonstructured manner to synthetic bypass grafts (e.g., Gore-Tex).

Aneurysm

The accepted criterion for the ultrasonographic diagnosis of an aneurysm is an increase in diameter of 50% or greater than the native artery. See pages 43–44 for more discussion of aneurysms; see figure 13-10 for an

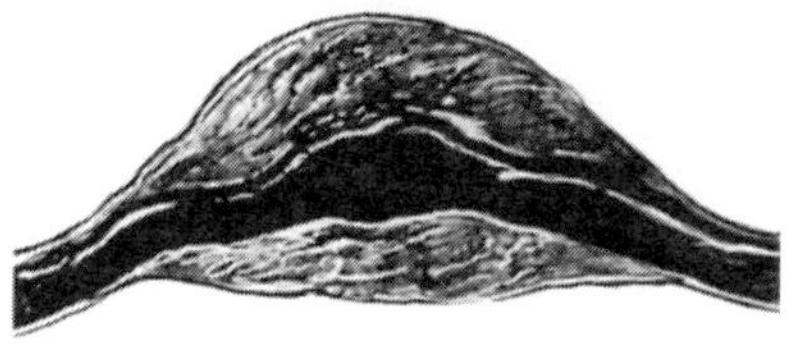

Figure 13-10. Popliteal aneurysm. From Salles-Cunha SX, Andros G: *Atlas of Duplex Ultrasonography.* Pasadena, Appleton Davies, 1988.

example of a true aneurysm. Pseudoaneurysm can also occur in native arteries subsequent to a catheter procedure such as angiography, heart catheterization, or trauma. See figure 13-11 for an example of a pseudoaneurysm.

8 Intraoperative monitoring:

A major intraoperative application of duplex and color flow imaging is to check the patency of the anastomotic sites, as well as to identify and evaluate any suspicious stenotic or turbulent areas in the vein bypass grafts (e.g., valve sites or suspected branch sites).

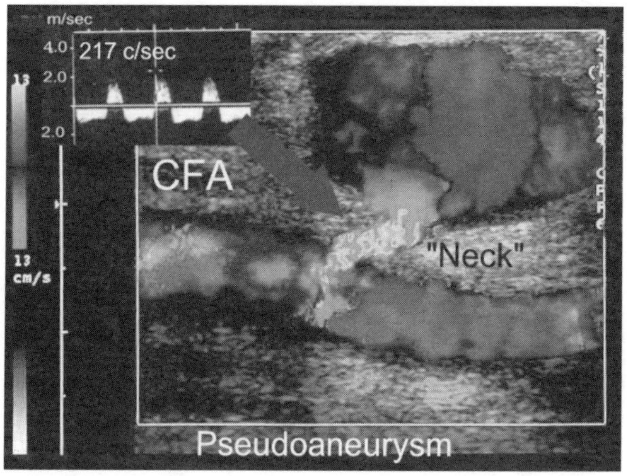

Figure 13-11. Pseudoaneurysm of the common femoral artery (CFA). The "neck" and unique Doppler waveforms (to and fro) are diagnostic hallmarks of a pseudoaneurysm. See color plate 3.

Duplex Scanning and Color Flow Imaging of the Abdominal Vessels

CHAPTER 14

1 Capabilities:

- Aortoiliac vessels: To determine the presence/absence of significant stenosis, follow up bypass grafts, and evaluate aneurysms.
- Renal artery: To determine if significant stenosis of greater than 60% diameter reduction is present.
- Kidney: To help determine presence/absence of nephrosclerotic disease and to evaluate and follow up transplants.
- Mesenteric arteries: To determine presence/absence of significant stenosis, which may account for or cause mesenteric bowel ischemia.
- Liver: To evaluate for suspected portal hypertension and to evaluate pre/post liver transplants.

2 Limitations:

- Depth of ultrasound penetration (size of patient)
- Bowel gas
- Previous abdominal surgery (scar tissue)
- Shortness of breath and rapid respiration
- Nonfasting patient

3 Patient positioning:

- Supine, with minimal head elevation (a pillow is acceptable) as the procedure begins.
- Left lateral decubitus (LLD) for access to the right flank and the right lateral decubitus (RLD) for access to the left flank. These positions allow the bowel to fall to the other side, providing better access to the renal arteries and kidneys.

- The above positions are also useful for accessing the aorta and superior mesenteric artery.
- Evaluation of the abdominal vessels requires creative patient positioning and scanning approaches in order to optimize the study.

4 Physical principles:

See the brief descriptions of physical principles on page 118 of chapter 12 and on page 126 of chapter 13.

5 Technique and interpretation:

General Remarks

- Using a 5, 3, or 2.25 MHz transducer (a higher-frequency transducer, e.g., 7 MHz, may be needed for children or thin elderly patients) and duplex ultrasonography with or without color flow imaging, the examiner begins scanning for the vessels of interest.
- Combining longitudinal (sagittal) and transverse (see Note) approaches, and other approaches, as necessary, the examiner evaluates the gray-scale and color flow patterns, observing for aneurysm, plaque, and other pathologies and findings. *Note:* The transverse approach includes both the orthogonal 90-degree plane relative to the body itself and the short axis view, which is transverse in relation to the structure under examination.
- Using a longitudinal approach in order to accurately and appropriately set the Doppler angle (≤60°), the examiner also assesses the Doppler waveform qualities, peak systolic velocity, and, when appropriate, the end diastolic velocity of the arteries under examination.

Aortoiliac Arteries—Technique

See *General Remarks,* and examine:

- Proximal, mid, and distal aorta to the bifurcation
- Common iliac artery bilaterally
- Proximal, mid, and distal external iliac artery bilaterally
- Internal iliac artery bilaterally
- Brief observation is made of the celiac artery, SMA and renal artery branches

Note: It is important, especially when evaluating an AAA, that the measurements are made of the maximum diameter of the aneurysm. These measurements must be made perpendicular to the aorta, *not* transverse to the body. Many patients have an angulated aorta. In order to obtain accurate measurements, the transducer must be adjusted so that it is perpendicular or orthogonal to the aorta, even if this means the plane of insonation is oblique. Some authors recommend also measuring the diameter of an AAA in the sagittal and coronal planes for reproducibility and to avoid errors based on oblique transverse measurements.

See also the discussion of technique on page 128 of chapter 13.

Aortoiliac Arteries—Interpretation

- Stenosis:

  The criteria applied to the lower extremities are utilized here as well. See *Interpretation* on pages 131–132 of chapter 13.

- Aneurysm:

  For the aorta, a dilation of greater than 3 cm qualifies for designation as an aneurysm. In general, an increase in diameter of 50% or more qualifies an artery as aneurysmal. See figure 14-1.

  The majority of abdominal aortic aneurysms (AAA) are atherosclerotic and infrarenal, i.e., below the renal artery branches.

  The examiner should note the type of aneurysm (true [i.e., fusiform or saccular], false, or dissecting) and the presence of thrombus, if any. See the example of a thrombotic aneurysm in figure 14-2.

  The most frequent complication and danger of an abdominal aortic aneurysm is rupture, but it can also embolize. Both embolization and thrombosis are considered primary complications of peripheral arterial aneurysms. It is not uncommon for both abdominal and peripheral aneurysms to contain varying amounts of thrombus.

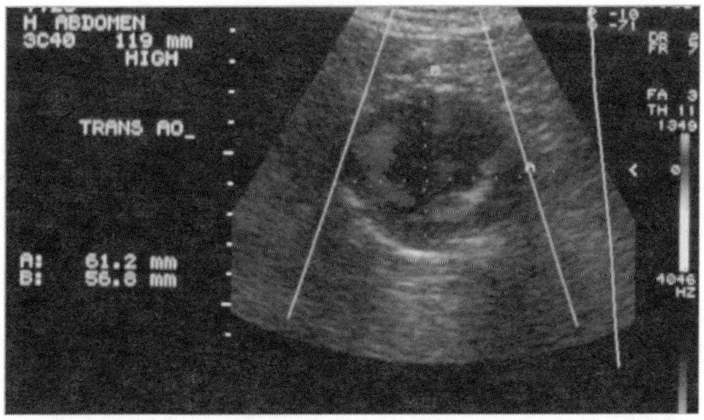

Figure 14-1. In this transverse image, the A/P and lateral measurements range from nearly 5.7 cm to 6.1 cm. Note also the presence of thrombus within the vessel.

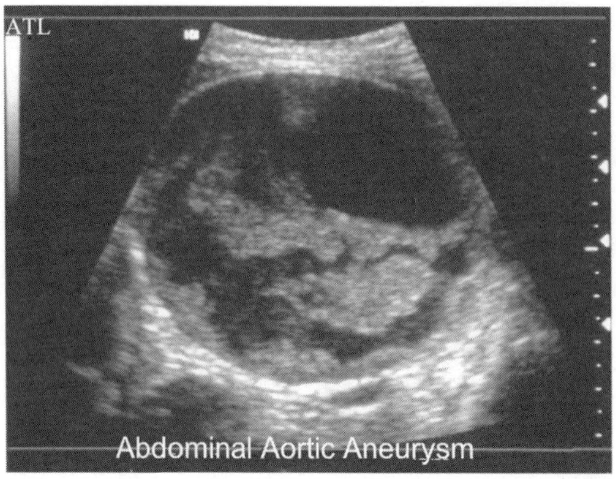

Figure 14-2. Transverse image of an abdominal aortic aneurysm at least 7 cm in diameter. It contains a mixture of echoes that are more heterogeneous than homogeneous. Peak systolic velocities vary according to the extent of thrombus. Large lumen (less thrombus) = lower velocities; small lumen (more thrombus) = higher velocities; preocclusive = lower velocities.

Renal Artery and Kidney—Technique

See *General Remarks* on page 138. It is helpful to know that many of the patients undergoing this study present with hypertension (controlled or not well controlled) and that many of these hypertensive patients have "renovascular" hypertension. Renovascular hypertension is often caused by renal artery stenosis (which can be secondary to atherosclerosis or fibromuscular dysplasia) or occlusion.

- The examiner begins in the longitudinal plane, acquiring velocity data from the celiac and superior mesenteric arteries.
- The peak systolic velocity of the aorta is obtained just distal to the superior mesenteric artery. The aorta is evaluated proximally as needed.
- Using a transverse approach, the examiner locates the renal arteries. The left renal vein is a good landmark for identifying the left renal artery, as depicted in figure 14-3. See also figures 14-4 and 14-5, which show not only the left renal vein, but also the renal artery branches.

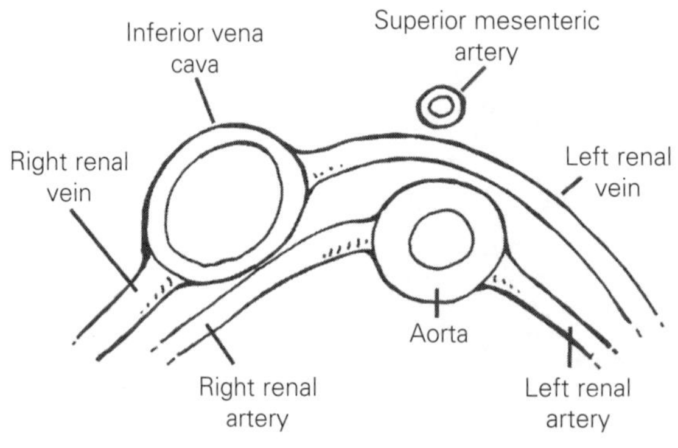

Figure 14-3. Transverse view of the abdomen showing the position of the left renal vein in relation to the left renal artery.

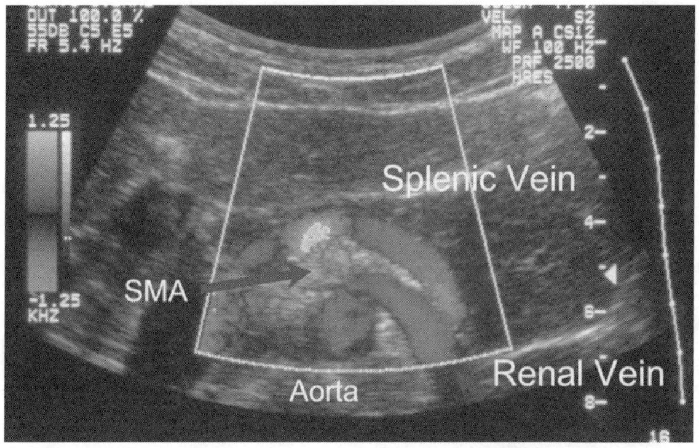

Figure 14-4. Transverse view of the abdominal aorta also showing the SMA, left renal vein, and splenic vein. This image is a good reminder to watch the color bars for flow direction. The veins, in this example, are not blue. See color plate 8.

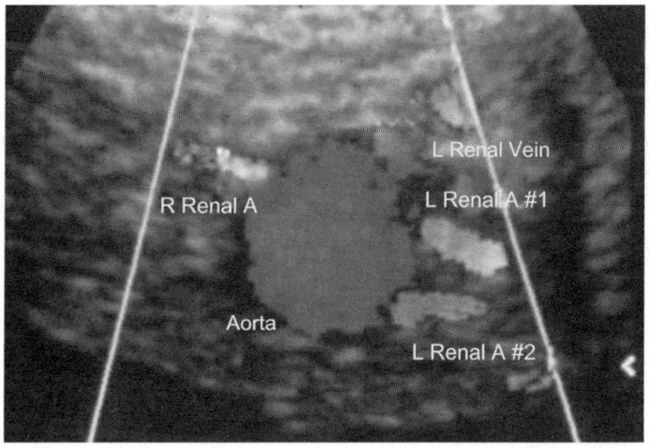

Figure 14-5. Transverse image of the abdominal aorta, showing a portion of the left renal vein, as well as two left renal arteries and a single right renal artery. (Note: In this particular view, it would be impossible to insonate the right renal artery with a Doppler angle of 60 degrees or less; therefore, the peak velocities would be falsely elevated.)

- Kidney size and morphology are evaluated bilaterally. Figure 14-6 on the following page shows a normal-sized kidney with color flow clearly demonstrating the segmental arteries.

- The peak systolic and end diastolic velocities are obtained from the following vessels:

  Renal artery—proximal, mid, and distal levels bilaterally

  The segmental arteries (located in the pelvis of the kidney) and interlobar arteries if necessary (located more distal and outward beyond the pelvis but not yet to the edge) in the upper and lower poles of the kidney (see figure 14-25 on page 156)

- The examiner looks for and notes any secondary or accessory renal arteries. See figures 14-5 and 14-7 and color plate 9.

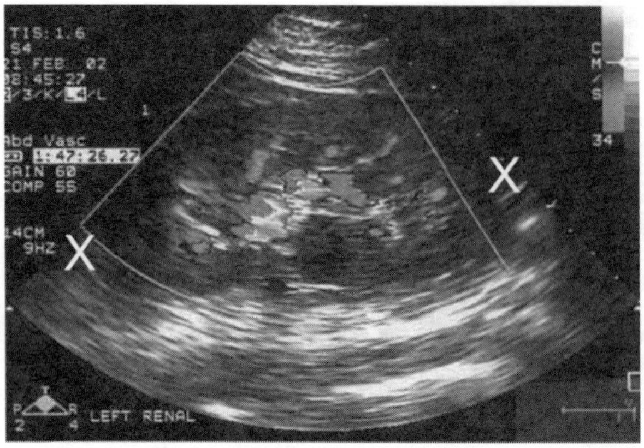

Figure 14-6. Normal-sized kidney with color flow clearly showing the segmental arteries.

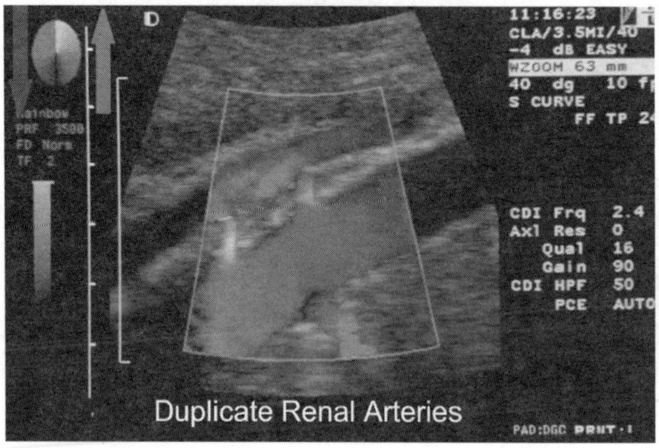

Figure 14-7. Longitudinal (sagittal) view of the abdominal aorta with duplicate right renal arteries and a single left renal artery. This image is often obtained by using an oblique approach from the right side of the abdomen. Because of the configuration of the image, it is often referred to as the "banana peel" image.

Renal Artery and Kidney—Interpretation

- The renal arteries and kidney arteries (i.e., segmental, interlobar) are normally characterized by their low resistance (see figure 14-8). Other arteries with normally low-resistance flow patterns include the celiac, hepatic, and splenic arteries. All of these arteries deliver blood to organs that have vasodilated vascular beds.

- Aortic flow is usually high-resistance, as reflected in figure 14-9A. However, if the Doppler signals are obtained proximal to the renal artery branches, the aortic flow pattern will have a lower-resistance quality (reversal disappears and diastole increases somewhat), as depicted in figure 14-9B, because it is providing flow to a low-resistance renal vascular bed, as well as to the normally high-resistance peripheral vascular bed (iliac and extremity arteries). Other arteries with normally high-resistance flow patterns include the fasting superior and inferior mesenteric arteries.

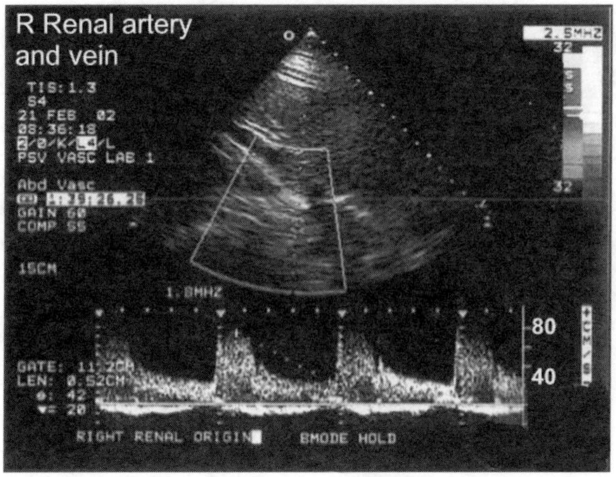

Figure 14-8. Transverse image of the right kidney with a long view of the distal renal artery. The Doppler spectral waveforms, both arterial and renal vein, are normal. Note the compliance peak in the renal artery Doppler waveform, which may or may not be seen in normal waveforms: An initial systolic peak (the "compliance peak") is followed by another peak that is similar in amplitude or slightly higher than the first.

- The renal-to-aortic ratio (RAR) is calculated by dividing the highest peak systolic velocity of the renal artery by the peak systolic velocity of the aorta and interpreted as follows:

$$\frac{\text{Renal Artery PSV}}{\text{Aortic PSV}}$$

Normal < 3.5

Abnormal ≥ 3.5 (this indicates a 60% or greater diameter reduction) *, **

See figures 14-10A and B for examples of these measurements.

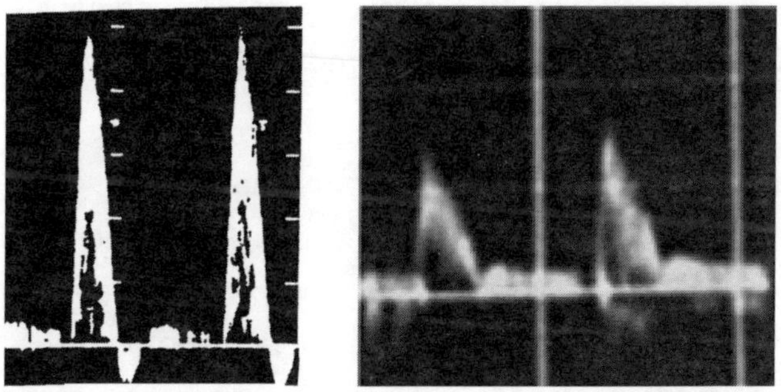

Figure 14-9. Doppler spectral analysis of the aorta. **A** Spectral analysis of normal (pulsatile) high-resistance (triphasic) aortic Doppler signals obtained distal to the renal artery branches. **B** Spectral analysis of a normal, but less pulsatile, aortic Doppler signal obtained proximal to the renal artery branches.

* Kohler TR, Zierler RE, Martin RL, et al: Noninvasive diagnosis of renal artery stenosis by ultrasonic duplex scanning. J Vasc Surg 4:450–456, 1986.

** Neumyer MM: Native renal artery and kidney parenchymal blood flow duplex evaluation with/without color flow imaging. In Rumwell CB, McPharlin MM, Strandness DE, et al (eds): *Vascular Laboratory Policies & Procedures Manual.* Pasadena, Davies Publishing, 2000.

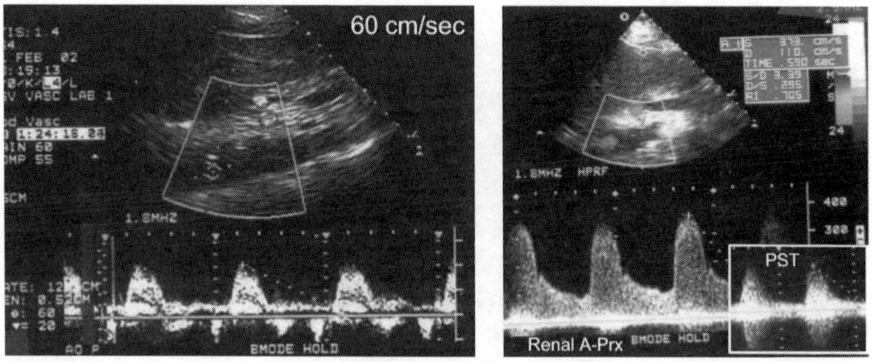

Figure 14-10. **A** Doppler spectral waveforms obtained from the aorta (proximal to the renal artery branches), with peak systolic velocities measuring 60 cm/sec. **B** Longitudinal image of the abdominal aorta and the proximal portion of the right renal artery. Doppler waveforms show severe spectral broadening, abnormally elevated PSVs of 373 cm/sec, and end diastolic velocities of 110 cm/sec. **Inset:** Poststenotic turbulence (PST) following the hemodynamically significant stenosis. Applying the RAR (renal to aortic ratio), 373 cm/sec is divided by the aorta PSVs of 60 cm/sec. The RAR in this example is calculated as > 6.0, which is abnormal. If the aortic PSVs had been unreliable so as not to be able to use the RAR calculation, the fact that the renal artery PSVs are ≥ 200 cm/sec, followed by poststenotic turbulence, strongly suggests a ≥ 60% stenosis.

- The RAR may not be accurate in the presence of an aortic aneurysm or when the peak systolic velocities of the aorta are < 40 cm/sec or > 90–100 cm/sec. In either scenario, finding peak systolic velocities of ≥180–200 cm/sec, along with poststenotic turbulence, is considered by some to be abnormal and suggestive of a ≥60% diameter reduction.*

- A significant stenosis—in this case a reduction in diameter of 60% or more—should conform to the stenosis profile (refer to chapter 2, pages 35–37) and exhibit poststenotic turbulence. There will be spectral broadening in the abnormal waveform. (See figure 14-10B.)

- The resistance of the renal arteries is also evaluated. The ratios used to evaluate kidney arterial flow have also been utilized for the renal arteries. Since renal artery flow is normally of low resistance, the absence or reduction of end diastole suggests distal disease of either the distal renal artery or within the kidney itself. Use of the ratios provides additional information, especially as the values are compared from one side to the other.

- The kidney is examined for morphologic abnormalities (cyst, cortex thinning, other defects). Normal pole-to-pole measurements of length are usually in the range of 10–12 cm. Some kidney lengths can be as small as 8 cm and still be considered normal, however, depending on the patient's body habitus. It is important to compare the measurements of both kidneys. See figure 14-11.

- Kidney arterial flow: The arteries of the kidney (i.e., segmental, interlobar, arcuate) are normally characterized by their low resistance (figure 14-12). Ratios such as the end diastolic ratio (EDR)—also known as

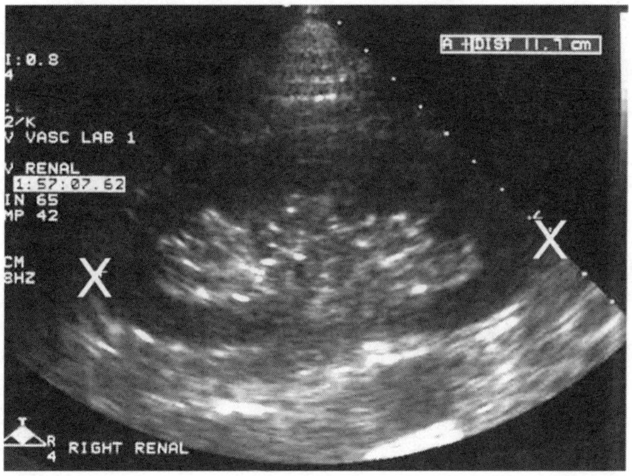

Figure 14-11. Longitudinal view of a normal kidney. The "X" located at both ends of the kidney show how the kidney length is measured.

the parenchymal resistance ratio (PRR)—can be applied to determine if flow resistance has increased (which is considered abnormal):

$$\frac{\text{End Diastolic Velocity}}{\text{Peak Systolic Velocity}}$$

Normal > 0.2

Abnormal < 0.2

An EDR or PRR of less than 0.2 indicates an increase in resistance within the kidney parenchyma. An example of the type of waveform obtained in this situation appears in figure 14-12.*

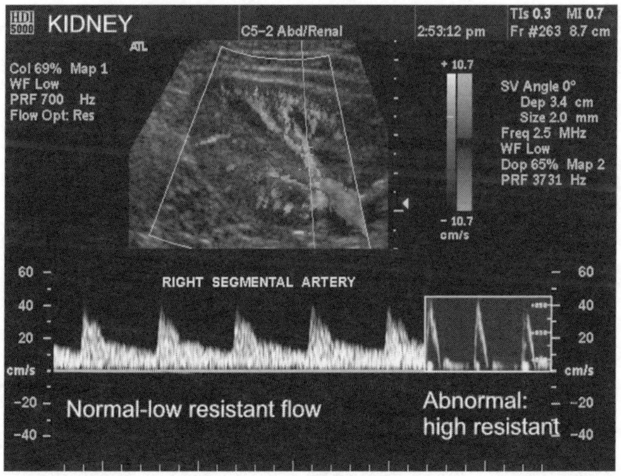

Figure 14-12. Transverse view of the right kidney, using a coronal approach. Doppler signals are obtained from a segmental artery within the kidney. Normally, the signals are low-resistance, as noted on the image beneath the first set of waveforms. In an abnormal (e.g., nephrosclerotic) kidney, the Doppler signal is high-resistance.

* Neumyer MM: Native renal artery and kidney parenchymal blood flow duplex evaluation with/without color flow imaging. In Rumwell CB, McPharlin MM, Strandness DE, et al (eds): *Vascular Laboratory Policies & Procedures Manual.* Pasadena, Davies Publishing, 2000.

- Another ratio used (to a lesser degree) to determine whether resistance is increasing in the kidney (especially the transplanted kidney) is Pourcelot's ratio/resistivity index (RI), which is calculated as follows:

$$\frac{PSV - EDV}{PSV}$$

Normal < 0.7

Abnormal (increased resistance) ≥ 0.7 *

Example of velocity findings in a kidney with abnormal arterial flow patterns:

Doppler findings: PSV = 70, EDV = 10

End diastolic ratio (EDR) = 10/70 = 0.14

$$\text{Resistivity index (RI)} = \frac{70 - 10}{70} = 0.85$$

- The renal resistance-index value (RRIV)** also uses the peak systolic velocity and the end diastolic velocity obtained from the kidneys' segmental arteries. The purpose of using this value is to help prospectively identify patients whose renal function or blood pressure will improve after correction of the renal artery stenosis. A lower-resistance RRIV value of < 80 is associated with improvement in both blood pressure and renal function after the correction of renal artery stenosis. Conversely, an RRIV of ≥ 80 is a strong predictor of worsening renal function and lack of blood pressure improvement, despite the correction of renal artery stenosis. Calculation of the RRIV is as follows: $1 - EDV/PSV \times 100$. When you compare this to the EDR or PRR, described above, it is the inverse of that number. For instance, if the PRR is 0.3, then the RRIV is $0.70 \times 100 = 70$.

- Proximal high-grade stenosis or occlusion of the renal artery may result in a dampened (prolonged upstroke) but still low-resistance waveform (see figure 14-13A). The angiographic result (see figure 14-13B) is an example of what can cause the poor flow depicted in figure 14-13A.

- Some references describe the usefulness of evaluating the kidney arterial flow by determining the acceleration time (AT) and acceleration index (AI).

* Rifkin MD, Needleman L, Pasto ME, et al: Evaluation of renal transplant rejection by duplex Doppler examination: value of the Resistive Index. AJR 148:759–762, 1987.

** Radermacher J, Chavan A, Bleck J, et al: Use of Doppler ultrasonography to predict the outcome of therapy for renal artery stenosis. N Eng J Med 344:410–417, 2001.

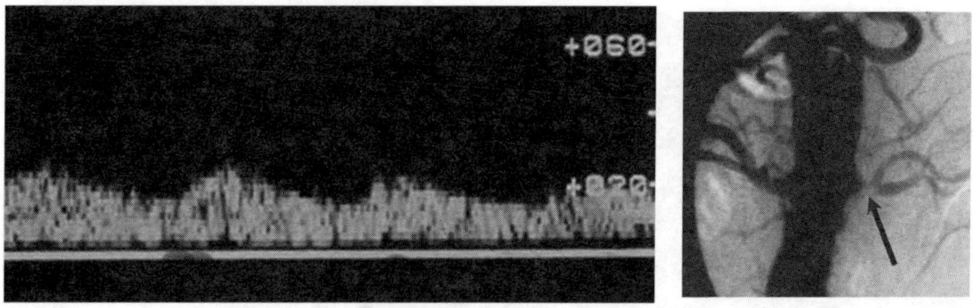

A B

Figure 14-13.
A Doppler spectral analytic arterial signal obtained from the kidney in the presence of a proximal high-grade stenosis or occlusion of the renal artery. This waveform pattern is termed tardus parvus. **B** Angiogram showing an occlusion of the left renal artery. Collateral vessels are providing poor quality arterial flow to the kidney.

The AT is the time interval from the onset of systole to the initial peak and is reported in milliseconds (msec). A proximal stenosis of $\geq 60\%$ diameter reduction is most likely to produce an AT of ≥ 100 msec.

The AI describes the slope of the Doppler velocity waveform. It is calculated as the change in velocity between the onset of systole and the systolic peak (cm/sec) divided by the AT. The units of measure for AI are cm/sec^2. A positive result (consistent with a $\geq 60\%$ diameter reduction) is defined as AI ≤ 291 cm/sec^2.[*, **]

Mesenteric Arteries—Technique

Patient history: Patients who present with a history of dull, achy, or crampy abdominal pain 15 to 30 minutes after meals may suffer from *mesenteric ischemia.* Mesenteric ischemic pain is also known as *mesenteric angina.* This condition may be caused by a stenosis or occlusion of vessels such as the superior mesenteric, celiac, or inferior mesenteric arteries.

Mesenteric ischemia may be acute or chronic. Both conditions are difficult to diagnose. The study described here assists with the diagnostic process, but an arteriogram is essential for diagnosis.

- It is essential that this study always be performed on a fasting patient.
- Peak systolic and end diastolic velocities are obtained in the longitudinal plane from the following arteries:

  Celiac trunk (artery), hepatic artery, and splenic artery, all of which often require a transverse approach to the aorta (see figure 14-14)

* Issacson JA, Zierler RE, Spittell PC, et al: Noninvasive screening for renal artery stenosis: comparison of renal artery and renal hilar duplex scanning. J Vasc Technol 19:105–110, 1995.

** Martin RL, Nanra RS, Wlodarczyk J, et al: Renal hilar Doppler analysis in the detection of renal artery stenosis. J Vasc Technol 15:173–180, 1991.

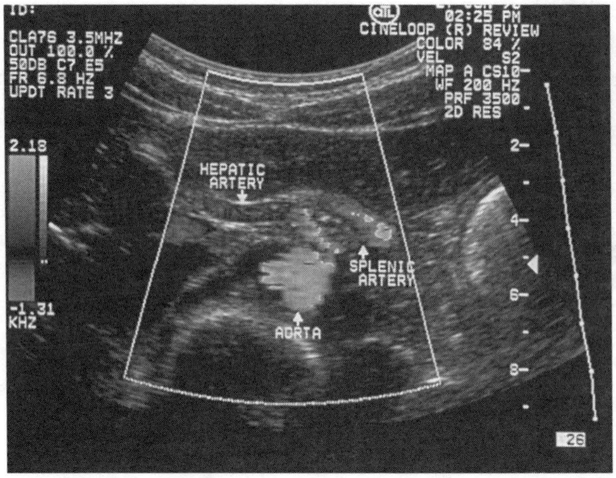

Figure 14-14. Celiac artery (celiac trunk) with branches. Hepatic artery takes blood to the liver; the splenic artery takes blood to the spleen. All Doppler flow is low-resistance because organs have a constant need for blood supply.

Proximal, mid, and distal superior mesenteric artery (see figure 14-15)

Inferior mesenteric artery, if possible

Aorta (to rule out aortic abnormalities)

- After the fasting study (i.e., *postprandially*), in selected cases (e.g., presence of a mesenteric bypass graft) the patient is given a high-calorie liquid meal (e.g., Sustacal, Ensure). The examiner documents the amount of high-calorie liquid ingested (usually 1 or 2 cans) and the time when the patient has completed drinking the high-calorie liquid. The examiner notes the time that the postprandial testing begins; the number of minutes between the ingestion of the liquid and the onset of any symptoms is important. If the patient experiences no symptoms, the examiner notes that fact.

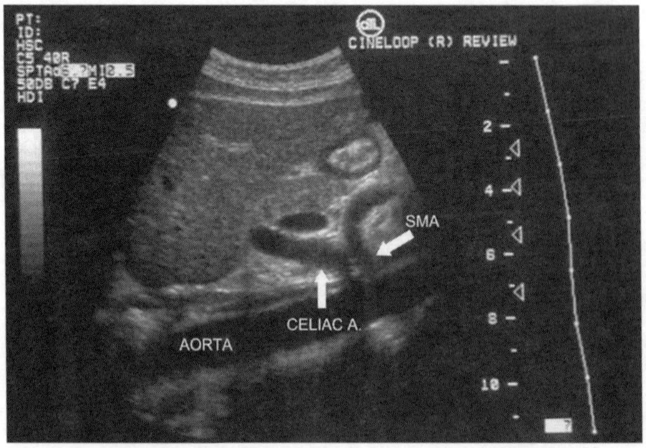

Figure 14-15. B-mode image using a 3 MHz transducer in the longitudinal (sagittal) approach demonstrates the celiac artery followed by the superior mesenteric artery branching off the aorta.

The study is repeated approximately 20–30 minutes after the meal (or sooner if the patient becomes symptomatic), and the peak systolic and end diastolic velocities are obtained as they were before the meal (i.e., *preprandially*).

- The postprandial part of the examination depends on stressing the bowel. In the fasting state, no major mesenteric blood flow is required, but more blood flow is required on a continuous basis after eating. Testing the patient both before and after ingestion of a known amount of calories makes it possible to compare the hemodynamic responses to both states.

- This exam is usually performed in the fasting state to determine the presence of a significant proximal stenosis. Postprandial testing is performed more often to evaluate the function of a mesenteric bypass graft. Vascular ultrasound plays a limited role in the evaluation of mesenteric disease.

Mesenteric Arteries—Interpretation

- In the fasting state, the normal pattern of superior mesenteric artery (SMA) blood flow is characterized by high resistance (figure 14-16A). After eating (or after a high-calorie meal in the vascular testing setting), the flow pattern changes to one of lower resistance (figure 14-16B), as evidenced by the increased diastolic quality. Simply stated, once an individual has eaten, there is digestive work to do, and the arterial blood vessels feeding the mesentery vasodilate in order to provide the additional blood flow necessary for that work. Normally, the end diastolic velocity has been found to at least double postprandially.

- Abnormally, the postprandial pattern of blood flow in the superior and inferior mesenteric arteries remains high-resistance. This finding,

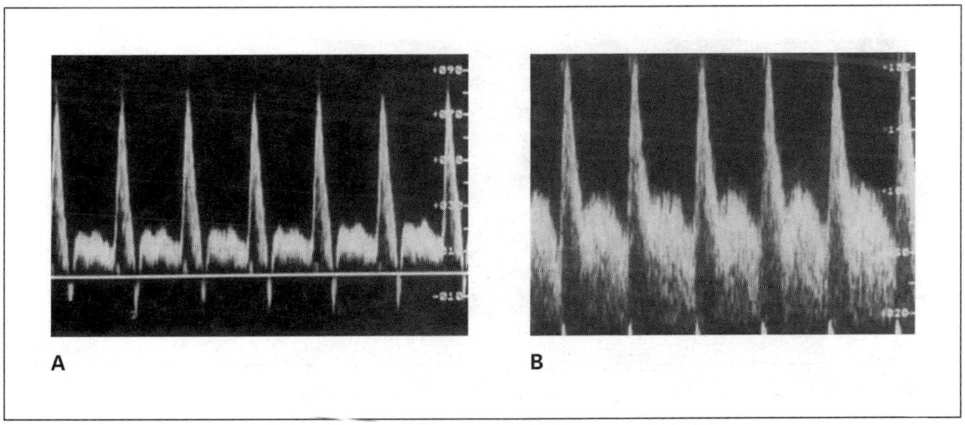

Figure 14-16. Superior mesenteric arterial signals. **A** Normal preprandial signal. Note the flow reversal component and the fairly high-resistance quality. **B** Normal postprandial signal. Note the absence of the flow reversal and the lower-resistance quality.

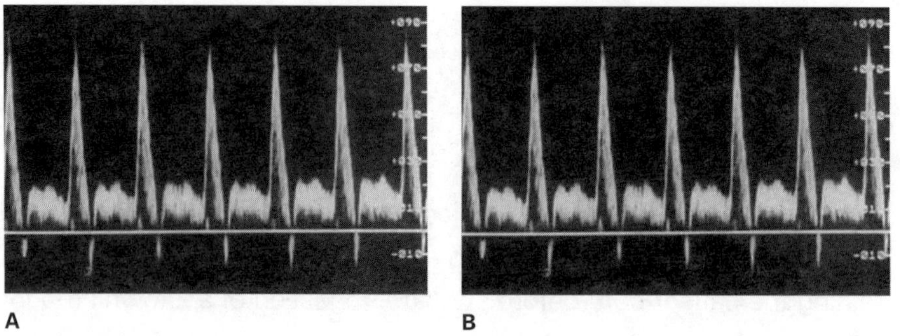

Figure 14-17. Superior mesenteric arterial signals. **A** Normal preprandial high-resistance signal. **B** Abnormal postprandial Doppler signals continue to have high-resistance flow quality.

illustrated by figure 14-17, suggests distal mesenteric ischemia. The high-resistance (i.e., increased pulsatility) blood flow pattern occurs due to the distal obstruction.

- The pattern of blood flow in the celiac artery is not significantly affected postprandially, and so the previous criteria do not apply. Rather, the superior mesenteric and celiac arteries are examined for proximal stenosis.

- Published variances of peak systolic velocity in the celiac artery vary from 50 to 160 cm/sec with end diastolic velocities of ≤ 55 cm/sec. An abnormal fasting peak systolic velocity of ≥ 200 cm/sec is predictive of a ≥ 70–99% diameter reduction.* Figure 14-18 shows a spectral Doppler signal from a normal celiac artery. Figure 14-19 shows a stenotic celiac artery with poststenotic turbulence.

- Published variances of peak systolic velocity in the superior mesenteric artery (SMA) vary from 110 to 177 cm/sec. Abnormal fasting peak systolic velocities of ≥ 275 cm/sec are predictive of a ≥ 70–99% diameter reduction.* See figure 14-20.

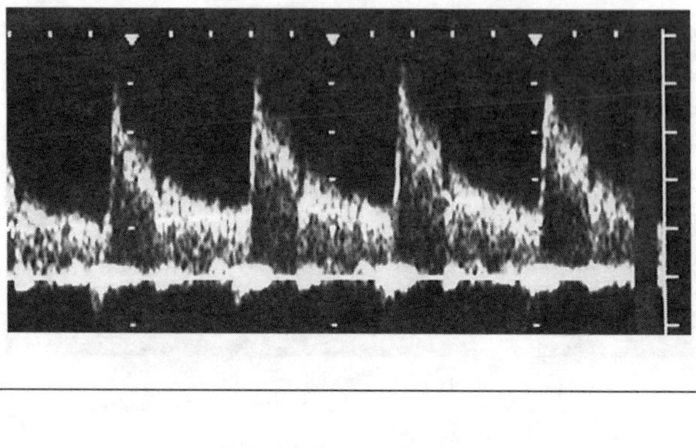

Figure 14-18. The spectral Doppler signal from a normal celiac artery.

* Moneta GL, Yeager RA, et al: Duplex ultrasound criteria for diagnosis of splanchnic artery stenosis or occlusion. J Vasc Surg 14:511–518, 1991.

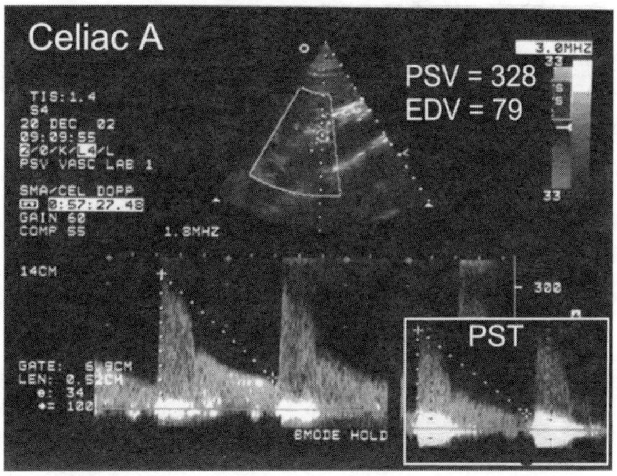

Figure 14-19. Longitudinal image of the abdominal aorta and celiac artery. The Doppler spectral waveforms are abnormal, showing spectral broadening and elevated peak systolic velocities (328 cm/sec). **Inset:** Doppler signals obtained distal to the stenosis described as poststenotic turbulence (PST). Note its bidirectionality and spectral broadening. The PSVs strongly suggest ≥ 70% stenosis of the celiac artery.

- Additional criteria utilizing the SMA end diastolic velocities has also been validated. In this study, an EDV of > 45 cm/sec was the best indicator of severe stenosis in the fasting SMA. Also used along with other useful thresholds in identifying patients with severe stenosis: the absence of a triphasic SMA waveform and a fasting PSV of > 300 cm/sec.*
- Hyperemia reflects decreased arterial resistance in superior mesenteric circulation in response to eating.**
- The increased diastolic flow in the celiac artery reflects decreased end-organ resistance of hepatic and splenic circulation. There is little change in response to food challenge.**

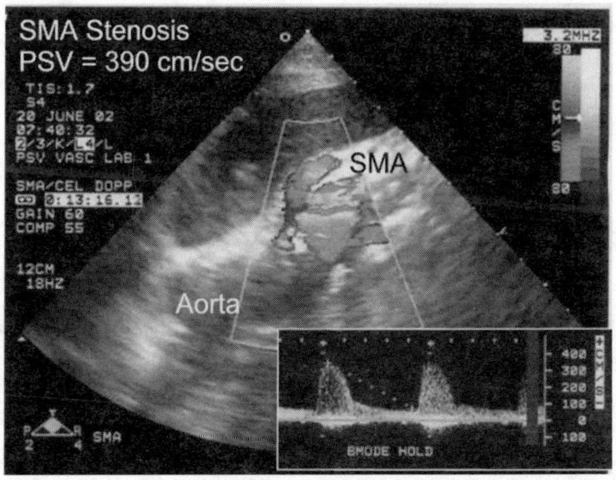

Figure 14-20. Longitudinal image of the abdominal aorta and superior mesenteric artery (SMA) in a fasting patient. The Doppler spectral waveforms show spectral broadening and elevated peak systolic velocities (390 cm/sec) at the proximal portion of the SMA. Poststenotic turbulence is present but not shown here. The findings suggest ≥ 70% stenosis.

* Bowersox JC, Zwolak RM, et al: Duplex ultrasonography in the diagnosis of celiac and mesenteric artery occlusive disease. J Vasc Surg 14:780–788, 1991.

**Criteria and other information from original articles published by Gregory L. Moneta, MD, Oregon Health & Science University Hospital, Portland, Oregon.

Table 14-1. Summary criteria for interpreting pre- and postprandial findings in the superior mesenteric and celiac arteries.

Description	***SMA***	***Celiac***
Preprandial (fasting)		
PSV	High	High
EDV	Low	High
Flow reversal	Yes	No
Postprandial (food challenge)		
PSV	Marked increase	No change
EDV	Marked increase	No change
Loss of flow reversal	Yes	N/A
Velocity criteria		
Normal PSV*	110–177 cm/sec	50–160 cm/sec
Stenosis criteria* (includes poststenotic turbulence)	PSV ≥ 275 cm/sec predicts ≥ 70% diameter reduction	PSV ≥ 200 cm/sec predicts ≥ 70% diameter reduction

*Criteria and other information from various articles by Gregory L. Moneta, MD, Oregon Health & Science University Hospital, Portland, Oregon.

- As always, poststenotic turbulence (PST), an example of which appears in figure 14-19, is a characteristic of the stenotic profile. See table 14-1.

- Extrinsic compression of the celiac artery origin by the median arcuate ligament of the diaphragm is a fairly frequent anatomic finding, especially among young, athletic women. A reversible celiac artery stenosis occurs and is rarely the cause of clinical symptoms. In such cases the high-velocity signals indicative of stenosis are substantially improved with deep inspiration and return with expiration. See figure 14-21. An audible bruit can be auscultated, and during duplex/color flow imaging a "color bruit" may appear during exhalation.

- The celiac artery branches into the hepatic and splenic arteries. The liver and spleen have fixed metabolic requirements and are not likely to be influenced by the postprandial state. Therefore, high peak systolic and end diastolic velocities are normally evident in these vessels at all times.

- The hepatic artery and splenic artery Doppler signals should be low-resistance in quality because each provides arterial inflow to a particular organ. There are no specific peak systolic velocity criteria for what

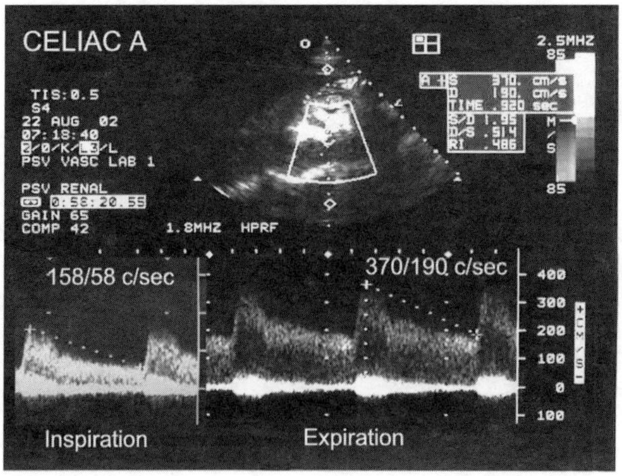

Figure 14-21. Longitudinal image of the abdominal aorta and celiac artery. The Doppler spectral waveforms were recorded at both phases of respiration. Extrinsic compression of the celiac artery during expiration is a frequent finding. In this example, it causes peak systolic (and end diastolic) velocities to increase (370/190 cm/sec). During inspiration, velocities decrease to 158/58 cm/sec.

suggests stenosis in either of these vessels (other than the "stenosis profile" described previously), but if the celiac artery is occluded, retrograde flow may occur in the hepatic artery and, if so, should be identified during the duplex evaluation.

● Because of its small caliber, it is quite difficult to locate the inferior mesenteric artery (IMA), which branches off the distal aorta. Easy detection and ultrasonographic dominance of the inferior mesenteric artery may suggest occlusion of the superior mesenteric artery. See figure 14-22.

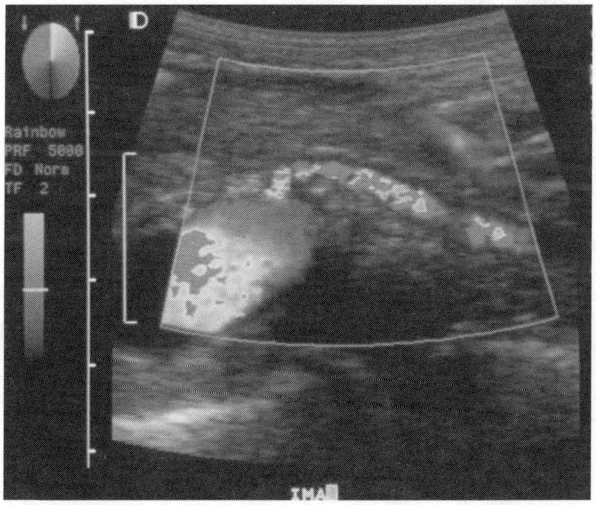

Figure 14-22. Inferior mesenteric artery (IMA) easily observed in this image because of a superior mesenteric artery (SMA) occlusion.

- Studies indicate that duplex evaluation of the mesenteric arteries suggests chronic mesenteric ischemia when there are abnormal findings for at least two of the three mesenteric vessels (celiac, superior mesenteric, and inferior mesenteric arteries) in symptomatic patients.

6 Organ transplants:

Liver Transplant (Allograft)

Liver transplantation is becoming a more frequently utilized procedure to treat patients with end-stage liver disease. In the pre- and postoperative assessment of these patients, the examiner uses a 3 or 5 MHz transducer.

- The preoperative duplex evaluation of candidates for liver transplantation includes documenting the patency of the portal vein, splenic vein, superior mesenteric vein, hepatic veins, inferior vena cava, and hepatic artery while observing for abnormalities (i.e., tumors in the liver and surrounding tissue) and/or other vasculature. Doppler waveforms are evaluated according to the appropriate criteria. It is also important to determine the status of the biliary tree. Abnormalities such as thrombosis of the portal vein can make transplantation difficult, if not impossible. See figure 14-23.

- Postoperatively, duplex ultrasonography is used to document patency of the portal vein (figure 14-24), splenic vein, superior mesenteric vein, hepatic veins, inferior vena cava, and hepatic artery, as well as portal vein flow direction and vessel size. Ideally, portal vein size should be ≤ 1–1.5 cm.

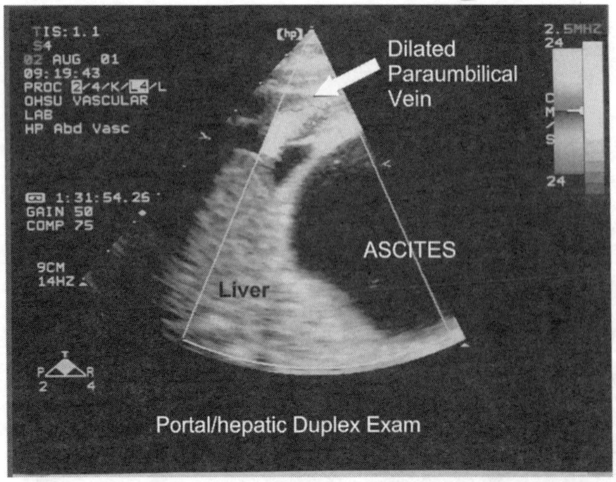

Figure 14-23. Sagittal/oblique view of the liver revealing ascites and a dilated paraumbilical vein. It is not unusual to observe one or both of these findings in patients being evaluated for liver transplantation. A sudden increase of ascites often precedes portal vein thrombosis, especially in cirrhotic patients. Dilation of the paraumbilical vein is a specific finding for portal hypertension, in which normally hepatopetal flow becomes hepatofugal as blood flow is diverted to the systemic venous circulation through a collateral pathway.

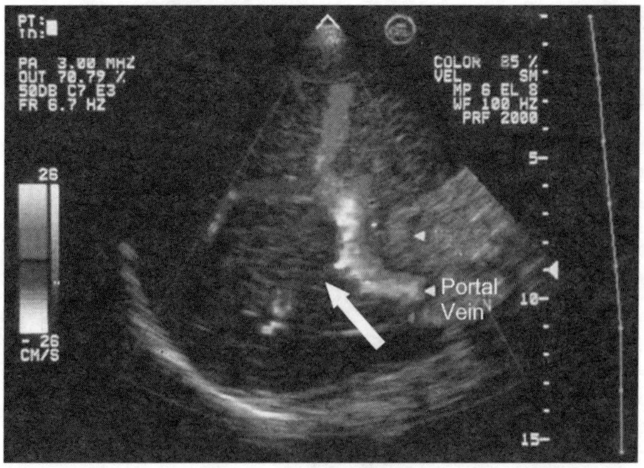

Figure 14-24. Duplex scan of the liver with a patent, normal portal vein. Flow direction (arrow) is toward the liver (hepatopetal).

- Postoperative complications include allograft rejection, pseudoaneurysm, hepatic infarction, and thrombosis of the portal vein, inferior vena cava, and/or hepatic artery. Obviously, the hepatic artery inflow becomes crucial to the viability of the liver.

  The location of this artery provides a challenge to obtaining a proper Doppler angle. If the Doppler angle is high (i.e., 80–90 degrees), no flow may be detected because of the poor or nonexistent Doppler shift. Before concluding that the hepatic artery is occluded, the examiner must be certain that an appropriate Doppler angle has been used. If possible, the proximal hepatic artery is examined where it branches off the celiac artery to determine the presence or absence of flow.

  Acute rejection will cause liver dysfunction. It has been thought by some that rejection would cause arterial resistance (measured by the resistivity index) to increase. Others, however, suggest that rejection is a cellular process which does not affect the peripheral vessels.

Renal Transplant (Allograft)

- Both living related–donor and cadaveric kidneys are implanted most commonly in the right iliac fossa.

- Postoperative examination is facilitated if the examiner knows what type of anastomosis was used. The transplanted renal artery vessels are anastomosed either end-to-side with the external iliac artery or end-to-end with the internal iliac artery. The venous anastomosis connects the renal vein to the external iliac vein in an end-to-side fashion. Most often, the donor ureter is implanted into the bladder directly.

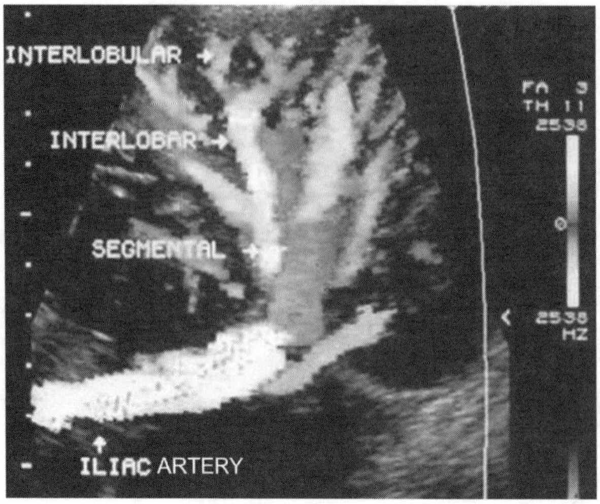

Figure 14-25. Transplanted kidney located in the iliac fossa, which makes scanning easier because of the more shallow location of the kidney. The kidney branches are clearly seen in the kidney pelvis (segmental arteries) to the outer region (interlobular arteries).

Technique

- Duplex evaluation is usually performed with a 5 MHz transducer. See figure 14-25.
- Longitudinal and transverse approaches are utilized.
- The length and A/P diameter of the allograft are measured for subsequent follow-up.
- Careful B-mode observation is made of parenchymal echogenicity, including the renal cortex, renal sinus, and pyramids.
- B-mode observation continues for the detection of fluid collections (e.g., hematoma, abscess). If there are regions of hypoechogenicity within the allograft, Doppler ultrasound is used to differentiate between a vascular abnormality or a fluid collection.
- Postoperative follow-up also includes Doppler spectral analysis of the aorta, internal or external iliac artery, donor renal artery, external iliac vein, donor renal vein, and allograft vessels. Peak systolic and end diastolic velocities are obtained and flow qualities assessed as to whether the flow resistance is normal or abnormal.

Interpretation

- B-mode signs of rejection include:

  Increased renal transplant size

  Increased cortical echogenicity

  Hypoechoic regions in the parenchyma

- The promise of Doppler analysis as a method for diagnosing acute rejection continues to be a source of discussion. There are those groups who have found the resistivity index to be predictive of acute rejection; others, however, have found that pulsatility abnormalities are not reliable. What is interesting is that conditions other than acute rejection—renal vein thrombosis, infection, cyclosporin toxicity, and tubular necrosis, to name a few—can increase pulsatility.
- According to many sources, biopsy continues to be the most reliable method for rejection diagnosis.
- With respect to other vascular complications, duplex scanning also has diagnostic value. Its applications include:

  Renal artery stenosis

  Renal vein thrombosis

  Arteriovenous fistula

  Pseudoaneurysm

Preoperative Mapping Procedures

CHAPTER 15

Preoperative Epigastric Artery Mapping

The *deep superior epigastric artery* (EA) is the terminal branch of the internal mammary artery. The diameter at its origin is about 1.6 mm. The *deep inferior epigastric artery* arises from the external iliac artery in the lower abdomen. Smaller branches of both arteries anastomose in a region known as the *watershed area* located approximately midway between the superior and inferior epigastric arteries. See figure 15-1.

Both arteries and their perforators contribute to the blood supply of the rectus abdominis muscle. Vertically positioned and lying on either side of midline, this long strap muscle, along with the subcutaneous fat, arteries, perforators, and overlying skin, is referred to as the *transverse rectus abdominis myocutaneous (TRAM) flap*.

1 Indications and purpose:

- To identify the location and determine the patency of the epigastric arteries and their perforators.
- The muscle with the best arterial blood supply is used for the TRAM flap in autogenous breast reconstruction.

Figure 15-1.
Rectus abdominis muscle is positioned vertically. Note the locations of the superior and inferior epigastric arteries and branches.

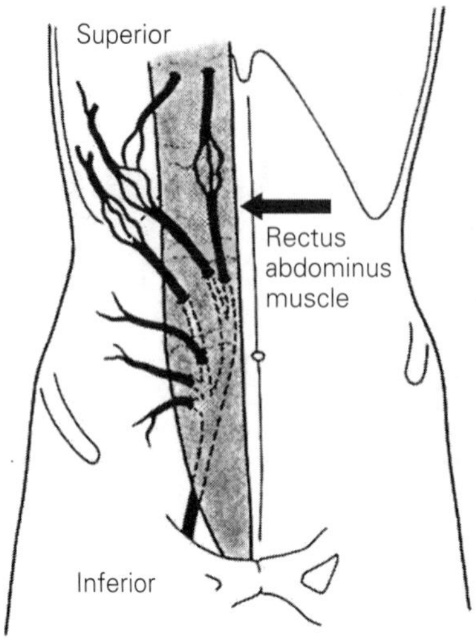

2 Technique and interpretation:

- Imaging the epigastric arteries requires considerable skill and is time-consuming.
- A 7.5 MHz linear array or 5 MHz phased array transducer is used.
- Color flow Doppler (in a low-flow setting) or power Doppler is used to identify location and obtain velocities of the epigastric arteries and their perforators.
- The epigastric arteries are approximately 3–4 cm below the skin surface. The diameter of the perforators is less than 1 mm.
- During the preoperative mapping, the location of the main perforators are marked on the skin with an indelible pen.

3 Limitations:

The success of this examination may be limited by:

- Depth
- Scarring
- Effects of radiation treatment

Preoperative Internal Mammary Artery Mapping

The *internal mammary artery*—also called the *internal thoracic artery*—arises from the arch of the subclavian artery. It descends along the posterior sides of the cartilage of the upper 6 ribs, about 1 cm from the sternum.

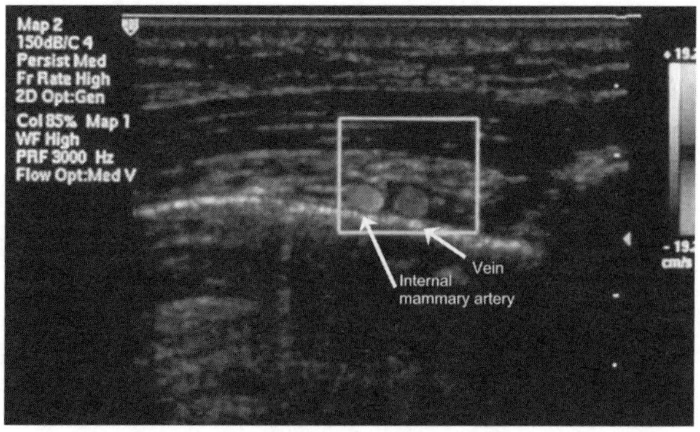

Figure 15-2. Transverse image showing the internal mammary artery and companion vein.

1 Indications and purpose:

- To determine if the artery can be used as a recipient site for a TRAM flap in breast reconstruction or as a graft to the left anterior descending (LAD) coronary artery.

2 Technique and interpretation:

See the technique and limitations of epigastric artery and perforator mapping above. Additional notes:

- Sonographically, the internal mammary artery appears as a tubular structure about 2 mm in diameter, and it is characterized by low-resistance flow. See figure 15-2.
- The supraclavicular fossa is used for accessing the left subclavian artery to make certain it is patent.

Preoperative Radial Artery Mapping

The *radial artery* originates from the brachial artery and travels down the lateral side of the forearm (in anatomic position) into the hand. In the hand it gives off a branch to form the superficial palmar arch. The radial artery then terminates in the deep palmar arch of the hand where it joins the deep branch of the ulnar artery.

1 Indications and purpose:

- To determine the suitability of the radial artery as a graft for coronary artery bypass grafting.

2 Technique and interpretation:

- Patient is supine with arms resting at his or her sides.
- Patency of the palmar arch is assessed by the modified Allen test (see pages 79–81). If the results of the modified Allen test are negative, the

examiner may proceed with the mapping exam. If the Allen test is positive, there is no need to proceed, because removal of the radial artery would compromise perfusion to the hand.

- Duplex ultrasonography (using a 7.5 or 5 MHz transducer) is used to evaluate the brachial, ulnar, and radial arteries for gray-scale and Doppler flow qualities. Peak systolic velocities (PSVs) are obtained at various increments.
- Observe for abnormalities such as abnormal Doppler quality, increased PSVs, and calcification of the artery wall.
- Measure artery diameter. Acceptable diameter may vary according to the criteria of the particular vascular diagnostic facility or surgeon.

Preoperative Vein Mapping

Preoperative vein mapping may be performed in the upper and/or lower extremities. In the lower extremity, the great and/or small saphenous veins are mapped to assess their acceptability for use in grafting procedures. (The *great* and *small saphenous veins* are the currently acceptable terms for what formerly were called the *greater* and *lesser saphenous veins.*) In the upper extremity, the cephalic and/or basilic veins are mapped.

1 Indications and purpose:

- To determine suitability of the vein for use as an extremity or coronary artery bypass and/or dialysis access graft.

2 Technique—lower extremities:

- Duplex ultrasonography (using a 7.5 MHz transducer or higher) is used for gray-scale evaluation.
- The great saphenous vein (GSV) and sometimes the small saphenous vein (SSV) are mapped from proximal to distal in a transverse view. See figure 15-3.
- The entire length of the vein is evaluated at increments for compressibility (coaptation of vein walls), diameter, and continuity. Typically, the increments for the GSV are:

  High, mid, distal thigh

  Knee and below knee

  Proximal, mid and distal calf

  Ankle

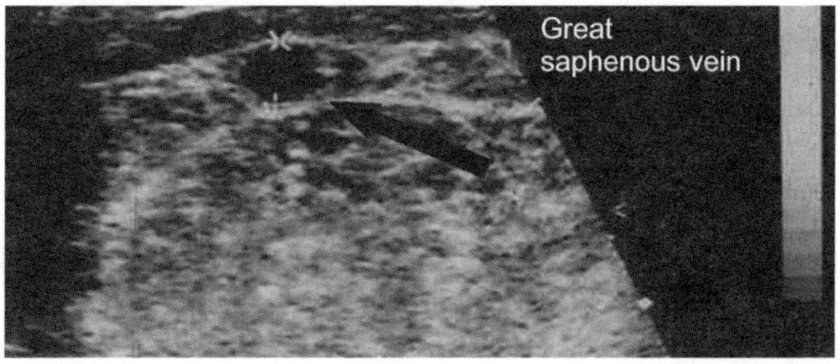

Figure 15-3.
Transverse view of the great saphenous vein in the thigh. Diameter measurement is 4 mm, which is acceptable.

3 Technique—upper extremities:

- Duplex ultrasonography (using a 7.5 MHz transducer or higher) is used for gray-scale evaluation.
- The entire length of the cephalic and basilic veins is mapped from proximal to distal in a transverse view. See figure 15-4.
- The entire length of the vein is evaluated at increments for compressibility (coaptation of vein walls), diameter, and continuity. Typically, the increments are:

  Upper arm: proximal, biceps, above elbow, elbow

  Forearm: proximal, mid and distal levels

  Wrist

Note: The basilic vein usually branches at the elbow level, making it difficult to follow it more distally.

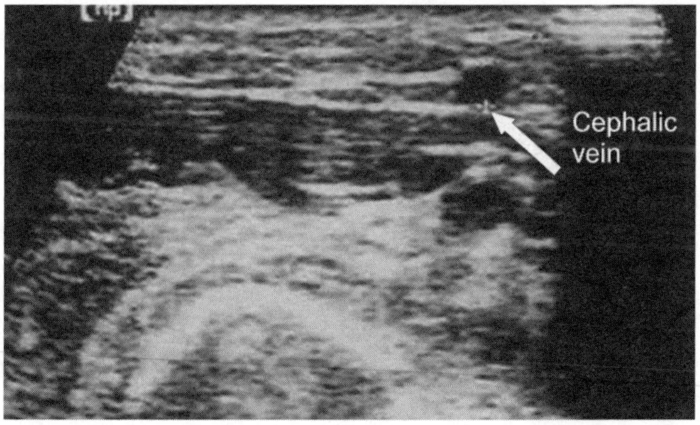

Figure 15-4.
Cephalic vein located above the elbow. Wall does not appear to be thickened; diameter measurement is approximately 3 mm, which is acceptable.

4 General techniques:

- Proximal tourniquet often assists with expansion of the vein size.
- Vein wall compressibility (coaptation of vein walls) is evaluated.
- Outer edge to outer edge diameter measurements (mm) are obtained incrementally along the extremity vein.
- Vein length (cm) is also measured and evaluated for continuity.
- Vein wall thickening and mural calcifications are watched for and, if present, noted, because these findings could rule out using the vein as a bypass graft.

5 Interpretation:

- Normal findings include:

  Complete vein wall compressibility without evidence of thrombosis.

  Acceptable diameter measurement criteria vary but venous diameter should be at least 2–3 mm. The basilic vein is often larger.

- Abnormal findings include:

  Incompressibility or other evidence of thrombosis.

  Inadequate diameter.

  Inadequate length.

  Excessive vein wall thickening or mural calcification.

CHAPTER 16 Other Conditions

Arteriovenous Fistulae

1 Arteriovenous fistulae can be congenital or traumatic. They set up an abnormal connection between the high-pressure arterial system and the low-pressure venous system, causing rather marked anatomic and hemodynamic changes.

2 When a fistula is located close to the heart, the potential for cardiac failure increases; peripherally located fistulae are less likely to cause congestive heart failure but more likely to cause ischemia.

3 An arteriovenous fistula may involve proximal and distal arteries and veins as well as collateral arteries and veins. Its diameter and length predict the resistance it offers.

- Compared to a normal artery, flow in the artery proximal to the fistula is greatly increased, especially during diastole, because the fistula markedly reduces resistance. See figure 16-1 on the following page.

- Depending on the size of the fistula, once blood flows past it there may be normal arterial flow beyond in the distal arteries, and flow resistance often becomes more pulsatile (e.g., triphasic). See figure 16-2. Blood pressure in the distal artery is always reduced somewhat. The direction of blood flow, on the other hand, is normal *if* the fistula resistance exceeds that of the distal vascular bed.

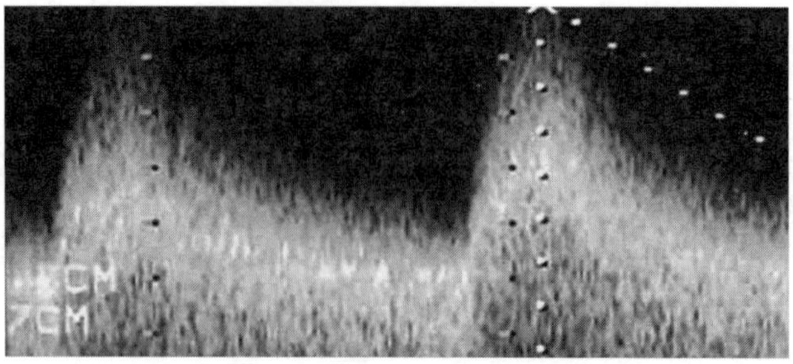

Figure 16-1. Arterial flow proximal to the fistula.

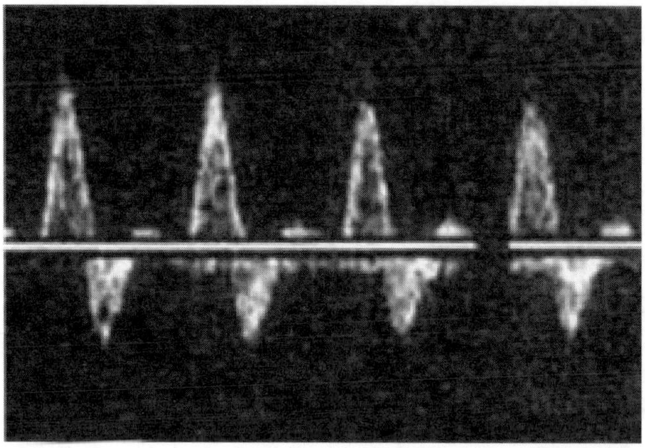

Figure 16-2. Arterial flow distal to the fistula.

- Flow through the fistula (connection between the artery and the vein) is usually characterized by elevated peak and end diastolic velocities associated with spectral broadening. The flow pattern is often bidirectional (turbulent) as depicted in figure 16-3. See also figure 16-6.
- Venous flow approaching the AVF has somewhat elevated velocities. Phasicity is present, but the venous signal has somewhat increased pulsatility because of the connection with the artery via the AVF. See figure 16-4.
- Venous flow moving away from the AVF has elevated velocities throughout. Phasicity is present, as well as increased pulsatility because of the mixed arterial and venous flow. See figure 16-5.
- A large, chronic fistula tends to elevate venous pressure, and blood flow may be retrograde in the distal vein. Most often, the valves that otherwise would prevent this retrograde flow are incompetent.

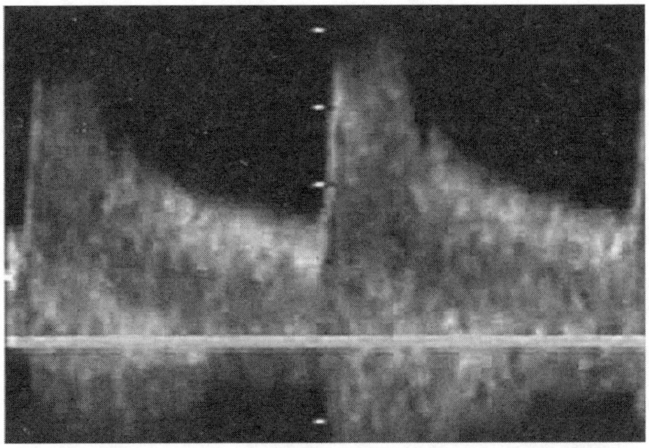

Figure 16-3. Arteriovenous fistula with typical high peak systolic and end diastolic velocities, coupled with spectral broadening and a bidirectional flow pattern.

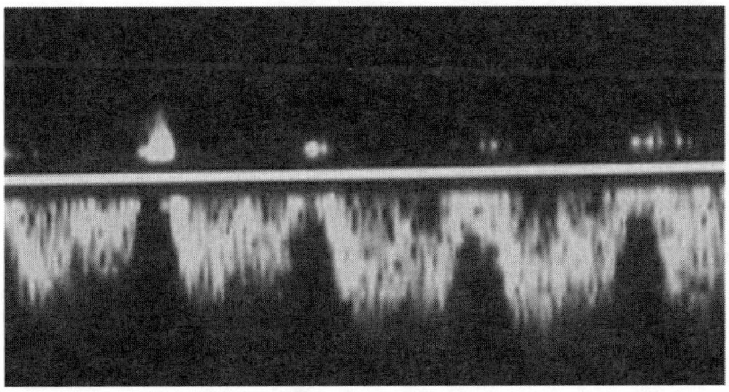

Figure 16-4. Venous flow pattern as it moves toward the arteriovenous fistula.

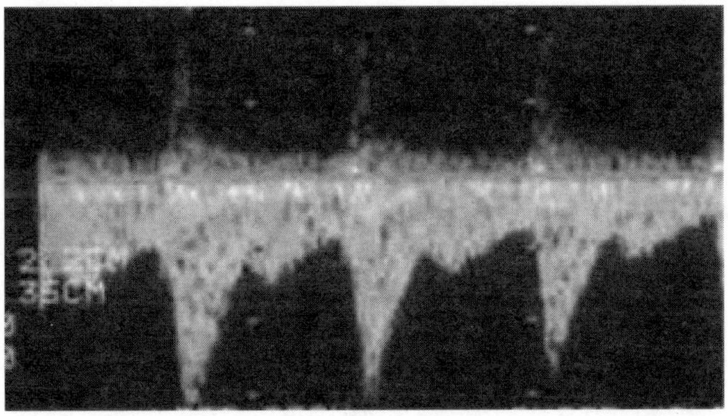

Figure 16-5. Venous flow pattern as it moves away from the arteriovenous fistula.

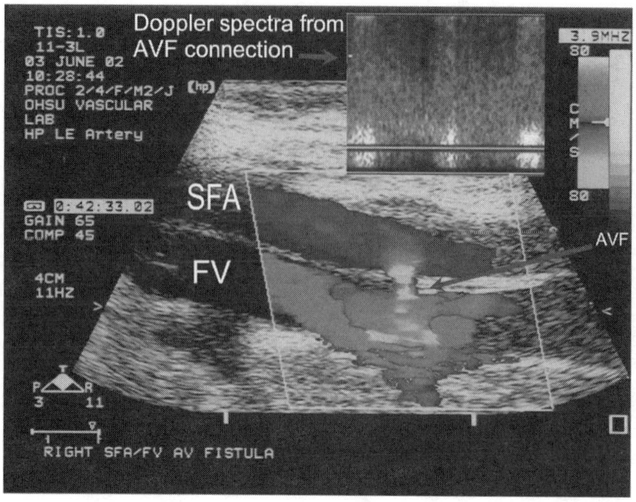

Figure 16-6. Longitudinal view of an arteriovenous fistula between the superficial femoral artery and the femoral vein. Note the characteristically elevated peak systolic and end diastolic velocities along with a bidirectional flow pattern.

Compartment Syndromes

1 Compartment syndromes are caused by swelling within the osteofascial compartments of the leg or arm. Intercompartmental pressure increases to the point where it exceeds capillary perfusion pressure, thereby decreasing vascular perfusion and compromising nutritive blood flow to the tissue. Nerves are the most susceptible to ischemia, with muscle necrosis occurring later.

2 Compartment syndromes most commonly occur following revascularization to correct prolonged ischemia caused by embolic, traumatic, or other less common conditions. Compartment syndromes also can develop spontaneously or as a result of bleeding within a compartment.

3 Clinical findings include paresthesias, pain (especially with passive motion of the muscle), weakness of involved muscle, and tension of the compartment. A late sign is loss of pulses.

4 Diagnosis is made on the basis of the clinical assessment and may be confirmed by measuring compartment pressures.

5 *Example:* Following repair of an occluded or traumatized tibial artery, swelling may become evident. Because the compartment is bound by bone, fascia, and interosseous membrane, outward swelling is impossible. Swelling therefore may compress the tibial arteries, causing necrosis of the muscles and severe pain, tenderness, foot drop, and other neurological changes.

6 Therapy: Fasciotomy is the appropriate treatment.

Cystic Adventitial Disease

1 With this disease process, cystic fluid accumulates in the wall of the peripheral artery, which can cause narrowing or occlusion of the vessel. Symptoms can include claudication, with bruit evident in some cases. The loss of distal pulses during flexion of the foot or hand may be documented.

2 The etiology of cystic adventitial disease—most common in males 40 to 50 years of age—is uncertain, but can include a congenital abnormality exacerbated by repetitive trauma associated with regions of joint connection or be related to development of a ganglion.

3 The sonographic presentation of cystic adventitial disease is that of a soft tissue mass. This condition may sometimes but not often be discovered as an incidental finding during the peripheral arterial sonographic examination.

Popliteal Artery Entrapment Syndrome

1 Popliteal artery entrapment syndrome is thought to be caused by compression of the popliteal artery by the medial head of the gastrocnemius muscle or fibrous bands. *Examples:* The popliteal artery may have an abnormal location, e.g., deviating medially around a normally positioned gastrocnemius muscle, *or* the medial head of the gastrocnemius muscle may have an anomalous origin. See figure 16-7.

2 This syndrome is most commonly found in young men and is bilateral in about one-third of the cases. Repeated trauma to the popliteal artery may result in the development of aneurysm, thrombosis, atherosclerosis, and emboli.

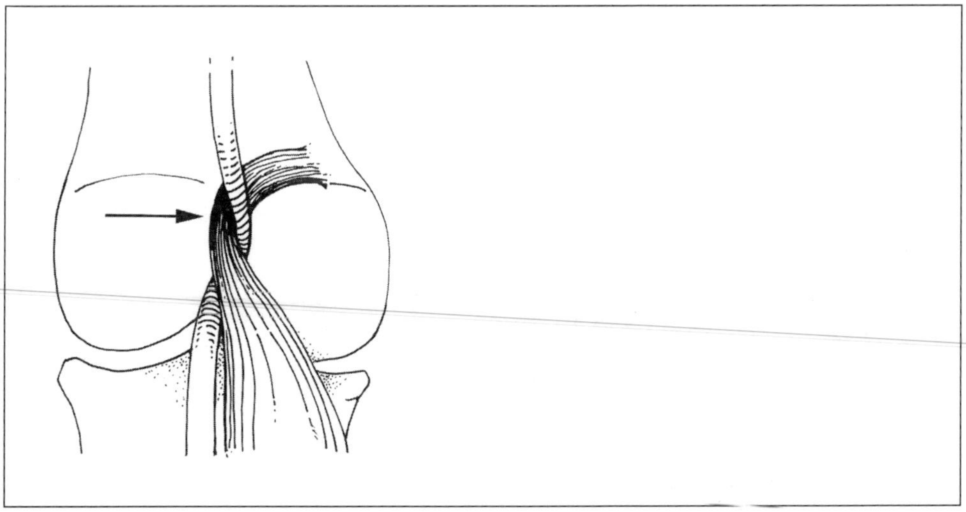

Figure 16-7. Popliteal entrapment with medial displacement of the popliteal artery by the gastrocnemius muscle.

3 Patients present with symptomatic arterial occlusion or intermittent claudication. With the knee extended and with active plantar flexion or passive dorsiflexion of the foot, the pulses may diminish or waveforms (i.e., Doppler, plethysmographic) may be altered. *Example:* Young male complains of pain in the calf muscle following exercise. The vascular laboratory findings reveal normal pressures, pulses, and plethysmographic waveforms at rest, but appreciably abnormal decreases during active plantar flexion or passive dorsiflexion of the foot.

Thoracic Outlet Syndrome (TOS)—Arterial Component

1 Thoracic outlet syndrome occurs when there is neurovascular bundle compression by the shoulder structures (cervical rib, costoclavicular, scalene muscle).

- The etiology of thoracic outlet syndrome is not well understood, but many investigators attribute the symptoms (at least in the majority of cases) to neurogenic compression of the brachial plexus when the arm is in certain positions, while a small percentage of cases are the result of venous (subclavian vein) or arterial (subclavian artery) compression.

- Symptoms include numbness or tingling of arm and pain or aching of shoulder and forearm. Exercise and upward arm positions can increase discomfort and exacerbate symptoms. Of note is data indicating that approximately 25–30% of the population have asymptomatic compression.

2 Technique:

- Plethysmographic techniques and/or Doppler waveform analysis are utilized to detect arterial changes:

  The PPG is attached to the index finger, or the continuous-wave Doppler is used to monitor the radial artery, or a brachial cuff is applied to monitor plethysmographic pulse volume waveforms.

  Resting waveforms are obtained, and then the patient's arm is placed in various positions as the pulsations are monitored at each position.

- Arm positions (patient sits erect, legs dangling from exam table):

  Resting position—hand in lap

  Arm at 90 degree angle in same plane as torso

  Arm at 180 degree angle in same plane as torso

  Exaggerated military stance (patient still sitting)

Adson maneuver positioning: Exaggerated military stance (e.g., shoulders back) with the head turned sharply toward the arm being tested (waveforms obtained); then turned sharply away from the arm being tested (waveforms obtained). Called *right Adson* when head is turned right, *left Adson* when head is turned left. Additionally, many examiners also ask the patient to lift the head up and take in a deep breath during each Adson maneuver while recording the waveforms.

Causative position as described by patient

3 Interpretation:

- Normally, the resting waveform obtained by PPG (or continuous-wave Doppler) is maintained throughout the various positions. See figure 16-8A.
- If the waveforms are partially reduced (attenuated) by one or more of the positional changes, this is considered borderline abnormal by some, completely abnormal by others. See figure 16-8B.
- If there is complete flattening of the waveforms, which is due to extrinsic compression of the artery at the thoracic outlet (see figure 16-8C), patient history is coupled with findings for interpretation (e.g., which positions produced the abnormal findings, what symptoms did the patient experience during the exam, etc.). It is also useful to note that a large percentage of patients have abnormal findings but are asymptomatic.

4 Therapy:

- Treatment is usually conservative (shoulder exercises).
- Surgical treatment, when indicated, consists of resecting the first rib with/without scalene muscle splitting.

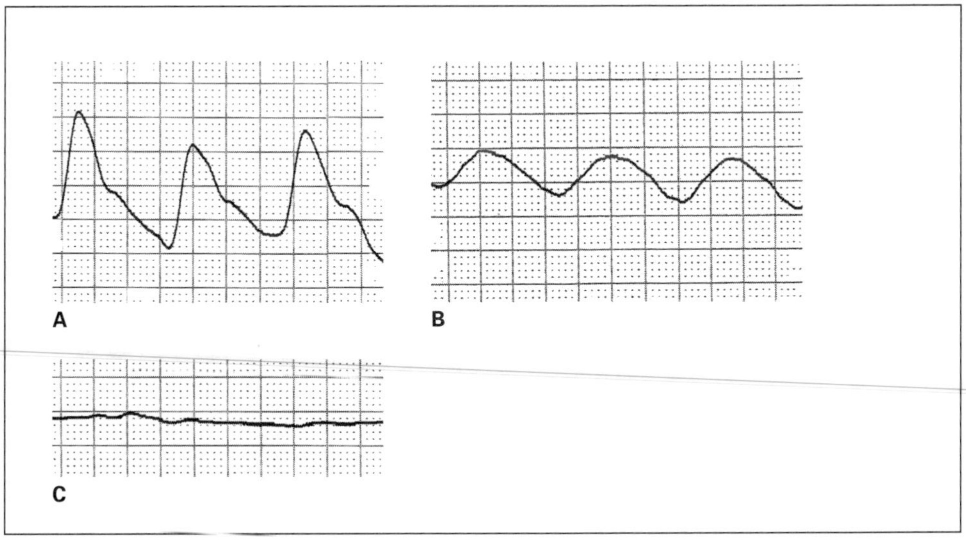

Figure 16-8. Photoplethysmographic waveforms. **A** Normal waveform with the arms in a resting position. Note the normal quality (e.g., reflection in the downslope) of the waveform. **B** Partial attenuation of the waveforms (borderline abnormal). **C** Complete flattening of the waveforms (abnormal).

Trauma

1. Arterial injuries can result from blunt trauma (e.g., long bone fractures that secondarily injure vessels) or penetrating trauma (e.g., stab or gunshot wounds and iatrogenic injuries from vessel catheterization).
2. Clinical presentation includes minimal to severe hemorrhage and/or hematoma and even the absence of distal pulses, all depending on severity of the injury.
3. *Example:* Intimal tear results from a gunshot wound or dislocation. Although the intima tears, the media and adventitia remain intact. Pulses remain intact initially, but thrombus forms at the site of injury and an occlusive intimal flap develops or dissection occurs.

Invasive Tests and Therapeutic Intervention

CHAPTER 17

Arteriography

For quick reference, see plates 1–7 on pages 418–420. Plate 1 shows an occluded superficial femoral artery with reconstituted flow via collateral vessels. Plate 2 is an image of a popliteal aneurysm. Plates 3 and 4 are aortograms of normal and abnormal conditions, while plate 5 demonstrates aortic occlusion with collateralization. Plates 6 and 7 reveal filling defects and other abnormalities of the aorta and renal arteries.

1 Types:

- Intraarterial injection
- Intravenous digital technology (not commonly employed)

2 Method:

- Intraarterial injection involves the percutaneous puncture of a superficial artery and the insertion of a very thin catheter (Seldinger technique).

- The most commonly used arteries are the common femoral, axillary, and brachial. The safest approach is the common femoral artery.
- Once the catheter is positioned properly, a liquid contrast agent is injected and flows with the moving blood, making it possible to image the arterial lumen but not the arterial wall itself.
- A rapid film changer technique is used to expose the films sequentially as the contrast agent moves through the vessel.
- Using fluoroscopy, the examiner may obtain and store digital information for later manipulation and interpretation.
- Following removal of the catheter, pressure is applied to the puncture site, and the patient is positioned supine for 6 to 8 hours of bedrest. A sandbag may be applied on top of the pressure dressing to prevent hemorrhage.

3 Interpretation:

- Arteriography is primarily an anatomic—not a functional—study.
- For many large muscular arteries, a hemodynamically significant stenosis is one that reduces the diameter of the arterial lumen by 50% and therefore its total cross-sectional area by 75%. Such a stenosis markedly reduces pressure and flow.
- Normal anatomy is expected to be visualized on the films as the contrast media completely fills the vessel. (See figures 17-1 and 17-2.)
- Interpretation is based on how much (if any) of the artery does not fill with blood containing the contrast agent. The extent and location of the *filling defect* is ascertained. (See figures 17-3, 17-4, and 17-5.)

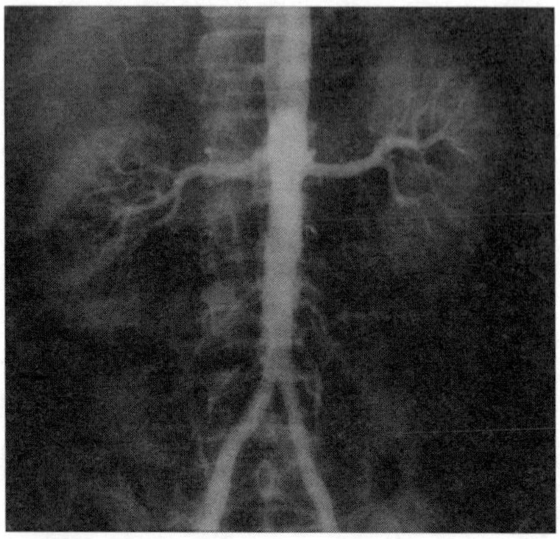

Figure 17-1. Normal aortogram, including the renal artery branches.

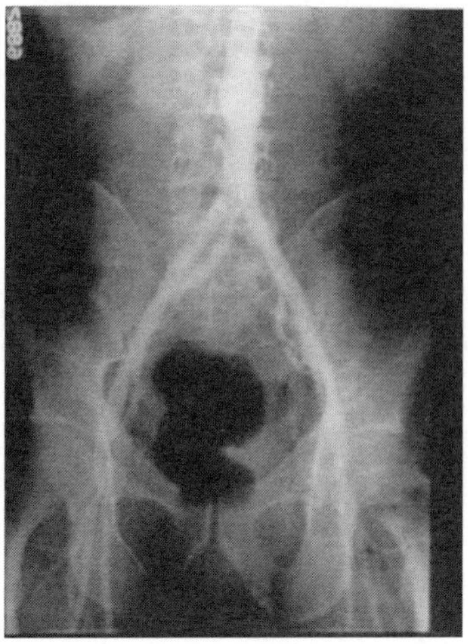

Figure 17-2.
Aortogram showing the normal abdominal aorta and right and left common iliac, external iliac, and internal iliac arteries.

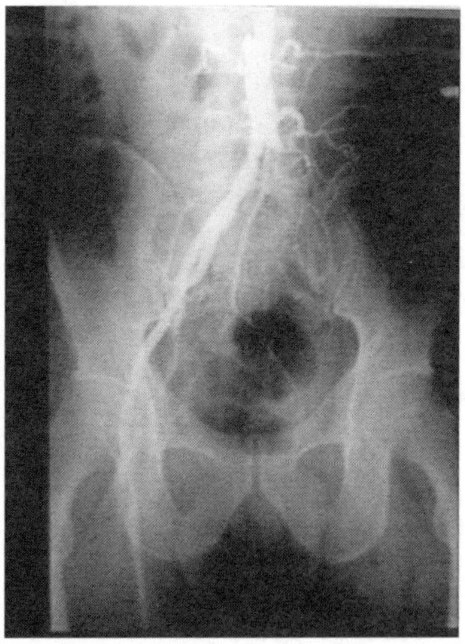

Figure 17-3.
Preoperative aortogram showing the abdominal aorta and bifurcation vessels. There is an occluded left common iliac artery with minimal collateralization at the origin of the left common iliac artery.

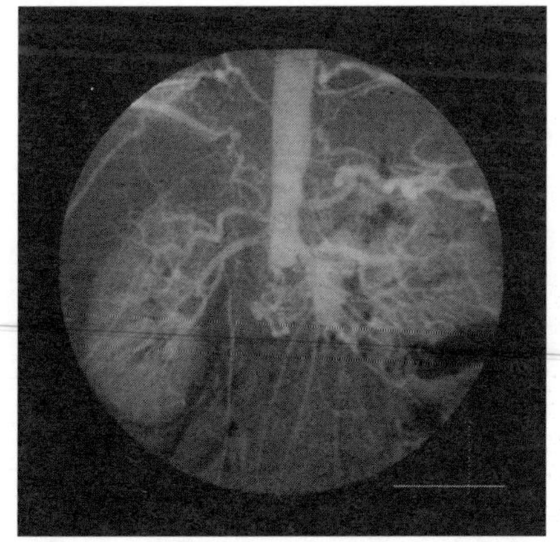

Figure 17-4.
Aortic occlusion with collateralization. Several collateral branches provide flow to the lower extremities. As the body attempts to maintain perfusion, the distal capillary beds vasodilate. The normally high-resistance patterns assume the low-resistance flow qualities of that vasodilated vascular bed.

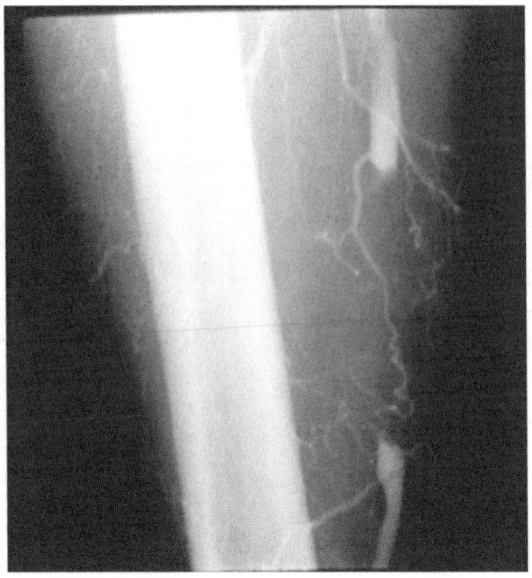

Figure 17-5. Left femoral arteriogram revealing an occluded superficial femoral artery (SFA) with reconstitution at the distal SFA/popliteal artery level via collateralization.

Atherosclerosis (Stenosis or Occlusion)

- Atherosclerotic plaque appears as irregular or smooth negative images on arteriograms wherever the plaque displaces the contrast media. The origins of the great vessels at the aortic arch, other bifurcations (carotid, aortic, femoral, etc.), and the adductor canal are common locations for atherosclerotic plaque. See figure 17-6.
- When there is extensive disease and patients present with signs and symptoms of embolization, it is difficult to identify the source of the emboli.

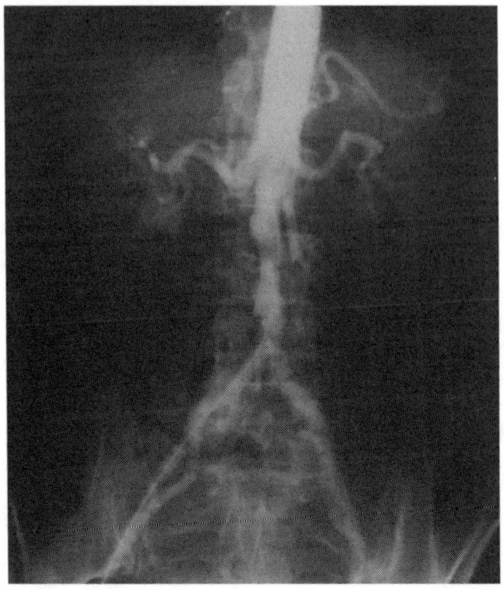

Figure 17-6. Aortogram showing several areas of filling defects, although the renal arteries appear normally patent.

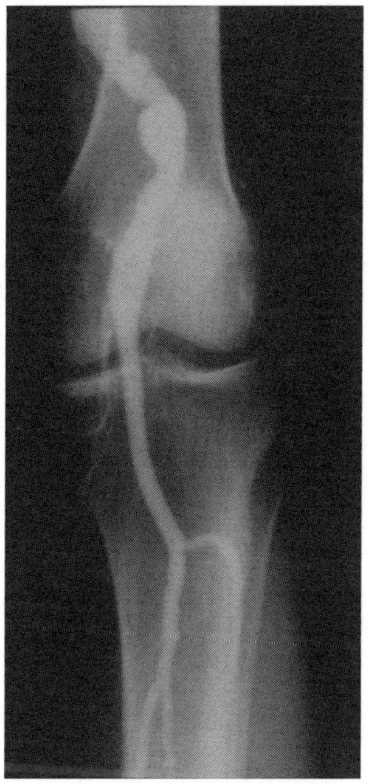

Figure 17-7. Popliteal aneurysm. The thrombus evident in this type of peripheral aneurysm is often the source of distal embolization. It is not unusual for a person with one aneurysm (e.g., aortic) to have aneurysmal disease at other levels (e.g., femoral, popliteal).

- Collateral vessels are usually present in cases of long-standing occlusion. (See figures 17-4 and 17-5.)

Aneurysm

- Dilation of the artery is a hallmark of aneurysm.
- Aneurysms most commonly develop in the infrarenal aorta, then next most commonly in the femoral, and then the popliteal arteries. (See figure 17-7.)
- Thrombus may develop along the dilated wall of the aneurysmal artery, creating the appearance of a straight tube on arteriography.
- Other signs are lack of arterial branches where you would expect them and elongation and tortuosity of the vessel.

Vasospasm

- Vasospasm appears as severe narrowing of the arterial lumen, usually without occlusion.

Fibromuscular Dysplasia

- Multiple arterial stenoses, which are caused by medial hyperplasia and which create the appearance of a "string of beads," typify fibromuscular dysplasia (FMD). Poststenotic dilatation also may be present. (See figure 17-8 on the following page.)

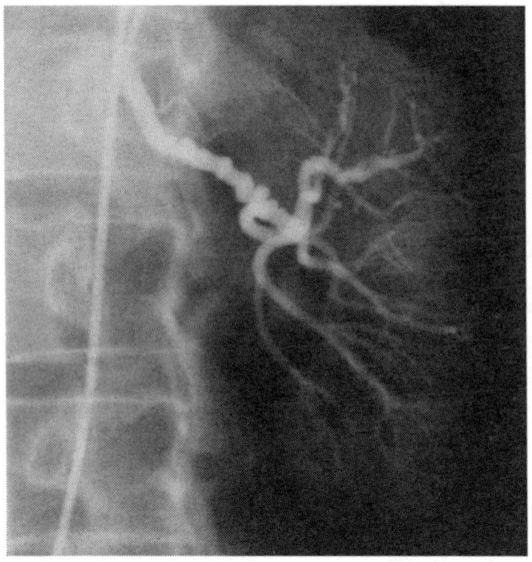

Figure 17-8. In this renal artery angiogram, note the "beaded" appearance of the renal artery. If duplex scanning were used to evaluate this artery, there would be variable peak velocities (more elevated at the points where the artery is more narrowed). In addition, spectral broadening would be consistently observed through the vessel.

4 Limitations:

- It may not be possible to perform arteriography on a patient who is allergic to the contrast agent (many times the allergy is to the iodine).
- Arteriography is inaccurate in its functional (hemodynamic) assessment. It also cannot provide multiple images in multiple planes in real time. A two-dimensional view is standard.
- Patients in kidney failure cannot undergo arteriography.

5 Complications:

- Puncture site hematoma
- Pseudoaneurysm
- Local arterial occlusion
- Neurologic complications

MR Angiography (MRA)

Magnetic resonance imaging (MRI) and angiography (MRA) use radio-frequency energy and a strong magnetic field to produce multiplanar images. MR instruments can quantitate blood flow and construct images that look like angiograms. Without the use of contrast agents, MRA is capable of distinguishing blood flow from soft tissue. See figure 17-9.

1 Purpose:

- To evaluate abdominal aortic aneurysm (AAA) and dissection.
- To evaluate peripheral arterial disease.

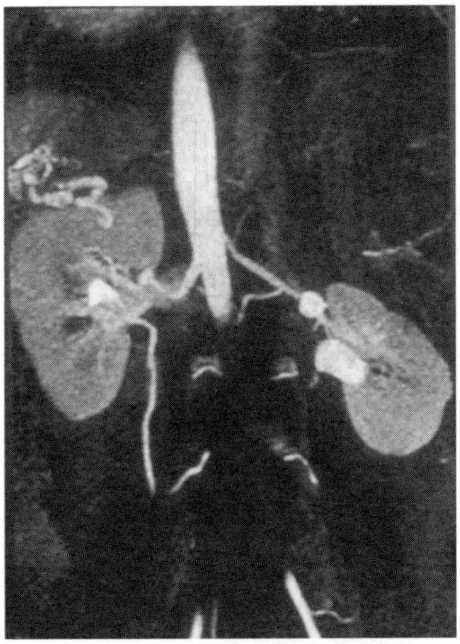

Figure 17-9. MRA of the aorta and branches. There is an aortic occlusion, with no definite collateral branches seen. (Some of the poor visualization may be caused by signal drop-out.) The renal arteries are patent.

2 Limitations:

- Patients with pacemakers, monitoring equipment, and other metallic objects, e.g., surgical clips, are impossible to assess.
- Degree of stenosis is difficult to assess accurately.
- Stenoses may be overestimated because of nonlaminar or slow-flow states, which can result in the loss of magnetic signals.
- MRA is more expensive than ultrasound.
- Patients who have claustrophobia have difficulty undergoing the exam.
- Interpretation requires considerable skill.

3 Interpretation:

- Similar to the interpretation of arteriograms.
- Excellent agreement with ultrasound and CT in determining aortic diameter.

Computed Tomography (CT)

Computed tomography (CT) uses ionizing radiation to obtain cross-sectional images of the aorta and other body structures. It can be performed with or without the use of contrast agents, but intravenous contrast allows for more discrete evaluation of the vasculature. See figure 17-10.

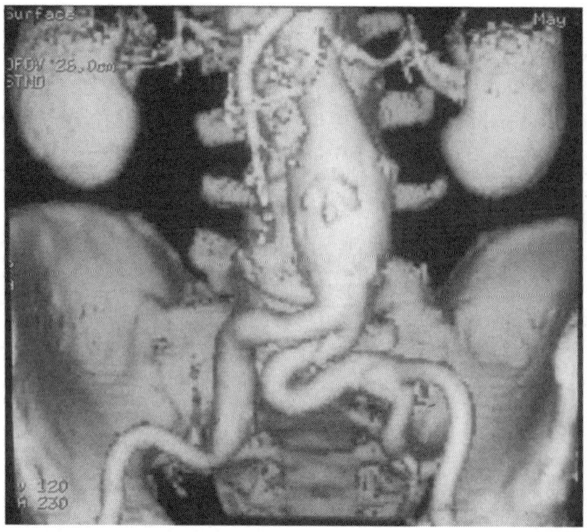

Figure 17-10.
An aortic aneurysm is seen in this CT image with contrast (also known as CT angiography) of the aorta and other branches.

1 Purpose:

- To help define relationship of aorta to renal artery origins.
- To evaluate various types of aneurysms.
- To assess for presence/location of dissection.
- To evaluate cerebral infarctions, intracranial aneurysms, hemorrhage, and arteriovenous malformation (AVM).

2 Limitations:

- Images degraded by patient motion and presence of surgical clips.
- Requires more time and is more expensive than ultrasound.
- Limited application in peripheral arterial disease (PAD) because of the small size of the vessels.

3 Interpretation:

- Determines dimension of aorta and size of aneurysms.
- Often used to evaluate infarctions after a stroke.

Medical Therapy

Medical management includes behavior modification and drug therapy, as follows:

- Stop smoking, as nicotine has been shown to cause vasoconstriction and has potentially harmful effects to the endothelium.
- Exercise regularly and consistently to promote the development of collateral circulation.

- Weight control and a low-cholesterol diet may enhance normal endothelial cell metabolism.
- Aspirin is an antiplatelet drug that decreases platelet aggregation and, thereby, thrombotic activity.
- Medications that decrease blood viscosity.
- Antihypertensive drugs may serve to decrease shear forces on the endothelial cells.
- Protection to prevent injury and/or infection to the extremity of interest.

Surgical Therapy

The decision to operate depends on the extent and severity of disease and on the patient's symptoms.

Endarterectomy

Endarterectomy is the surgical removal of atherosclerotic material—and usually a portion of the intimal lining—from the arterial lumen.

Bypass Grafts

1 Purpose:

Bypass grafts provide an alternate pathway for distal blood flow, i.e., around significant stenoses or occlusions. Bypass grafts are most successful when there is inflow obstruction and patent distal runoff. Many patients have obstructions at more than one level of the arterial tree. If the lesions are of equal severity, the proximal lesions may be bypassed first.

2 Graft materials:

- Synthetic (e.g., Gore-Tex)
- Reversed vein (saphenous, cephalic)
- In situ vein

3 Common bypass grafts:

- Aorta to both iliac arteries.
- Aorta to bifemoral. (See figure 17-11A.)
- Femoral to popliteal.
- Femoral to posterior tibial artery, to anterior tibial artery, or to peroneal artery. (See figure 17-11B.)

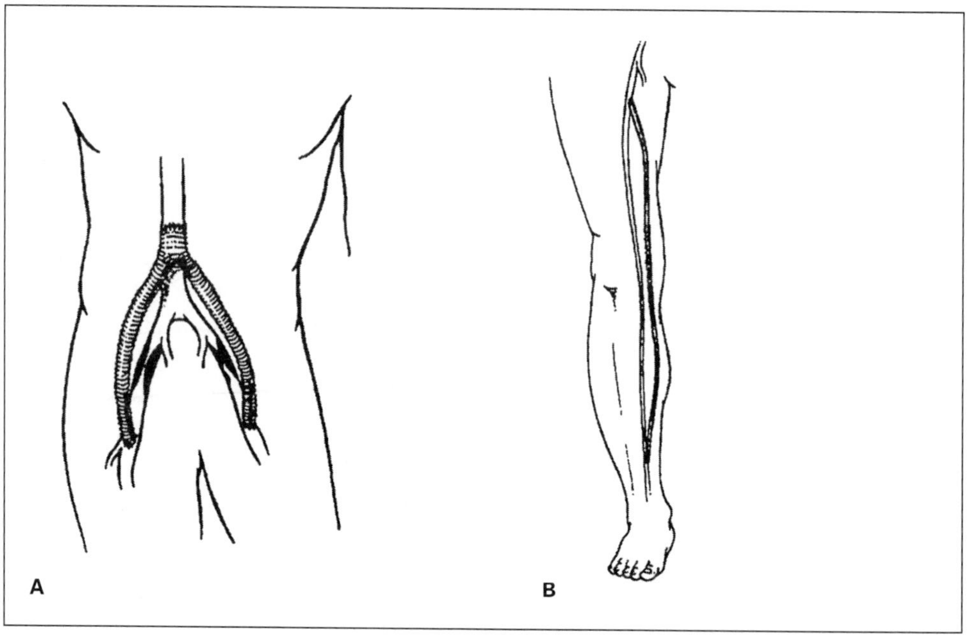

Figure 17-11. Common bypass grafts. **A** Aorta-to-bifemoral bypass graft. **B** Superficial femoral-to-posterior-tibial bypass graft. From Ridgway DP: *Introduction to Vascular Scanning: A Guide for the Complete Beginner,* 3rd edition. Pasadena, Davies Publishing, 2004.

- Axillary to femoral or axillary to bifemoral.
- Femoral to femoral.

Sympathectomy

Sympathectomy is the surgical interruption of part of the sympathetic nervous pathway to help treat severe arterial occlusive disease by producing vasodilation and thereby enhancing blood flow to an extremity. This procedure is no longer commonly performed.

Endovascular Procedures

Angioplasty

1 Purpose:

Angioplasty (percutaneous transluminal angioplasty [PTA]) is used to dilate precisely that segment of a vessel that contains a focal atherosclerotic lesion.

2 Technique:

- PTA is performed in much the same manner as angiography except that a balloon-tipped catheter is used.
- Under fluoroscopic guidance and with the injection of small amounts of contrast media, the tip of the catheter is positioned within the stenotic segment and the balloon slowly inflated, pushing the plaque against the walls of the vessel, dilating the artery, and restoring the arterial lumen to dimensions that are closer to normal.

3 Limitations:

Balloon angioplasty cannot be applied to all vessels or all types of lesions. The procedure is usually performed for focal lesions in large vessels such as the iliac, femoral, and popliteal arteries. Angioplasty has also been performed on the renal artery. Its application is limited in vessels of smaller diameter.

4 New applications and clinical developments:

- A newer application for ultrasound involves using duplex ultrasonography to guide balloon angioplasty as well as stent placement in the treatment of femoral-popliteal and infrapopliteal arterial occlusive disease (see below). Although ultrasound guidance is not routinely performed, it may hold promise as an alternative to the use of fluoroscopy and injection of contrast material.

- Subintimal angioplasty is a minimally invasive endovascular procedure that may be used in some patients who are not candidates for traditional surgical or endovascular treatment. Originating in the United Kingdom, subintimal angioplasty is used in the United States as a means to provide blood flow around a totally occluded artery. It is performed in the arterial wall rather than in the arterial lumen. Under duplex ultrasonographic guidance, a wire is passed within the wall of the blood vessel, into the subintimal space rather than through the occluded true lumen of the vessel, from above to below the occlusion. Once the wire is passed, a balloon is inflated to open the new pathway for blood to flow through. In some cases a stent is used to maintain patency of this new pathway in the subintimal space.

Stents and Stent Grafts

Because of the acknowledged problem of restenosis following transluminal angioplasty, intravascular stents are frequently used. A stent is a small device designed to maintain the intraluminal structure and patency of an artery. Some describe the stent as a kind of scaffold.

Several types of stents have been developed and used worldwide, but to date only a few have undergone significant investigation. The three major types of stents are *balloon-expandable* (which employ a balloon catheter for its insertion and deployment), *self-expanding* (which are compressed into a small-diameter introducer sheath and then deployed), and *thermal expanding* (which currently are not in wide use).

In addition, there is what is described as the *stent graft* or *covered stent*. This particular stent has been used to repair specific types of aneurysmal disease, especially in the aorta. The three basic types of aortic stent grafts are *tube grafts, bifurcated grafts,* and *aortoiliac grafts.* It is very important to know the details about the endovascular aortic repair prior to any ultrasound assessment, because potential complications vary according with graft configuration. A much more complex evaluation, as well as greater technical expertise, is required for evaluating the aortic stent grafts.*

1 Purpose:

- Stents are used to treat stenosis and maintain patency of vessels such as the renal, iliac, femoral, and carotid arteries.
- Stent grafts are most often used to repair aneurysmal disease of the aorta.

2 Interventional technique:

- Insertion and deployment techniques vary with type of stent. Insertion of the introducer sheath involves the use of techniques similar to those in arteriography (e.g., fluoroscopy, needles, and in some cases guide wire).

3 Limitations and complications:

- Depending on the stent type, limitations can include course and size of vessel, abdominal gas (aortic stent graft), and patients' inability to lie flat.
- Complications are similar to those of angiography. Other problems include intimal hyperplasia (which can cause restenosis), stent migration, twisting, dislodgment, and leaks.
- Complications of endovascular aortic stents that may be identified with duplex ultrasound assessment include endoleak, graft stenosis or thrombosis, dissection, problems at the attachment sites resulting in migration, and twisting of the limbs of a bifurcated graft.
- Other complications of endovascular aortic stents include lower extremity ischemia secondary to embolization, graft infection, delayed rupture of the aortic aneurysm, bleeding, pseudoaneurysm at the insertion site, and hematoma.*

*The majority of the information in this section on stents and stent grafts, including other details not mentioned here, can be found in the following references:

Carter KA: Color duplex ultrasound for the evaluation of endovascular stent grafts following endovascular repair of abdominal aortic aneurysm. J Vasc Ultrasound 29:137–141, 2005.

Sato DT, Gregory RT, Carter KA, et al: Endoleak after aortic stent graft repair: diagnosis by color duplex ultrasound versus CT scan. J Vasc Surg 28:657–663, 1998.

Nelms CR, Carter KA, et al: Color duplex ultrasound characteristics: can we predict aortic aneurysm expansion following endovascular repair? J Vasc Ultrasound 29:143–146, 2005.

4 Follow-up evaluation:

Postinterventional evaluation of aortic stent grafts is often routinely performed with computed tomography angiography (CTA). Nevertheless, in the hands of experienced vascular technologists who regulary follow the aortic stent grafts, duplex evaluation is more often used.

5 Ultrasonographic technique:

- The noninvasive technique used to evaluate stents (e.g., for renal, iliac, femoral arteries) or stent grafts (for aortic aneurysm) is duplex imaging (with or without Doppler color flow imaging), described elsewhere in this book by vessel. This technique includes B-mode evaluation using transverse and longitudinal approaches combined with Doppler spectral analysis, with the vessel in the longitudinal plane.

- Aortic endograft surveillance can also include ultrasound contrast, which may become a promising diagnostic tool for the detection of endoleaks following endovascular intervention of abdominal aortic aneurysms. Because color flow imaging has proven to be a reliable diagnostic tool for endograft surveillance, some believe that combining it with a continuous infusion of ultrasound contrast may improve the detection of any endoleaks that might exist.

6 Interpretation:

- Current practice involves using the normal/abnormal values for arteries without stents. The stenosis profile described on page 36 can be used to determine the presence or absence of abnormality. Studies are currently underway to determine if this practice of applying the established diagnostic criteria for unstented arteries to stented arteries is valid or if, on the other hand, those criteria should be revised for the evaluation of stents.

- Stent graft evaluation (see figure 17-12) requires information previously described. In addition, careful observation must be made of the B-mode ultrasound image to include proximal and distal attachment sites, entire graft and AAA sac, and native iliac arteries inferior to the attachment sites (if the graft is bifurcated or an aortoiliac graft). In addition to type I, II, III, or IV endoleaks as well as stent migration, the presence of a heterogeneous texture to the residual aneurysmal sac with some hypoechoic areas could suggest endotension or a type V endoleak. Endoleak types I–V are defined below.

- The stent graft is used to repair (exclude) an abdominal aortic aneurysm. A successful procedure should reduce the size of the

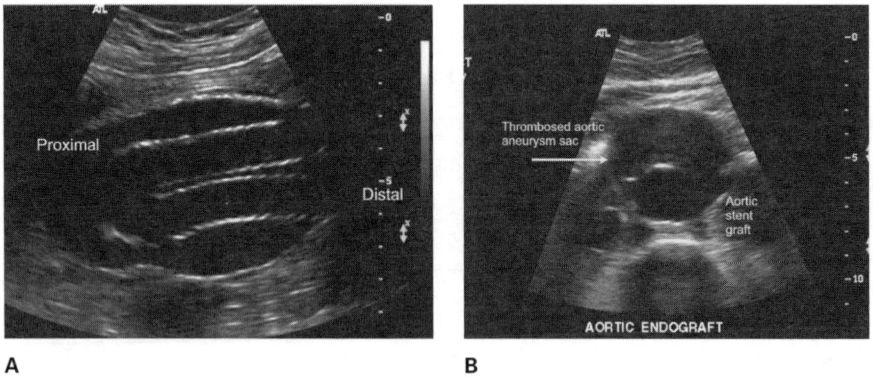

Figure 17-12. **A** Longitudinal B-mode view of an aortic stent graft. **B** Transverse B-mode view of the body of an aortic stent graft proximal to the bifurcation. Note the exclusion of what appears now to be a thrombosed aortic aneurysm sac.

aneurysmal sac. If blood flow is observed in the sac, or if the size of the sac fails to diminish or grows larger, there could be an endoleak as the result of one or more problems. The following types of leaks have been identified:

Type I: Attachment endoleak at proximal and/or distal attachment site(s).

Type II: Branch leaks, i.e., leaks that originate from native vessels normally branching off the aorta, allowing retrograde flow into the residual aneurysmal sac. The resulting blood flow outside of or around the endograft keeps the sac pressurized. Branches include lumbar artery and IMA.

Type III: Modular connect endoleak, i.e., leaks from the connections of modular components.

Type IV: Transgraft endoleaks—can occur from minute tears in the fabric or from changes in wall porosity.

Type V: Although not mentioned in the previous references, some other sources describe a Type V leak as endotension, a condition in which no visible blood flow is detected in the aneurysmal sac, but the sac contains anechoic areas and appears to expand or fails to retract over time.

All leaks are confirmed with spectral Doppler. True leaks will have reproducible arterial waveforms that differ from the graft flow waveform. The type of waveform (biphasic, monophasic, bidirectional) may help identify the type of endoleak and should be documented.

Once an endoleak is confirmed by spectral Doppler, color Doppler is used to learn more about the direction of flow (into or out of the sac) and whether or not the flow fills the entire sac. Even if no leak is suspected, most authors recommend that the aneurysmal sac be sampled throughout with spectral Doppler and/or evaluated with power Doppler.

Thrombectomy/embolectomy

1 Purpose:

- This procedure is used to extract embolus from the affected arterial segment.

2 Technique:

- In a procedure similar to balloon angioplasty, a catheter is inserted into the artery and passed down so that the tip is beyond the area of thromboembolism.
- The balloon is inflated, and the catheter is then withdrawn with the thromboembolus. This procedure may have to be repeated.

Atherectomy

1 Purpose:

- Atherectomy is an alternative to percutaneous transluminal angioplasty that cuts through or pulverizes the plaque with a rotational device at the end of the catheter.

2 Technique:

- The excised material is impacted into a storage compartment located in the catheter.

3 Limitations and advantages:

- Problems with this procedure include heat generation from some rotational devices, lack of guide wire in some devices (making a vessel more susceptible to puncture), and microembolization of pulverized particles.
- The advantages of atherectomy (with the directional type of atherectomy catheter) include reduced risk of peripheral emboli and intimal dissection.

The Value of Duplex Surveillance

- In addition to its use in guiding some endovascular procedures, vascular duplex ultrasound is also used to assess patency of the vessel following peripheral endovascular procedures. Protocols vary with regard to frequency of this postprocedure evaluation, depending on the method of endovascular intervention (e.g., PTA, atherectomy, stent).
- Whether the duplex ultrasound evaluation is performed intraoperatively and/or postoperatively, it can contribute to efficient and cost-effective care. For example, the finding of residual wall defects or restenosis

indicates the need for a secondary endovascular intervention. Many feel that duplex surveillance resulting in secondary interventions can increase the "assisted" patency rates in these vessels.

General Observations about Catheter Intervention

- The thrombotic component of any occlusion must be managed before angioplasty or atherectomy is performed. The main reason is that it may reduce the length of the occlusion and thus improve the initial success and late patency.
- Serious complications occur in less than 5% of cases. Among PTA patients, 2–3% have complications that require surgery or that alter or prolong the hospital course.
- Minor discomfort is associated with the procedure.
- Morbidity, mortality, and cost of angioplasty or atherectomy are low.

Pseudoaneurysm Treatment Options

Option 1: Do Nothing

The pseudoaneurysm spontaneously thromboses.

Option 2: Manual Compression

1 Purpose and indications:

- Under the guidance of B-mode ultrasonographic imaging and/or color flow imaging, transducer compression of particular types of pseudoaneurysms (e.g., femoral) is being accomplished.
- The size and location of the pseudoaneurysm and the location of its communicating channel to the native artery are important factors in determining whether a pseudoaneurysm is an appropriate candidate for ultrasound-guided compression.
- The key criterion in selecting a pseudoaneurysm for ultrasound-guided compression is the ability to uniformly and appropriately compress the "neck," i.e., the communicating channel between the native artery and the pseudoaneurysm (figure 17-13; see also color plate 3).

2 Technique:

- Although practices vary, many facilities obtain informed consent for the procedure and have a physician present as the technologist or sonographer provides the compression. During the compression procedure it is not uncommon for the patient to be somewhat sedated since the procedure is uncomfortable. Additionally, the patient may have

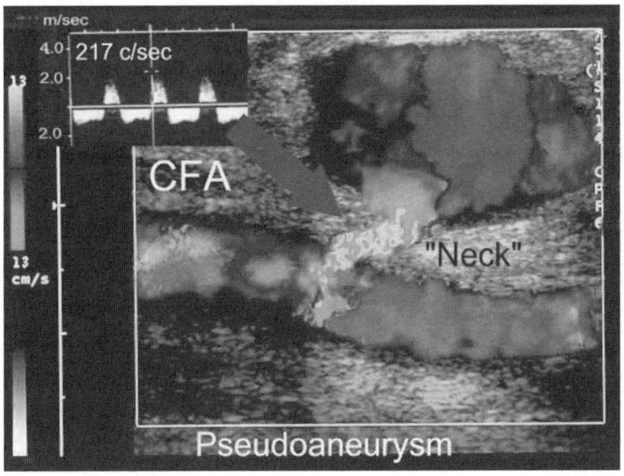

Figure 17-13. Longitudinal view of a pseudoaneurysm of the common femoral artery (CFA). Notice the clear view of the "neck."

an IV started, oxygen available, etc. This procedure should not be treated lightly. Although it may seem relatively safe and easy because ultrasound techniques are being used, complications can occur, including arterial occlusion and venous thrombosis.

- As yet, there is no standardized protocol. One method is to firmly compress the transducer against the skin, closing off the neck of the aneurysm for 10 minutes, then rest, and then compress again for 10 minutes. It is important to observe the duplex or color flow image during this procedure to determine whether the compressions are working. This process may take as long as 45–60 minutes, but its duration seems to vary widely. Because the compression is performed with the ultrasound transducer, the technologist alternately observes the native vessel and aneurysm (figure 17-14).

- It is important to maintain a therapeutic level of pressure without stopping arterial inflow to the extremity. It has been noted that photoplethysmographic monitoring of the great toe helps to ensure the maintenance of arterial inflow. Observation of the deep venous system is also an important component of the overall technique to make certain patency is maintained during the compression maneuver.

- The percentage of successful compressions varies widely. It is important to standardize a protocol for your site. With experience, one can determine with greater accuracy the overall effectiveness for the various types of pseudoaneurysms. Obviously, if the procedure is unsuccessful, it is likely the patient will undergo surgical intervention.

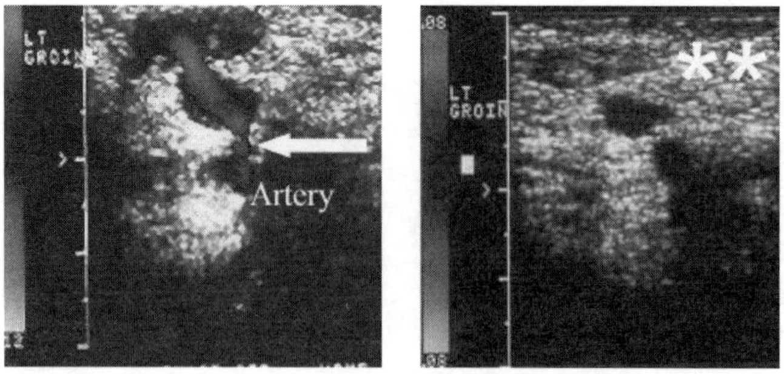

Figure 17-14.
A Transverse image of a common femoral artery (CFA) double pseudoaneurysm. The arrow is pointing to the "neck" of the pseudoaneurysm. **B** This image was obtained after three 10-minute transducer compressions of the pseudoaneurysm neck (each followed by a brief rest period in order to check the status of the pseudoaneurysm and allow the examiner's arm to rest as well). In this example there is successful thrombosis of the pseudoaneurysm (double asterisks).

Option 3: Thrombin Injection

- In addition to utilizing the ultrasound system to treat pseudoaneurysm by manual compression with the transducer, one can also use ultrasound to guide the injection of thrombin into the pseudoaneurysm. Thrombin injection is a promising alternative to manual compression techniques in most cases.
- Using standard scanning techniques, the pseudoaneurysm is located with the characteristic to-and-fro flow in its neck, and the body of the aneurysmal formation is identified. Under ultrasound guidance, thrombin is injected into the body of the pseudoaneurysm as blood flows into it. To minimize the possibility that thrombin will be introduced into the main artery, causing thrombosis, care is taken to avoid injecting thrombin where blood flow is directed away from the aneurysm and toward the main artery and also to avoid injecting thrombin into the neck of the structure.
- In some patients thrombin injection can adversely affect the clotting mechanism. The adverse effects of thrombin may also include those typical of allergic drug reactions, including anaphylaxis.

Option 4: Surgery

Operative repair by a vascular surgeon is indicated in the following situations:

- The neck of the pseudoaneurysm cannot be sufficiently compressed to cause the pseudoaneurysm to thrombose.
- The pseudoaneurysm is too large to be repaired (closed) by manual compression of the neck.
- Thrombin injection is too risky for the patient to undergo.

Operative repair can often be scheduled as a same-day surgery.

PART II

Cerebrovascular Evaluation

CHAPTER 18

Gross Anatomy, Physiology, and Fluid Dynamics

Anterior Circulation: Internal and External Carotid Arteries and Branches

1 The *internal carotid artery:*

- Originates from the *common carotid artery* and then travels into the base of the skull before giving off any branches.
- After passing into the skull, gives off the caroticotympanic branch and then its first major branch—the *ophthalmic artery*—at the carotid siphon, a significant curve of the distal internal carotid artery.
- Gives off the anterior choroidal and posterior communicating arteries before the ICA finally divides into the middle cerebral artery and anterior cerebral artery.
- Distributes blood to the anterior brain, the eyes, the forehead, and the nose, with the brain being a very low-resistance vascular bed.
- The majority of blood in the common carotid artery (70–80%) flows into the internal carotid artery to satisfy the tremendous metabolic demands of the brain.

2 The *external carotid artery:*

- Originates from the common carotid artery.
- Gives off eight branches supplying blood to the neck, face, and scalp, all of which are high-resistance vascular beds:

  Superior thyroid artery

  Ascending pharyngeal artery

  Lingual artery

  Facial artery

  Occipital artery

  Posterior auricular artery

  Superficial temporal artery

  Maxillary artery

- See figure 18-1.
- The superior thyroid artery is frequently visualized in carotid duplex evaluations. Because the internal carotid artery normally has no branches in the neck, the presence of the superior thyroid artery is one method to determine that the insonated vessel is the external carotid artery (ECA). Blood flow in the superior thyroid artery is usually retrograde compared to the common carotid artery because of the course of the superior thyroid artery. Although it is usually the first branch of the ECA, the superior thyroid artery sometimes originates from the distal common carotid artery.

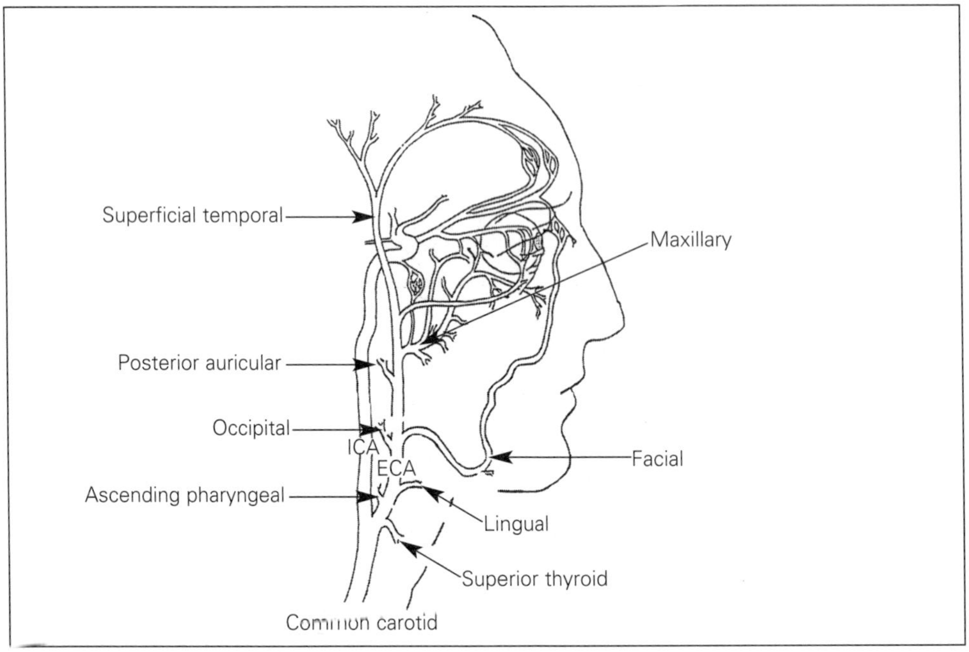

Figure 18-1. The internal carotid artery (ICA), the external carotid artery (ECA), and some of their main branches.

Posterior Circulation

1 The *right* and *left vertebral arteries:*

- Originate from the subclavian arteries (figure 18-2).
- Are asymmetrical in size with right usually smaller than left.
- Unite to form the basilar artery after they enter the skull through the foramen magnum (see figure 18-3 on the following page).

2 The *basilar artery:*

- Is formed by the confluence of the right and left vertebral arteries.
- Divides into the posterior cerebral arteries to form part of the circle of Willis.
- Supplies blood to the posterior structures of the cranial cavity.
- See figure 18-3 on the following page.

3 The *circle of Willis* is:

- An intracranial communication channel, approximately the size of a fifty-cent piece, which lies at the base of the brain connecting the anterior and posterior circulatory systems. When complete (some are not), it provides an important collateral pathway that maintains blood flow to the brain in cases of stenosis or occlusion.

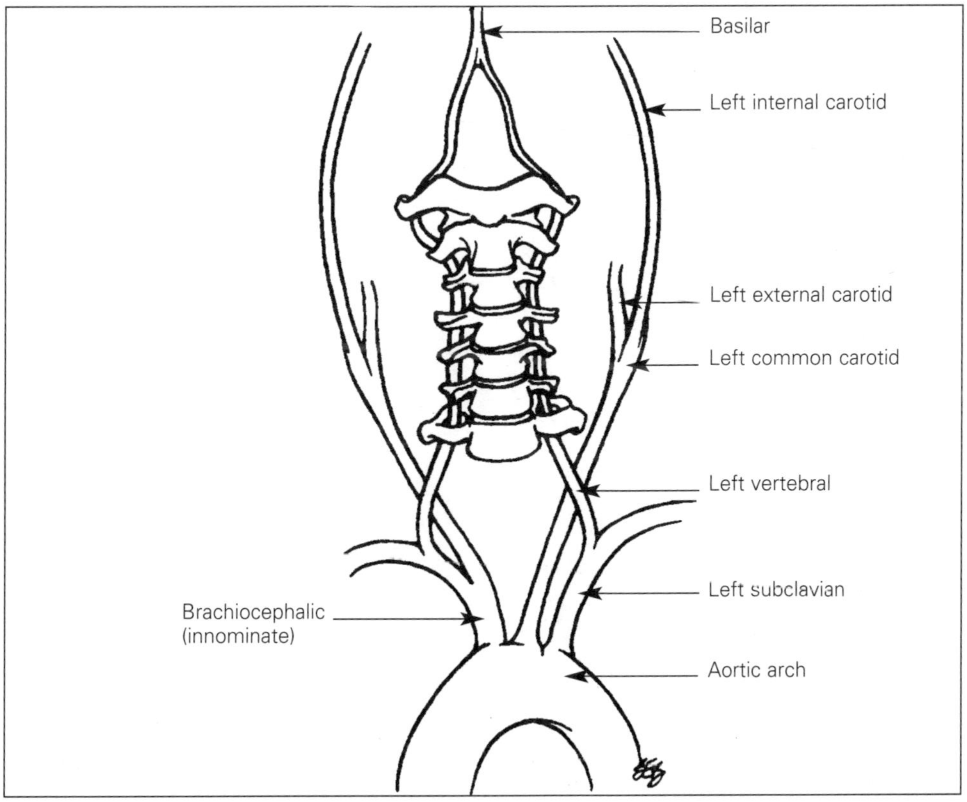

Figure 18-2. The subclavian artery is the origin of the right and left vertebral arteries, which then course through the transverse foramina of the cervical vertebrae. From Belanger AC: *Vascular Anatomy and Physiology: An Introductory Text.* Pasadena, Davies Publishing, 1999.

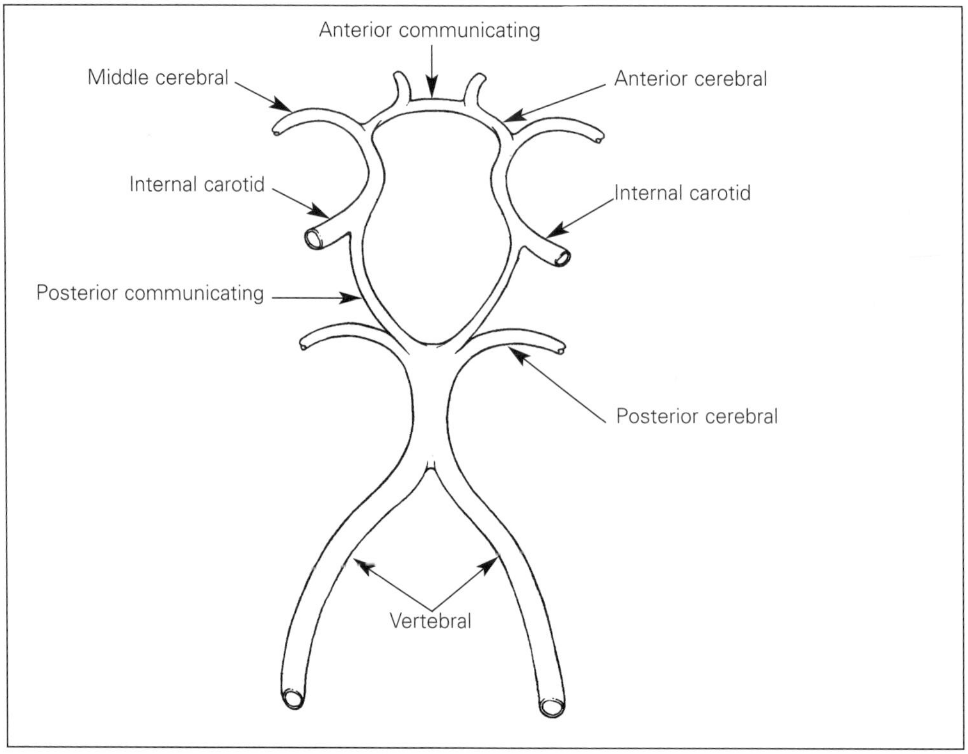

Figure 18-3. The right and left vertebrals unite to form the basilar artery after they enter the skull through the foramen magnum. The basilar artery then divides into the posterior cerebral arteries to form part of the circle of Willis. From Belanger AC: *Vascular Anatomy and Physiology: An Introductory Text.* Pasadena, Davies Publishing, 1999.

- A hexagonal arrangement of the distal internal carotid, anterior cerebral, and posterior cerebral arteries. Although these vessels are joined together by the *anterior* and *posterior communicating arteries,* blood flow through these communicators is usually evident only when they serve as collateral channels secondary to a hemodynamically significant lesion of the extracranial carotid circulation.
- There can be a number of variations to this structure, including vessel size and congenital anomalies.

Figure 18-4 illustrates the intracranial circulation.

Periorbital Circulation

The major branches of the ophthalmic artery that exit the orbit are illustrated in figure 18-4 and include the following vessels:

1. The *supraorbital artery,* which arises from the ophthalmic artery and travels anteriorly and superiorly to the globe. It joins the external carotid artery via branches of the *superficial temporal artery.*
2. The *frontal artery,* which also arises from the ophthalmic artery and exits the orbit medially to supply the mid-forehead. It too joins the external carotid artery via various branches, e.g., the superficial temporal artery.

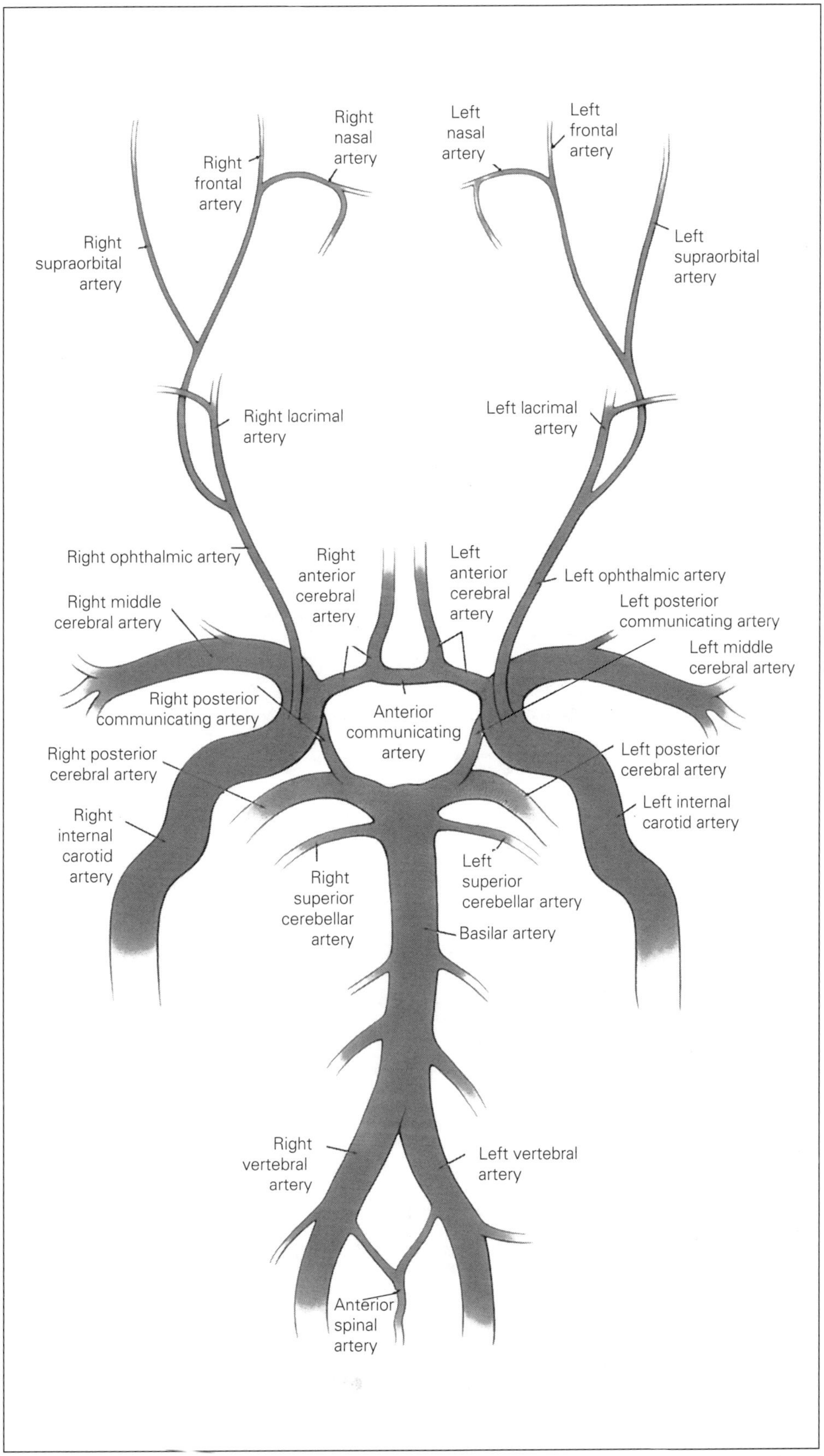

Figure 18-4.
Intracranial circulation. Courtesy of Meda-Sonics.

3 The *nasal artery,* which branches from the frontal artery to supply the nose. It becomes the *angular artery,* descends along the lateral border of the nose, and joins the external carotid artery via the *facial artery.*

Collateral Pathways

1 As illustrated in figures 18-3 and 18-4, the largest of the interarterial connections is the circle of Willis.

2 Intracranial-extracranial anastomoses include these:

- ICA–external carotid artery connections via the ophthalmic and orbital arteries, the meningohypophyseal branches, and the caroticotympanic branch.
- The occipital branch of the external carotid artery with the atlantic branch of the vertebral artery.
- The deep cervical and ascending cervical branches of the subclavian to the branches of the lower vertebral artery, the atlantic branch of the upper vertebral, and the occipital branch of the external carotid artery.
- The external carotid arteries across the midline.

3 Small intraarterial communications include the following:

- Intricate network of transdural anastomoses across the subdural space from the dural arteries to the arteries on the surface of the brain.
- Leptomeningeal collaterals form the meningeal border-zone network that connects the terminal cortical branches of the cerebral arteries.

Structural Anatomy of Arteries

1 The function of the artery is to carry blood away from the heart, transporting gases, nutrients, and other essential substances to the tissues. In addition, the arteries help to regulate blood flow by contracting and relaxing.

2 Microscopic anatomy:

The vessels of both the arterial and venous systems have three layers:

- The inner layer, the *tunica intima,* is a single layer of endothelial cells with a base membrane and connective tissue.
- The middle layer, the *tunica media,* is a thick layer of smooth muscles and collagenous fiber that is usually arranged in a circular fashion.
- The outermost layer, the *tunica adventitia,* is a thin fibrous layer of connective tissue and some smooth muscle fibers. This layer contains the

vasa vasorum, the "vessels of vessels" that supply blood to the walls of the larger arteries. In addition to transporting oxygen and nutrients, as well as removing waste products, the vasa vasorum can also function as a collateral pathway to maintain patency of the internal carotid artery in the presence of a focal occlusion.

Pressure/Flow Relationships

In addition to the discussion of arterial hemodynamics in chapter 2, this brief review clarifies certain important aspects and principles of cerebrovascular blood flow.

1 The Bernoulli principle states that the total fluid energy along a streamline of fluid flow is constant and that velocity energy and pressure energy are inversely proportional: The higher the velocity, the lower the pressure; the lower the velocity, the higher the pressure. Within any particular arterial segment, pressure gradients develop as a result of changes in the geometry and direction of the vessel, whether or not there is intraluminal disease. As figure 18-5 demonstrates, these pressure gradients, also known as areas of flow separation, are easily observed in systole as reversed flow. Regions of stagnation are evident in diastole.

2 Poiseuille's law describes flow through a rigid tube as being parabolic or laminar. The quantity of flow (Q) is related to the pressure gradient across an arterial segment (P), the radius of the vessel (r), the viscosity of the fluid (η), and the length of the vessel (L):

$$Q = (P)\pi r^4/8\eta L$$

- Flow is directly proportional to a pressure gradient and the size of the vessel.
- Flow is inversely proportional to the viscosity of the blood and length of the vessel, which are elements of resistance.
- Since the measurement of the radius is raised to the fourth power, i.e., multiplied by itself four times, small changes in the radius result in very large flow changes.

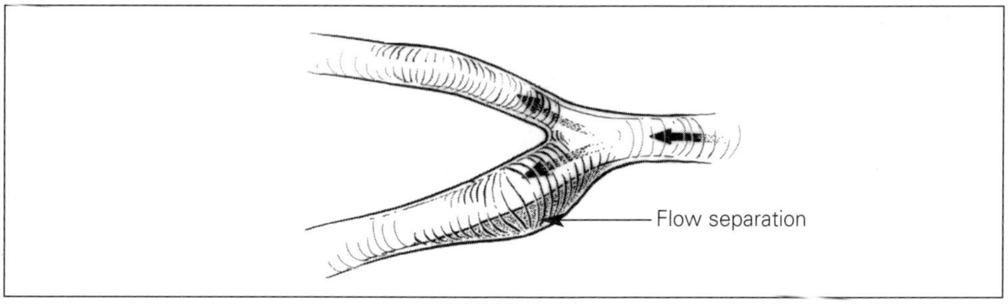

Figure 18-5. Normal area of flow separation (related to the Bernoulli principle) evident at the carotid bulb.

- Remember the law of conservation of mass: Velocity (V) = Volume flow (Q) divided by cross sectional area (A) [$V = Q / A$]. Considering that volume flow is constant, this law shows why the radius of a vessel is inversely proportional to the velocity.
- Compensatory mechanisms, e.g., increase in velocity, work to equalize flow across a lesion.

Effects of Stenosis on Flow

1. The blood must change direction as the flow stream narrows at the stenosis and enlarges as it exits the stenosis. The resulting eddy currents, turbulence, and vortices produced by changes in the flow stream cause an energy loss through inertia.
2. Velocity increases through a stenosis since velocity and area are inversely proportional. This acceleration increases energy losses.
3. Figure 18-6 illustrates the accelerated and turbulent stenotic flow evident on spectral analysis. Acceleration is reflected as increased frequency (F_d in the figure) or increased velocity measurement. Turbulence is detected by spectral broadening, which is a wide range of very low to very high frequency shifts. Spectral broadening results in the loss of the frequency or spectral window.

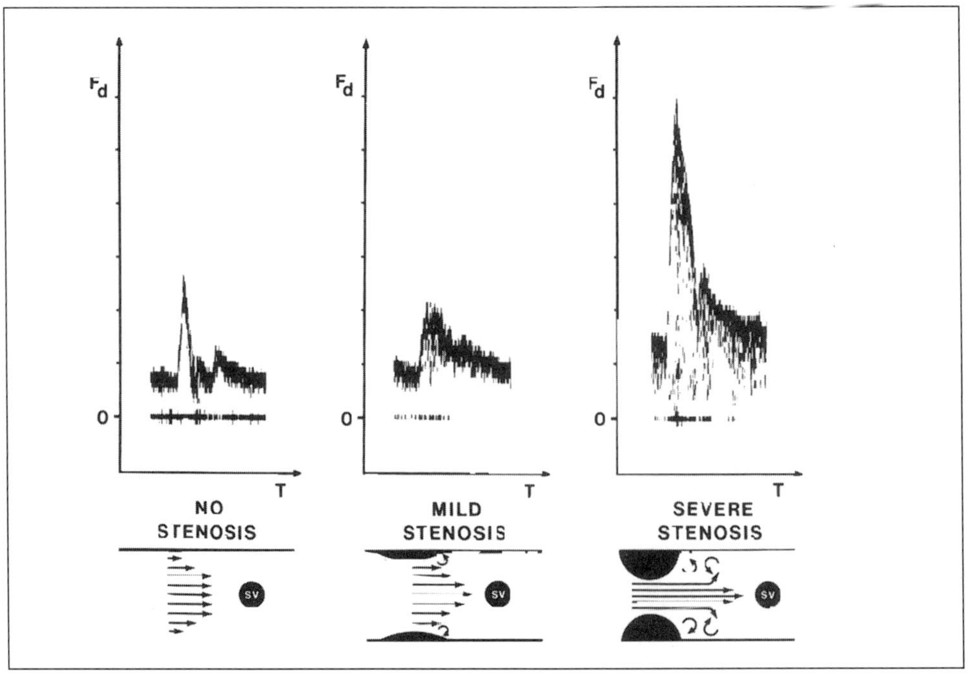

Figure 18-6. Laminar flow in normal artery evidenced by narrow band of frequencies during systole leaving a window beneath the envelope. Slight turbulence from a mild stenosis broadens the frequency band. Accelerated, turbulent flow caused by tight stenosis; note loss of frequency window. From Hershey FB, Barnes RW, Sumner DS [eds]: *Noninvasive Diagnosis of Vascular Disease*. Pasadena, Appleton Davies, 1984.

Patient History, Mechanisms of Disease, and Physical Examination

CHAPTER 19

Patient History: Signs and Symptoms

Note: It is very important to obtain a complete history and to confirm appropriate indications for testing before performing a study.

1 Symptoms of transient ischemia:

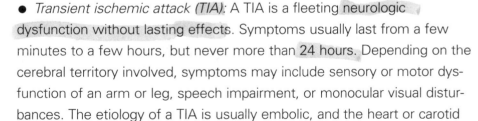

- *Transient ischemic attack (TIA):* A TIA is a fleeting neurologic dysfunction without lasting effects. Symptoms usually last from a few minutes to a few hours, but never more than 24 hours. Depending on the cerebral territory involved, symptoms may include sensory or motor dysfunction of an arm or leg, speech impairment, or monocular visual disturbances. The etiology of a TIA is usually embolic, and the heart or carotid artery is most often the source of the emboli.

- *Reversible ischemic neurologic deficit (RIND):* A RIND usually lasts longer than a TIA, but the deficit it creates resolves in time. That is, brain tissue is damaged but recovers completely.

- *Vertebrobasilar insufficiency (VBI):* Vertebrobasilar insufficiency usually causes bilateral symptoms of visual blurring or paresthesia and complaints of vertigo, ataxia, and drop attacks.

2 *Cerebrovascular accident (CVA) or stroke:* A stroke produces a permanent neurologic deficit. Strokes are classified as follows:

- *Acute:* Symptoms of sudden onset; considered unstable.
- *Stroke in evolution:* Symptoms come and go; considered unstable.
- *Completed stroke:* No progression or resolution of the symptoms; considered stable.

Patient History: Risk Factors and Contributing Diseases

1 Diabetes mellitus, the most common form of diabetes, is a chronic systemic disease characterized by disorders of the metabolism of insulin, carbohydrate, fat, and protein. As previously described on page 41, the chronic complications are primarily vascular.

2 Hypertension may be a causative factor in the development of atherosclerosis, or it may enhance the development of the atherosclerotic process. It almost certainly helps to precipitate clinical events associated with atherosclerosis, e.g., embolization and aneurysms.

3 Smoking is associated with irritation of the endothelial lining of the blood vessels.

4 Hyperlipidemia: Because lipids are insoluble in water, an excessive amount of plasma lipids is closely associated with the development of atherosclerosis.

Mechanisms of Disease

The two most common mechanisms of cerebrovascular insufficiency are ischemia and hemorrhage. The three leading causes of ischemia are atherothrombotic pathology, cardiogenic pathology, and lacunar strokes. Atherothrombotic pathologies include alterations in perfusion secondary to an atheromatous plaque, resulting in stenosis, occlusion, or a thromboembolic event. The cardiogenic pathologies usually responsible for cerebrovascular ischemia are altered cardiac function and, most often, embolism. A lacunar infarction is a small circumscribed loss of brain tissue caused by occlusion of one of the small penetrating arteries in the brain. Hemorrhage, the second most common mechanism of cerebrovascular insufficiency, may be the result of hypertension, ruptured aneurysm, or trauma (e.g., subarachnoid hemorrhage).

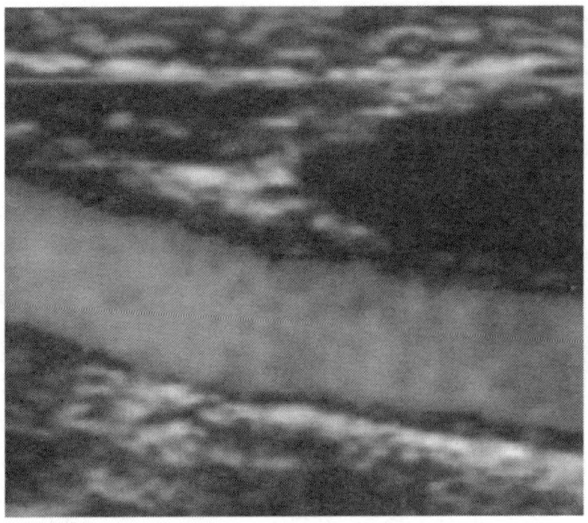

Figure 19-1. The thin layer of lipid material on the arterial wall can be difficult to visualize because of its low-level echoes. Color flow Doppler is helpful in determining that the vessel does not fill completely.

1 Stenosis: Atherosclerosis is a common pathological condition in which the arteries thicken, harden, and lose their elasticity, creating a narrowing that can reduce distal blood flow and adversely affect brain perfusion. Atherosclerosis is a form of arteriosclerosis that is characterized by a localized accumulation of lipid-containing material (atheroma), smooth muscle cells, collagen, fibrin, and platelets that form within or beneath the intimal surface of the blood vessel. This type of plaque usually occurs at the origin of the vessel, e.g., the proximal internal carotid artery. The distinguishing characteristics of plaque include the following:

- A *fatty streak* is a thin layer of lipid material on the intimal layer of the artery (figure 19-1; see also color plate 10).
- A *fibrous plaque* is an accumulation of lipids that is covered by more lipid material, collagen, and elastic fiber deposits (figure 19-2).

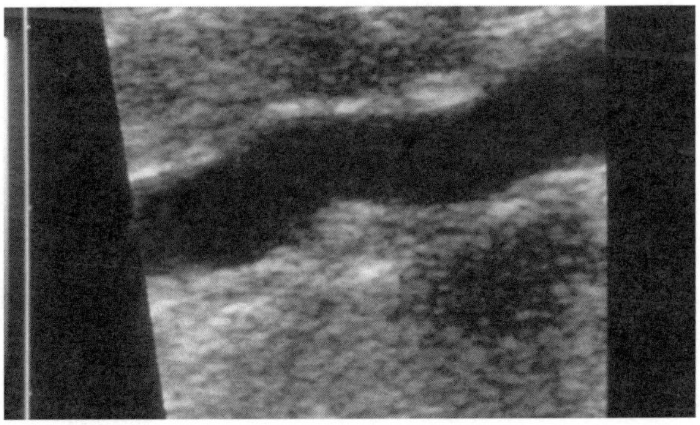

Figure 19-2. The increased amount of collagen in this fibrous plaque makes it easily visible on B-mode imaging.

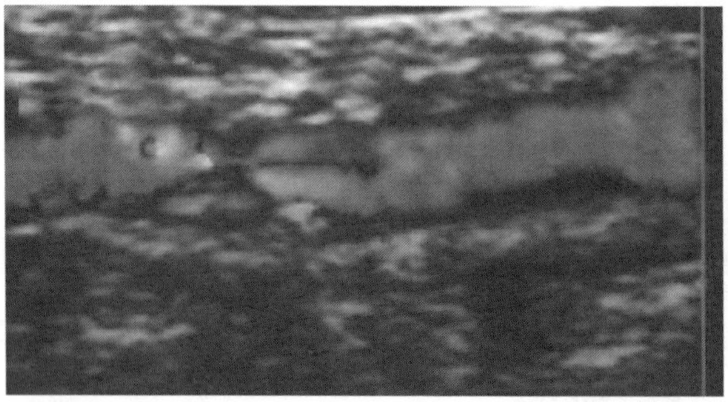

Figure 19-3. The heterogeneous quality of this plaque is consistent with the variety of cellular material it is composed of.

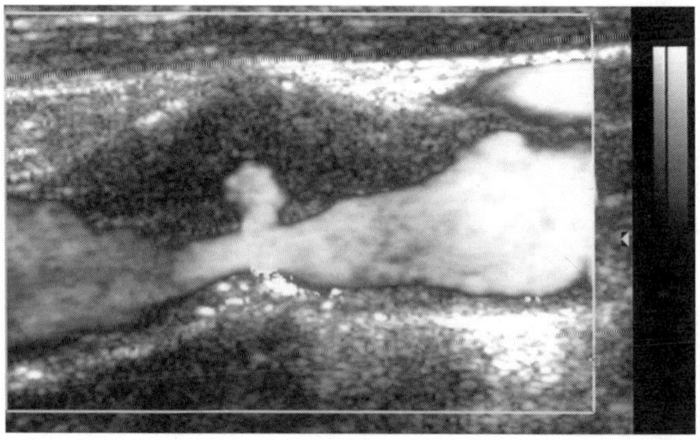

Figure 19-4. Although it is very difficult to accurately determine surface characteristics with B-mode, this scan, using power Doppler (other names: color-energy Doppler, color amplitude image), clearly illustrates the crater-like formation of the arterial wall consistent with an ulcerative lesion. Courtesy of Steve Bernhardt, BS, RDMS, RVT.

- A *complicated lesion* is a fibrous plaque that contains fibrous tissue, more collagen, calcium, and cellular debris (figure 19-3).
- An *ulcerative lesion* is characterized by the deterioration of the normally smooth surface of the fibrous cap. Ulcerative lesions are of particular importance because they may shed debris that embolizes distally (figure 19-4).
- *Intraplaque hemorrhage* is evident on duplex scans as a sonolucent area within the plaque (figure 19-5). This finding is also important as this plaque may be at higher risk for rupture, resulting in distal embolization.

2 *Embolism:* An embolus is a foreign substance or piece of a thrombus that moves through the circulatory system until it lodges in a distant blood vessel, resulting in complete or partial obstruction of that vessel. Emboli may be solid, liquid, or gaseous and may arise from the body or may enter from without. The heart is the most frequent source of emboli. Another common source is atherosclerotic plaque, which can break loose and travel distally until it lodges in a small or stenotic vessel. One

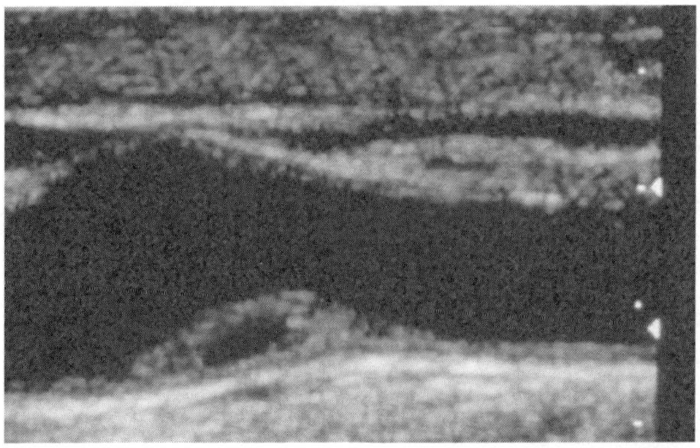

Figure 19-5. The well-defined area of low-level echoes in the center of the lesion suggests instability and the risk of plaque rupture with distal embolism.

example of this process is a *Hollenhorst plaque*, which is a bright plaque in the vessels of the retina. Detected by an ophthalmologist, Hollenhorst plaques are actually atheromatous emboli that contain cholesterol crystals and may have originated from plaque in either the carotid artery or more proximal vessels.

3 *Thrombosis:* This aggregation of blood factors is composed primarily of platelets trapped within a network of fibrin. One of the most common causes of a stroke, thrombosis alters cerebral perfusion secondary to a stenosis or embolism. As illustrated in figure 19-6, a stenosis, embolism, or thrombosis may exist individually or in combination.

4 *Aneurysm:* An aneurysm is an abnormal, localized dilatation of a blood vessel due to congenital defects or weakness of the vessel wall, which may be caused by trauma, infection, or atherosclerosis. Although aneurysms rarely occur in the cervical carotid artery, figure 19-7 documents one such case.

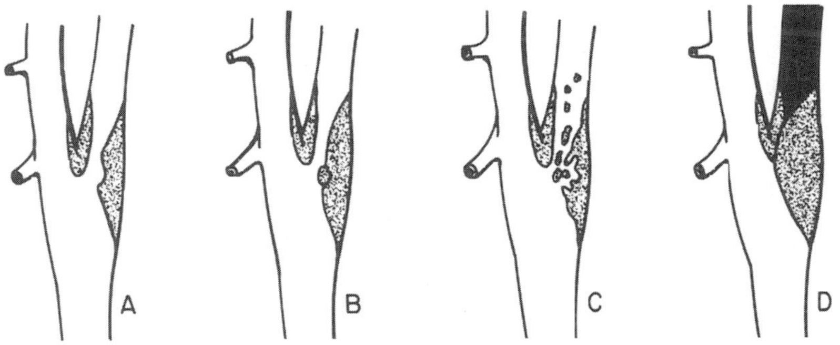

Figure 19-6. Atherosclerotic changes at the bifurcation of the carotid artery include (**A**) stenosis, (**B**) stenosis with surface platelets, (**C**) embolic process, and (**D**) thrombosis. From Roederer GO, Langlois Y, Strandness DE: Comprehensive noninvasive evaluation of extracranial cerebrovascular disease. In Hershey FB, Barnes RW, Sumner DS [eds]: *Noninvasive Diagnosis of Vascular Disease.* Pasadena, Appleton Davies, 1984.

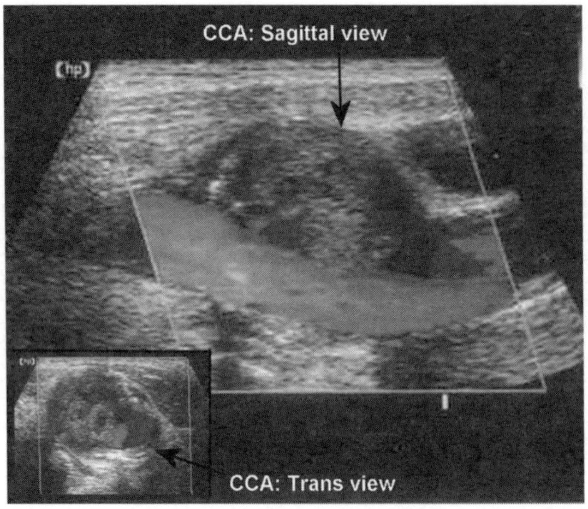

Figure 19-7. Sagittal view of aneurysmal disease of the cervical carotid artery documenting focal dilatation of the distal common carotid artery. **Inset:** Transverse view. Most patients evaluated for carotid aneurysms have very tortuous CCAs.

5 Nonatherosclerotic lesions include a variety of processes:

- Arteritis:

  Arteritis is an inflammation of an artery or arteries. Two types of arteritis are of note:

  Takayasu's arteritis: A chronic inflammation that results in narrowing of the arteries. The resulting arterial occlusion or near-occlusion explains why Takayasu's arteritis is also known as pulseless disease. Type I lesions involve the aortic arch and its branches; type II involves the thoracic and abdominal aorta; type III includes the aortic arch, its branches, and the thoracic and abdominal aorta; and type IV is related to pulmonary vessels.

  Temporal arteritis: The inflammation of the distal superficial temporal artery and/or its branches may cause a severe headache and/or sudden unilateral blindness.

- Carotid body tumor:

  The carotid body is a small mass of vascular tissue that adjoins the carotid sinus. It functions as a chemoreceptor sensitive to changes in the oxygen tension of the blood and signals necessary changes in respiratory activity to maintain homeostasis.

  Tumors are highly vascular structures that develop between the internal carotid artery and external carotid artery, and they are usually fed by the external carotid artery (figure 19-8; see also color plate 11).

  Excision of the tumor may also require ligation of the internal carotid artery and/or external carotid artery.

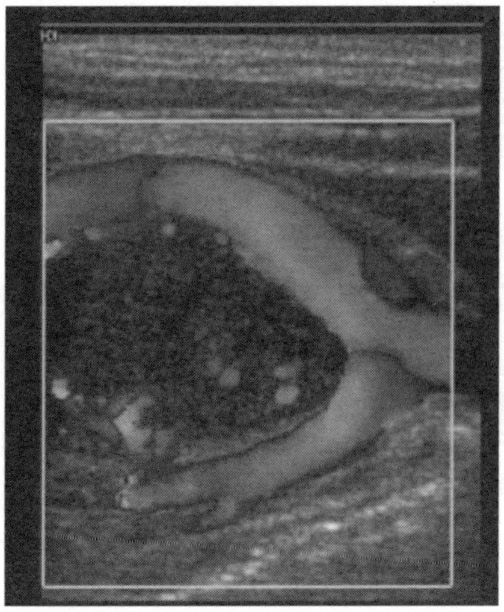

Figure 19-8. Sagittal view of a carotid body tumor, a highly vascular structure that develops between the internal carotid artery and external carotid artery.

- Collagen vascular connective tissue disorders
- Dissection:

  Dissection is most commonly caused by trauma that results in a sudden tear in the intimal wall.

  A false lumen is created, which may gradually extend proximally or distally (figure 19-9A). Figure 19-9B illustrates abnormal flow patterns detected in the false lumen. See also color plate 12.

  Blood in the false lumen may thrombose.

  Extension of the false lumen may result in complete obstruction to blood flow (see figure 19-10 on the following page).

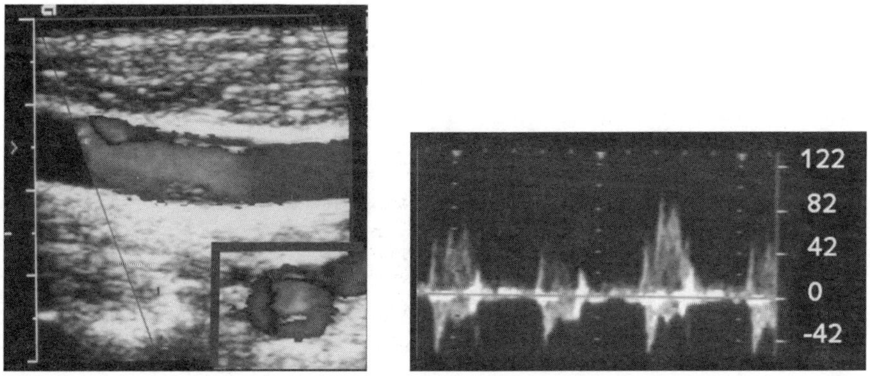

A B

Figure 19-9. **A** Longitudinal view of a common carotid artery dissection. The inset shows the wall defect in transverse view. Flow direction and velocities differ in each lumen. A distinguishing B-mode feature is the thin membrane dividing the main arterial lumen from the false lumen. **B** Abnormal flow patterns detected in this false lumen.

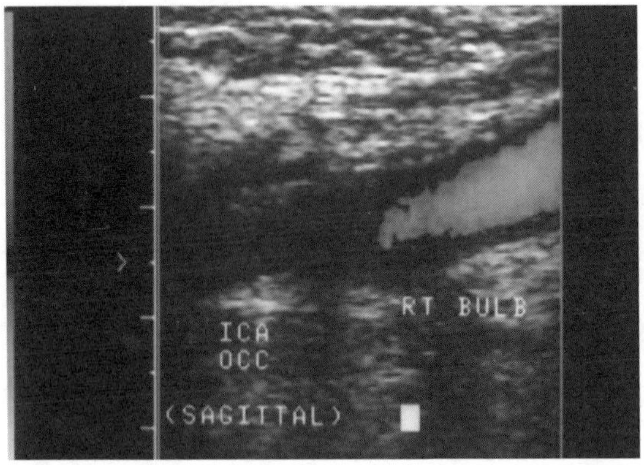

Figure 19-10. Extension of the false lumen has resulted in complete thrombosis of the distal portion of the vessel.

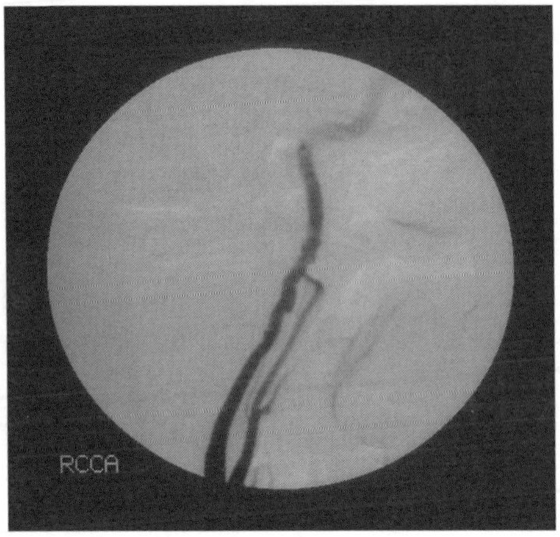

Figure 19-11. Angiogram showing internal carotid artery with fibromuscular dysplasia (FMD) of the distal portion of the vessel. While most carotid disease is atherosclerotic and occurs at the bifurcation, FMD is usually located quite distally. On angiography, the vessel with FMD typically appears to be beaded, a characteristic referred to as a "string of pearls." Doppler signals would show spectral broadening and increased velocities that vary with the location of the Doppler sample.

- Fibromuscular dysplasia (FMD):

  Most commonly caused by dysplasia of the media along with overgrowth of collagen. (Dysplasia of the intima is less frequently seen.) Dysplasia of the media occurs in the mid to distal segments of vessels, unlike atherosclerosis, which usually develops at the origin of vessels.

  Characteristic beadlike appearance on angiography (figure 19-11).

  More often seen in younger women.

- Neointimal hyperplasia:

  Intimal thickening results from rapid production of smooth muscle cells.

  Hyperplasia is a response to vascular injury, which may be the result of vascular reconstruction, e.g., carotid endarterectomy.

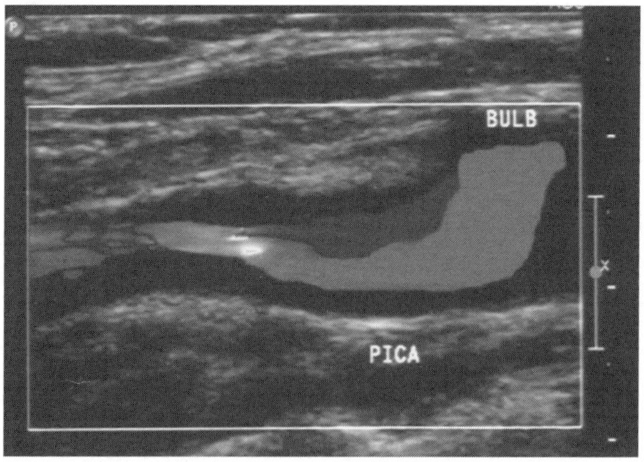

Figure 19-12. This sagittal view of the distal CCA and proximal ICA was obtained one year post-endarterectomy. The thickened arterial wall with very low-level echoes is consistent with intimal hyperplasia. This differs from a reoccurrence of atherosclerosis, which is usually not evident sooner than two years following carotid endarterectomy.

Denuding the endothelium surgically leads to platelet accumulation, endothelial regeneration, and proliferation of smooth muscle cells.

A hemodynamically significant stenosis may develop within 6 to 24 months following endarterectomy (figure 19-12; see also color plate 13).

- Trauma: Traumatic damage to vessels and surrounding tissue can produce a variety of conditions, including hemorrhage, occlusion, fistula, dissection, and pseudoaneurysm.

Physical Examination

1 Anterior circulation (carotid):

Normally, anterior vessels—internal carotid, anterior cerebral, middle cerebral, and anterior communicating arteries—supply blood to the brain's cerebral hemispheres. Since the left hemisphere of the brain controls the right side of the body, and vice versa, a left hemispheric stroke results in neurologic deficits to the right side of the face and body. The term "lateralizing" may be used to indicate which side or hemisphere of the brain has been affected. For example, left arm numbness is consistent with decreased perfusion of the right hemisphere by the anterior circulation. "Nonlateralizing" symptoms (e.g., ataxia, vertigo) are usually associated with problems of the brain stem or the posterior circulation. A monocular visual problem (i.e., amaurosis fugax, a temporary loss of vision in one eye) is related to disease of the ipsilateral internal carotid artery. As illustrated in figure 19-13A, neurologic deficits that may be attributable to problems with the anterior circulation include these:

- Unilateral *paresis* (weakness or slight paralysis on one side of the body).

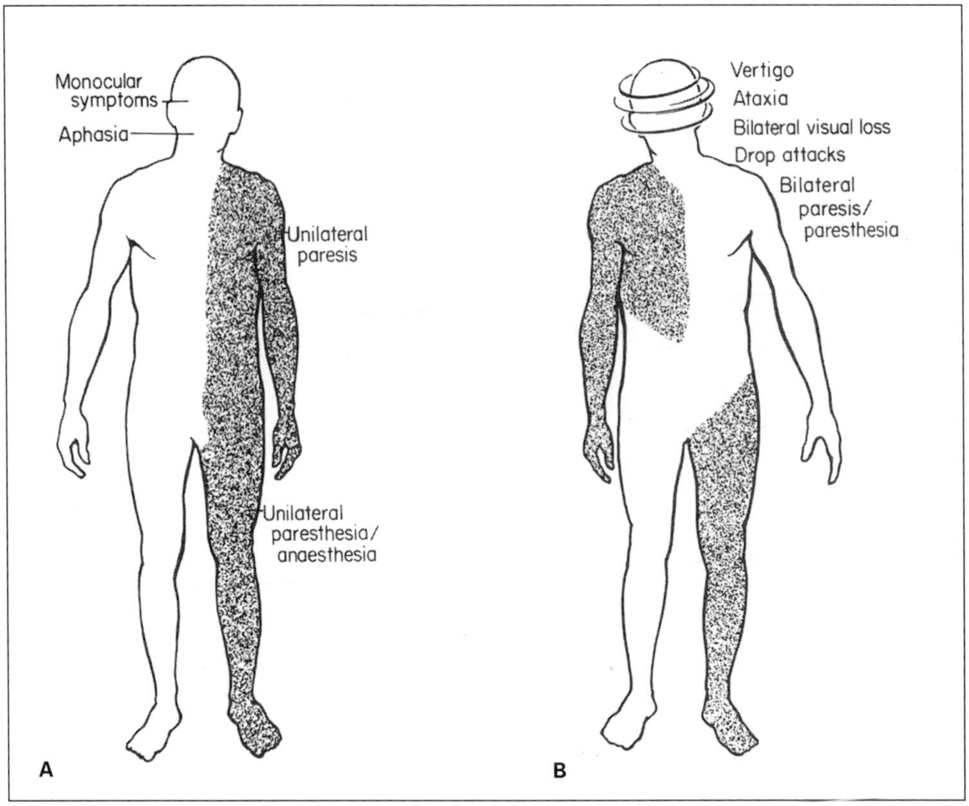

Figure 19-13. **A** Hemispheric signs and symptoms. **B** Vertebrobasilar signs and symptoms. From O'Connor AJ: The anatomy and pathophysiology of extracranial atherosclerotic cerebrovascular disease. In Hershey FB, Barnes RW, Sumner DS [eds]: *Noninvasive Diagnosis of Vascular Disease.* Pasadena, Appleton Davies, 1984.

- Unilateral *paresthesia* or *anesthesia* (tingling, numbness, or lack of feeling on one side of the body).
- *Dysphasia* (impaired speech) or *aphasia* (the inability to speak). Since the left hemisphere is dominant for speech, analytical capabilities, and verbal and auditory memory in a right-handed person, a lesion in the left middle cerebral artery may cause dysphasia or aphasia. The right hemisphere is dominant in a left-handed person.
- Behavioral abnormalities, which frequently accompany ischemia of the temporal lobe can be caused by infarction of the right middle cerebral artery.
- *Amaurosis fugax* (temporary partial or total blindness, usually of only one eye). Patients typically describe the temporary visual loss as "a shade coming down over the eye." Because the ophthalmic artery arises from the internal carotid artery, the origin of the embolic process causing the temporary visual loss may be a lesion in the ipsilateral carotid artery.
- *Homonymous hemianopia* (blindness in the corresponding, right and left, visual fields). This loss of vision in one-half of the visual field of both eyes may be evident after a stroke affecting the cerebral cortex.

2 Posterior circulation (vertebrobasilar):

Normally, posterior vessels—vertebral, basilar, posterior cerebral, and posterior communicating arteries—supply blood to the brain stem, cerebellum, and occipital lobes of the brain. As illustrated in figure 19-13B, symptoms of brain ischemia that may be attributable to problems of the posterior circulation include the following:

- *Vertigo* (difficulty in maintaining equilibrium: a sensation of either moving around in space or having objects move around the person). Vertigo is *not* dizziness or lightheadedness.
- *Ataxia* (muscular incoordination), usually described as the inability to control gait or to touch an article with the hand.
- Bilateral visual blurring.
- *Diplopia* (double vision).
- Bilateral *paresthesia* (the sensation of prickling or tingling of the skin) or *anesthesia* (loss of sensation).
- *Drop attack* (falling to the ground without other symptoms or loss of consciousness).

3 Nonlocalizing symptoms:

- Dizziness (a sensation of whirling within one's head, lightheadedness, and a tendency to fall).
- *Syncope* (a transient loss of consciousness).
- Difficulty with speech.
- Headache.
- Confusion.

Although infarction of the middle cerebral artery is the most common cause of a stroke, table 19-1 on the following page relates presenting signs and symptoms to the artery that may be responsible for them.

4 Auscultation:

- Normal flow patterns are evident when listening through a stethoscope.
- The normal lub-dub sound corresponds with closure of the atrioventricular valves followed by closure of the aortic and pulmonary valves.
- Bruits are abnormal sounds heard on auscultation caused by turbulent blood flow patterns that set up a vibratory response in the tissue, as happens with a hemodynamically significant stenosis. Because bruits may

Table 19-1.
Cerebrovascular signs and symptoms related to anatomy.

Affected artery	*Frequently seen signs and symptoms*
Internal carotid	Contralateral weakness, paralysis, numbness, and/or sensory changes; ipsilateral amaurosis fugax, and/or a bruit; may have aphasia if in the dominant hemisphere; occasionally altered level of consciousness.
Middle cerebral	Aphasia or dysphasia; severe contralateral hemiparesis or hemiplegia (more severe in face and arm rather than leg if upper-division MCA occluded); dysarthria; behavioral changes, confusional state, agitated delirium (if lower-division MCA occluded).
Anterior cerebral	Contralateral hemiparesis (especially in leg), incontinence, loss of coordination, impaired motor and sensory functions.
Vertebrobasilar	Numbness around lips and mouth, diplopia, poor coordination, dysphagia, vertigo, amnesia, and ataxia.
Posterior cerebral	Dyslexia, coma (paralysis usually absent).

arise from lesions other than those at the carotid bifurcation, and because they may be absent in severe stenosis or occlusion, cervical bruits are a nonspecific sign of hemodynamically significant carotid artery disease. Nevertheless, the absence of a bruit does not exclude the presence of stenosis because a preocclusive lesion may not produce a bruit. Also, a previously documented bruit that is no longer evident may indicate a lesion has progressed from < 90% diameter reduction to > 90% diameter reduction (preocclusive).

- Common sites for bruit evaluation include the low-to-mid neck for the common carotid artery, the mid-to-distal neck for the bifurcation, and just above and/or below the clavicle for the subclavian artery.
- The brachial artery in the antecubital fossa is a common site for auscultation when obtaining bilateral blood pressure measurements.
- Unequal brachial artery pressures may be related to subclavian or axillary disease. Please refer to chapter 6 on page 75.

5 Palpation:

- The pulse strength is evaluated manually.
- The strength of the pulse is graded and compared to other sites on the same side as well as to the same site on the contralateral side.
- Common sites for palpation of pulses include the common carotid, superficial temporal, subclavian, axillary, and brachial arteries.

Periorbital Doppler

CHAPTER 20

1 Capabilities:

The periorbital Doppler examination can detect hemodynamically significant lesions of the internal carotid artery by evaluating the flow to some of its terminal branches around the eye.

2 Limitations:

- It is only diagnostic in cases of hemodynamically significant lesions.
- It cannot differentiate an occlusion from a tight stenosis.
- It requires considerable skill.

3 Patient positioning:

The patient is positioned supine. The head of the bed can be elevated slightly, and the patient's head can rest on a pillow.

4 Physical principles:

When a sound wave is reflected from a moving target, the frequency of the received wave differs from that of the transmitted wave. This difference—the Doppler effect—occurs whenever there is relative motion between the source and the receiver of the sound wave. In the evaluation of blood flow, the red blood cell is the moving target and the transducer is the stationary source and receiver of the sound waves.

5 Technique:

- Patient is supine with eyes closed.
- Using an 8–10 MHz Doppler, the examiner locates the frontal artery at the inner canthus of the eye.
- The analog recorder is automatically calibrated when the equipment is turned on.
- Flow in the frontal artery should be antegrade. (The probe must be positioned carefully so that antegrade flow is not misrepresented as retrograde flow.)

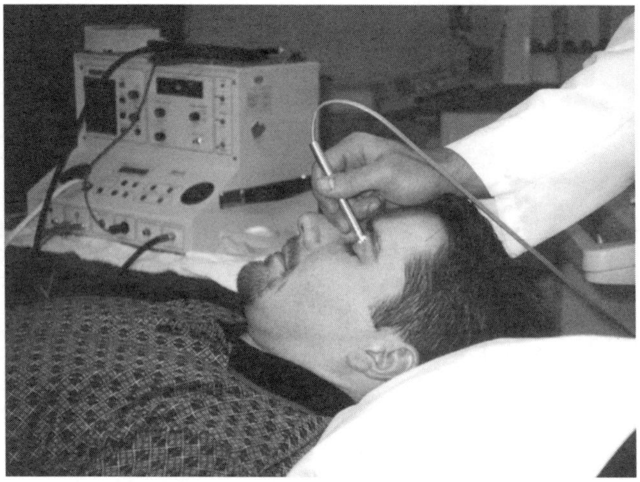

Figure 20-1. Doppler probe positioned to obtain frontal artery flow signals during the periorbital evaluation. The probe is placed lightly against the skin over an adequate amount of coupling gel with just a slight angle. Under normal circumstances, flow direction in the frontal artery is antegrade (toward the probe).

- A series of compression maneuvers are performed to detect the presence of abnormal collateral channels, which suggest disease. These compression maneuvers are performed both ipsilaterally and contralaterally to the following arteries:

  Facial artery

  Superficial temporal artery

  Infraorbital artery

  Common carotid artery

  Figure 20-1 and figure 20-2

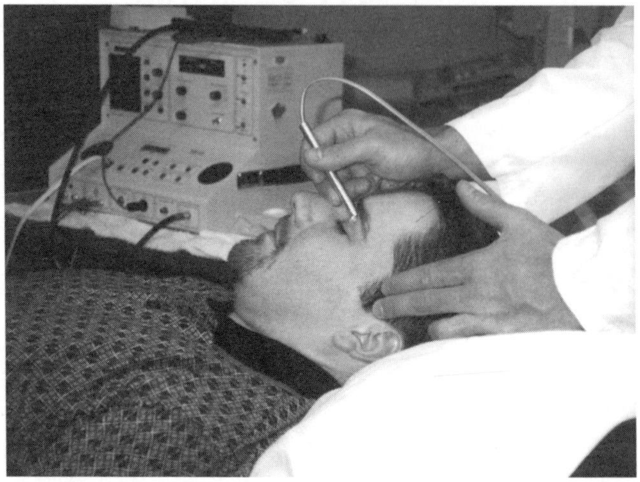

Figure 20-2. During the periorbital Doppler evaluation, compression maneuvers are performed to determine the presence or absence of collateralization. In this example the temporal artery, a branch of the external carotid artery, is manually compressed while the Doppler probe obtains arterial signals from the frontal artery, a branch of the ophthalmic artery. Under normal circumstances, compression of a distal branch of the external carotid artery does not result in flow reduction to a distal branch of the internal carotid artery. This compression maneuver either maintains signals or slightly augments the arterial Doppler signals.

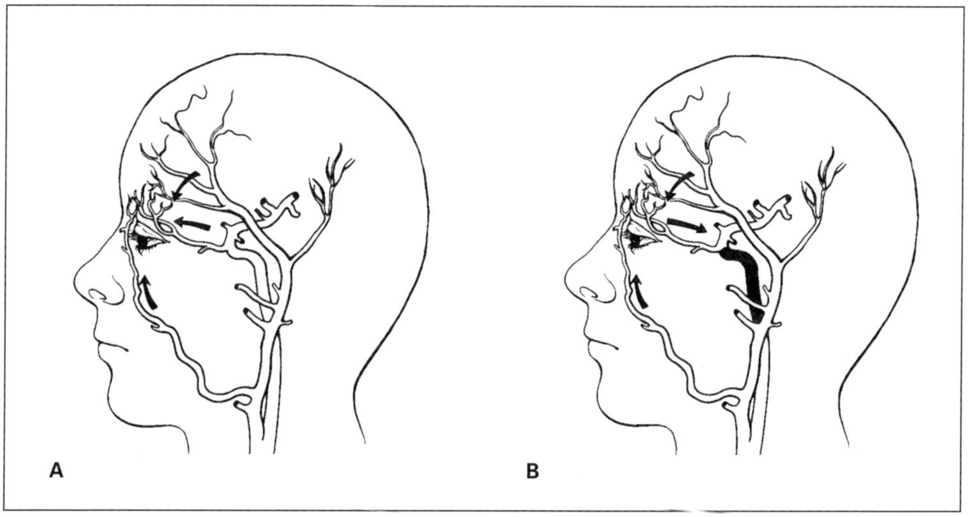

Figure 20-3. **A** As blood flows through its normal pathway, antegrade flow should be evident in the frontal artery. **B** Retrograde flow in the frontal artery is consistent with altered blood flow pathways secondary to a hemodynamically significant lesion in the internal carotid artery. From O'Connor AJ: The anatomy and pathophysiology of extracranial atherosclerotic cerebrovascular disease. In Hershey FB, Barnes RW, Sumner DS [eds]: *Noninvasive Diagnosis of Vascular Disease.* Pasadena, Appleton Davies, 1984.

The common carotid artery must be compressed with great care to avoid stimulating the carotid sinus and thereby altering heart rate and/or rhythm. Compressions are usually done low in the neck to avoid the bifurcation. Other dangers associated with compression of the carotid artery include decreased cerebral perfusion and distal embolization of dislodged plaque.

6 Interpretation:

Normal

- Antegrade flow should be evident in the frontal artery (figure 20-3A).
- Compression of the facial, superficial temporal, and infraorbital arteries should not diminish or reverse flow.
- Ipsilateral compression of the common carotid artery should diminish flow in the frontal artery secondary to decreased flow to the brain. Flow reversal may also be evident depending on collateral development.

Abnormal

- Retrograde flow in the frontal artery is consistent with a hemodynamically significant lesion of the ipsilateral internal carotid artery (figure 20-3B).
- Diminished or reversed flow during compression of the facial, superficial temporal, or infraorbital arteries suggests that flow into the frontal artery is from the vessel being compressed rather than the ipsilateral internal carotid artery. For example, decreased or absent flow in the frontal artery with compression of the superficial temporal artery signifies that flow in the frontal artery is abnormal. The superficial temporal artery is serving as a collateral pathway secondary to a hemodynamically significant lesion in the ipsilateral internal carotid artery.

Oculopneumoplethysmography (OPG-Gee)

CHAPTER 21

1 Capabilities:

- OPG-Gee is an indirect test that detects hemodynamically significant lesions of the internal carotid artery by evaluating flow in one of its terminal branches.
- OPG-Gee also provides indirect information about the development of collateral channels that maintain adequate blood flow to the brain.

2 Limitations:

- OPG-Gee cannot ascertain the exact location of a stenosis, which may be at any point from the origin of the carotid artery to the ophthalmic artery.
- It is not diagnostic in cases of non–hemodynamically significant or well-collateralized lesions.
- It cannot differentiate occlusion from a tight stenosis.
- It is not useful in documenting the progression of disease.
- It may be incapable of determining the ocular systolic pressure (OSP) in patients with severe hypertension.

3 Patient positioning:

- The patient is positioned supine.
- The head of the bed can be elevated slightly, and the patient's head can rest on a pillow.

4 Physical principles:

- Plethysmography is a technique that records changes in volume. In this vascular application, fluctuations in volume are related to changes in blood flow during systole and diastole, e.g., when arterial inflow exceeds venous outflow.
- As represented in figure 21-1, OPG-Gee measures ophthalmic arterial systolic pressure by applying a vacuum to the eye. As the vacuum

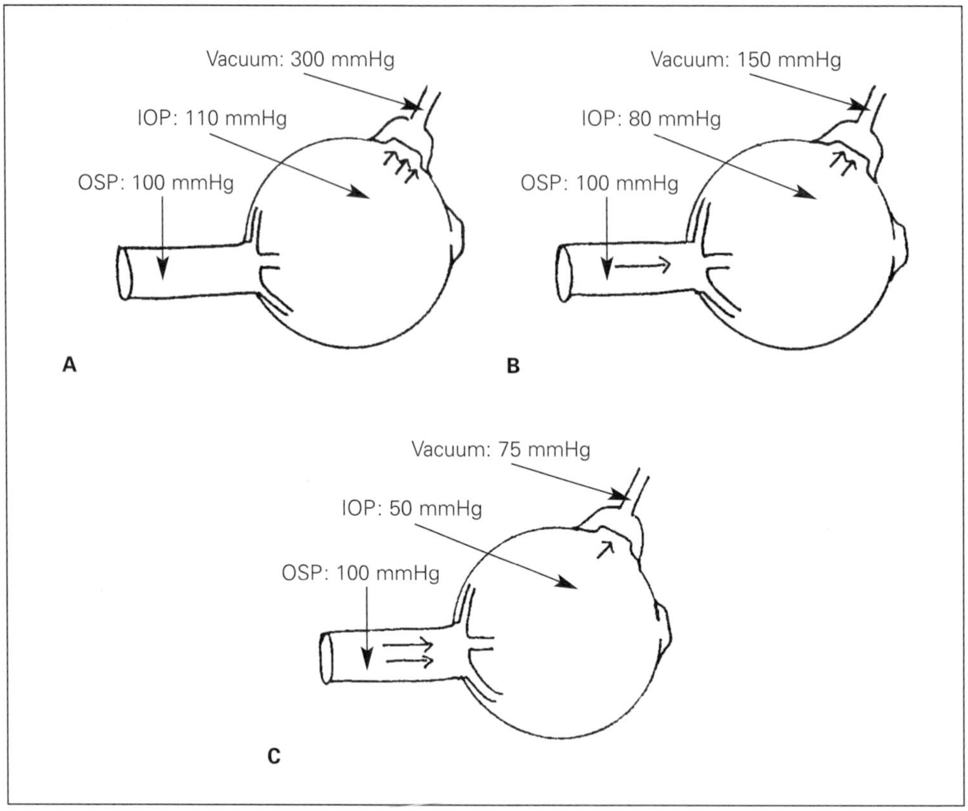

Figure 21-1. **A** Intraocular pressure exceeds the pressure in the ophthalmic artery. There is no flow into the globe and no detectable change in volume. **B** When intraocular pressure is slightly less than the pressure in the ophthalmic artery, there is some flow into the globe and a slight volume change is detected. **C** A large volume change occurs because there is a big difference between intraocular and ophthalmic arterial pressures.

distorts the shape of the globe, intraocular pressure increases to the point at which it obliterates arterial inflow (figure 21-1A). Strip-chart recordings are made as the vacuum is slowly decreased. The pulse waveform reappears when the ophthalmic arterial pressure exceeds the intraocular pressure (figures 21-1B and C).

- Since the pressure in the ophthalmic artery reflects the pressure in the distal internal carotid artery, the measurement of ophthalmic arterial pressure can be useful in detecting hemodynamically significant lesions.

5 Technique:

- A thorough patient history must be obtained to rule out the following contraindications to this test:

  Allergies to local anesthetics

  Eye surgery within last 6 months

  A past spontaneous retinal detachment

  Acute or unstable glaucoma

- Other ophthalmic disorders, e.g., myopia or conjunctivitis, are not considered contraindications to OPG-Gee.

- If there are no contraindications, the patient is supine, ECG leads are applied, bilateral brachial pressures are obtained using a stethoscope (noting the side of the higher pressure), and the local anesthetic is applied to the eyes.
- Calibration or standardization is performed by pressing the STD button. Standardized deflections should have an amplitude of approximately 10 mm on the chart (figure 21-2). If not, the examiner presses REC and adjusts GAIN controls for the eye channels, alternately pressing and releasing STD buttons until a 10 mm deflection is evident.
- Calibration is critical to the usefulness of this plethysmographic technique since differences in the size of the tracing are of diagnostic value.

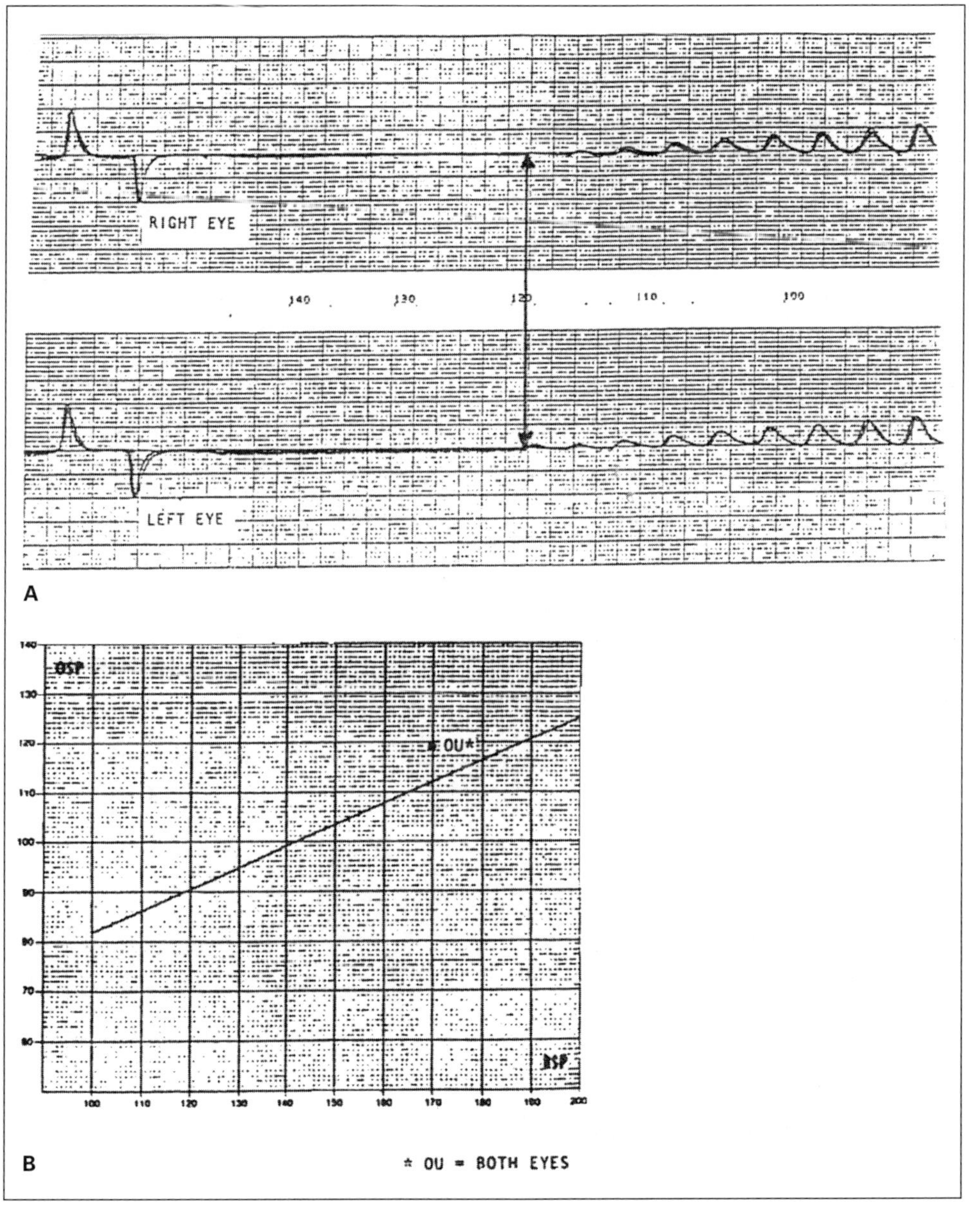

Figure 21-2.
A The first significant eye pulse on the right and left appear at the same OSP (approximately 119 mmHg) and are within normal limits. The initial deflections represent the calibration or standardization procedure. Standardization is achieved when the tracings demonstrate positive and negative deflections of approximately 10 mm.
B Plotting the values on a scoring grid demonstrates that both OSPs fall above the line discriminating normal from abnormal ratios.

For example, a significant reduction in the size of the tracing should reflect a significant reduction in blood volume.

- Eye cups are placed on the lateral sclera (whites of the eye), not on the cornea. After the patient has been informed that he or she will experience a loss of vision for a few seconds, a foot switch is depressed to activate the vacuum.
- The vacuum, measured in mmHg, increases intraocular pressure to obliterate arterial inflow:

  A patient with a systemic systolic pressure of < 140 mmHg may require only the 300 mmHg to occlude inflow.

  A patient with a systemic systolic pressure of > 140 mmHg usually requires a maximum vacuum of 500 mmHg to occlude inflow.
- After the vacuum is applied, the recording of ocular pulsations and ECG commences. As the vacuum is reduced, the decrease in intraocular pressure results in increased inflow and the reappearance of the waveforms. When optimum waveforms are evident, the vacuum is released, the eye cups are removed, and the brachial pressure is measured again from the arm with the higher pressure.
- If ocular pulsations are evident at the onset of the study using the 300 mmHg setting, blood flow was not obliterated, i.e., systolic pressure was not exceeded. The vacuum is deflated, and the examiner selects the 500 mmHg setting, reactivates the vacuum, and repeats the study.
- If the tracings are very irregular and suggest artifact, interpretation may be impossible. The examiner asks the patient to gently close the eyes and to relax the eye muscles as much as possible, reactivates the vacuum, and repeats the study.
- When OPG-Gee is required to assess the adequacy of collateral channels, carotid compression maneuvers are necessary.

  The examiner must *never* compress both common carotid arteries simultaneously.

  Compression is never performed in the area of the carotid sinus because stimulation of the carotid sinus can alter heart rate and/or rhythm, decrease cerebral perfusion, or cause distal embolization from dislodged plaque.

  The first phase consists of a 3–5 second compression while the OPG maintains an intraocular pressure of 60 mmHg.

  The second phase—which is performed *only* if pulsations are noted during the first phase—consists of a ≤15 second compression while the OPG decreases intraocular pressure from 110 mmHg to the level at which pulsations reappear.

Carotid compression should be released gradually to prevent the sudden return of turbulent blood flow.

- The patient is reminded not to rub the eyes for a specific length of time following the study.

6 Interpretation:

Normal

- Since the eye pulse should coincide with the upstroke of the T-wave on the ECG, the T-wave is used to differentiate true eye pulses from artifact, e.g., blinking.
- Identify the first significant eye pulse, i.e., the one followed by pulsations of increasing amplitude.
- Overlay the calibrated ruler on the right eye tracing, aligning the beginning of the upstroke with the appropriate OSP reading. Repeat this process with the left eye tracing.
- Ophthalmic systolic pressures should not differ by $\geq$ 5 mmHg. As illustrated in figure 21-2A on page 219, the first significant eye pulse on the right appears at the identical OSP as the left. Therefore, there is $<$ 5 mm difference in OSPs.
- A normal ratio of ophthalmic-to-systemic pressure should exist:

$$\text{OSP} - 39 \div \text{brachial systolic pressure (BSP)} \geq 0.430$$

As illustrated in figure 21-2B, this ratio appears as a discriminating line on a scoring grid with values above the line demonstrating a normal ratio.

- If the OSPs are greater than 140 mmHg, the examiner measures the amplitude of the first pulses. The difference in amplitude of the tracings should be $\leq$ 2 mm.

Abnormal

- OSPs that differ by $\geq$ 5 mmHg and/or an abnormal ratio of ophthalmic to systemic pressure:

$$\text{OSP} - 39 \div \text{BSP} \leq 0.429$$

- As represented in figure 21-3, the presence of a hemodynamically significant lesion is evidenced by either a difference in OSPs $\geq$ 5 mmHg and/or an abnormal ratio of ophthalmic-to-systemic pressure.

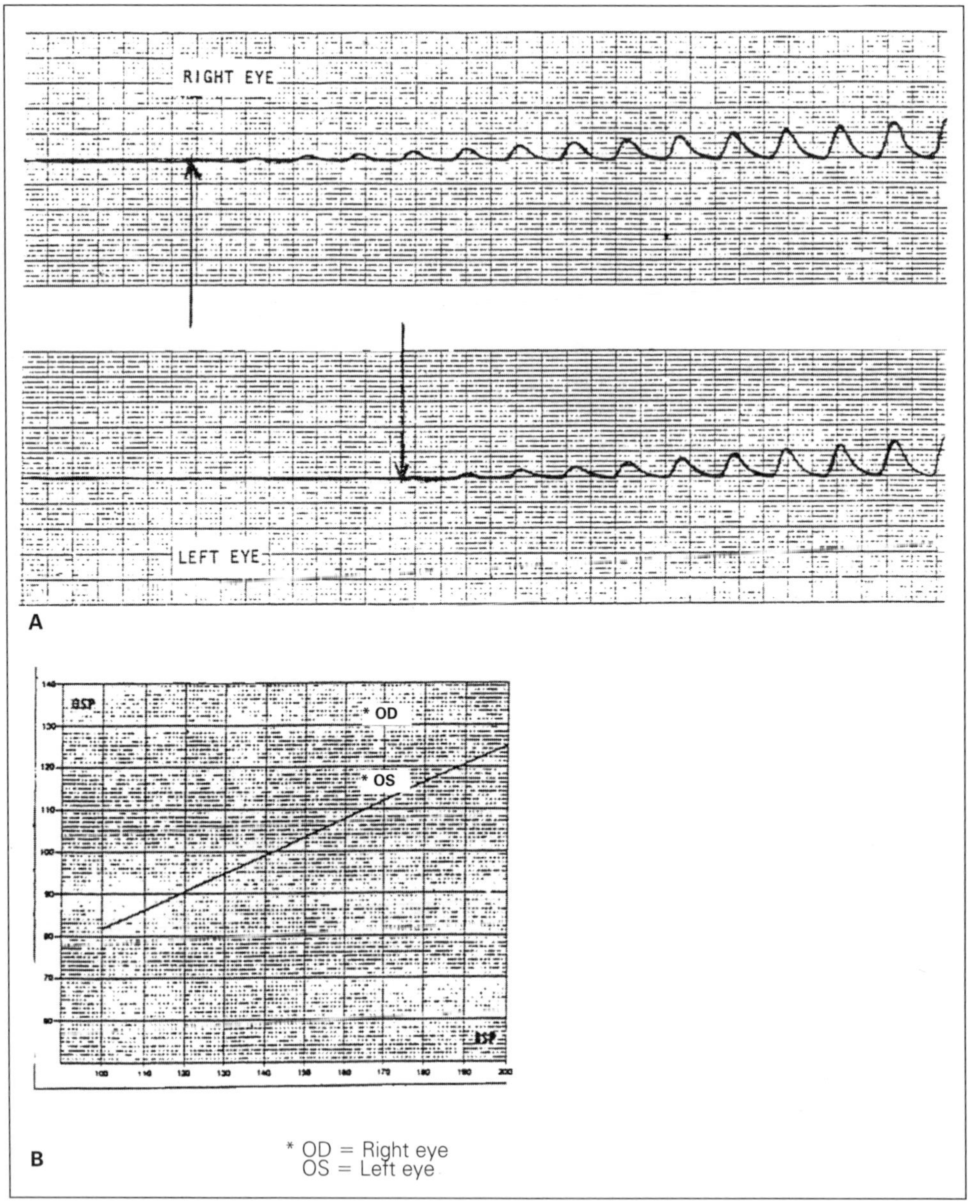

Figure 21-3.
A The first significant eye pulse on the right appears at approximately 133 mmHg pressure, while the one on the left appears at approximately 116 mmHg. Therefore there is ≥ 5 mmHg difference in OSPs. **B** Although both ratios fall above the discriminating line on the scoring grid, the difference between the left and right OSPs of ≥ 5 mmHg makes the left side abnormal.

- If the OSPs exceed 140 mmHg, the size of the first pulses are measured. A difference in amplitude of > 2 mm is abnormal, and the side with the least amplitude is considered positive.

7 Troubleshooting:

- Blink artifact (very irregular deflections not associated with the cardiac cycle) may be eliminated by reapplying the topical anesthetic.
- Cardiac arrhythmias may make the identification of the first significant ocular pulse difficult.
- Very diminished tracing bilaterally may suggest significant cardiac dysfunction, e.g., cardiomyopathy.

CHAPTER 22

Carotid Duplex Scanning and Color Flow Imaging

1 Capabilities:

- Duplex scanning (with or without color flow imaging) is an accurate means of localizing the presence of arterial disease in the extracranial carotid arteries. In addition, it can differentiate an occlusion from a tight stenosis with a high degree of accuracy.
- Duplex scanning is capable of documenting and following the progression of non–hemodynamically significant lesions.
- Duplex scanning provides information about the surface characteristics (e.g., smooth or irregular) of plaque and may help to identify the source of embolic phenomena.
- Duplex scanning can be used to evaluate pulsatile masses in the carotid or subclavian regions.

2 Limitations:

- Duplex ultrasonography may be adversely affected by:

  The presence of dressings, skin staples, or sutures.

  The size or contour of the neck.

  Patient movement.

  The depth or course of vessel.

  Rapid respiratory pattern.

  Acoustic shadowing from calcification.

- Duplex ultrasonography may overestimate disease when:

  Artifact is mistaken for plaque.

  Accelerated flow is mistakenly attributed to a stenosis. Other causes of increased velocity measurements include increased cardiac output, vessel tortuosity, compensatory flow for ipsilateral or contralateral disease, hyperemia, and/or a hypoplastic vessel.

  An inappropriate Doppler angle is used.

- Duplex ultrasonography may underestimate disease when:

  It fails to appreciate very low-level echoes of soft plaque.

  The examiner does not carefully interrogate the vessel and misses accelerated flow.

  A long, smooth plaque formation may not have the accelerated, turbulent flow patterns usually associated with a hemodynamically significant lesion.

  A high bifurcation precludes thorough evaluation of the internal carotid artery.

  An inappropriate Doppler angle is used.

3 Patient positioning:

The patient should be in a comfortable reclining or supine position with the head slightly hyperextended and turned slightly away from the side being interrogated.

4 Physical principles:

- *Duplex ultrasound:* Duplex ultrasonography combines physiologic information based on Doppler-shifted frequencies (spectral analysis) with the anatomic information of real-time, high-resolution, B-mode (gray-scale) ultrasonographic imaging. See also the brief description of physical principles on page 118.

- *Color flow imaging:* In addition to the information provided by conventional duplex ultrasonography, color flow imaging evaluates the Doppler flow information for its *phase* (direction toward or away from the transducer, on which basis color is assigned) and its *frequency content* (which determines the hue or shade of the assigned color).

- *Continuous wave and pulsed Doppler ultrasound:* Continuous wave (CW) Doppler uses two piezoelectric crystals, one to emit ultrasound continuously and the other to receive the reflected waves continuously. This results in a fixed sample size and no range resolution or ability to place the sample volume at a specific depth. Nor can continuous wave ultrasound create anatomic images.

Continuous wave Doppler has limited use in carotid evaluations. It may provide a simple method to determine patency of the carotid arteries and/or to provide limited blood flow characterization in very specific settings, such as the operating room following a carotid endarterectomy. It may also be used when severely elevated peak systolic velocities are evident. In some cases, CW Doppler may be the only means available to accurately display these very high velocities.

Unlike continuous wave ultrasound, pulsed Doppler does not continuously transmit and receive ultrasound. Rather, the multiple crystals in the pulsed Doppler transducer are excited in a quick burst producing ultrasound waves. This burst of transmission is followed by a "listening" period during which the crystals detect the reflected signals. Signals from more superficial structures are received almost immediately after transmission, while signals from deeper structures are detected after a longer listening period. This feature of pulsed Doppler, called range-gating, allows signals only from specific depths to be processed, thereby controlling sample size and range resolution. In this way, two vessels located directly above one another can be evaluated separately, and vessels also can be followed as their course changes.

Pulsed Doppler with spectral analysis is the primary tool utilized for the evaluation of blood flow: it provides very accurate blood flow characterization in conjunction with an image of the anatomy. This combination of physiologic (blood flow) and anatomic information is the basis of duplex ultrasonography.

- *Angle of insonation:* Reproducible and consistent velocity measurements require an angle of 45–60 degrees. Although a zero angle of insonation provides the greatest Doppler shift, it is not used in clinical applications for two reasons: (1) Most vessels of interest are parallel to the skin surface, so a zero angle of insonation is not possible. (2) More importantly, the criteria used in interpreting the significance of frequency or velocity measurements were established using a 60 degree angle of insonation. To rely on established diagnostic criteria, one must duplicate the conditions under which those criteria were established, i.e., with a 60 degree angle of insonation. However, a 60 degree angle is not always attainable. This may be due to the anatomic location of the arteries, or, even under normal circumstances, the normal curvature of the arteries. In such cases, the angle of insonation is more often less than 60 degrees and usually in the 45–60 degree range. Measurement variability (error production) can occur with angles greater than 60 degrees more frequently than with angles less than 60 degrees. It is very important that each vascular center independently validate their criteria for diagnosing carotid stenosis in order to determine their overall accuracy.

- *Spectral analysis:* Spectral analysis is a method of displaying the variety of frequencies of the flowing blood during systole and diastole. The scanner's built-in FFT (*fast Fourier transform*) technology automatically analyzes and displays the individual frequencies of the returned signals, creating a velocity profile that consists of time on the horizontal axis, frequency shifts on the vertical axis, and amplitude (intensity of the signals) as brightness.

Continuous wave Doppler produces a very broad spectrum because all of the layers of blood flow are displayed, from slow flow near the wall to faster flow in center stream. With CW Doppler, spectral broadening is an expected finding even with laminar flow. Pulsed Doppler, on the other hand, produces a narrow, well-defined spectrum when a limited number of frequencies or velocities are evident in laminar flow. With pulsed Doppler, spectral broadening is often associated with turbulent flow. (See figure 22-10 on page 233.)

- *Color Doppler:* Color flow Doppler information is displayed on the image after it is evaluated for its *frequency content* (which determines the hue or shade of the color) and its *phase* (i.e., direction toward or away from the transducer). The position of the color box, as well as the flow direction depicted on the color bar, assist in determining whether the blood is moving toward or away from the ultrasound beam.

5 Technique:

General Principles

- A 10, 7.5, or 5 MHz transducer is used.

- The sample volume of the pulsed Doppler is usually kept as small as possible, i.e., 1–1.5 mm, and placed in the center of the vessel and/or the flow channel.

- Measurements of a stenosis in the sagittal view can be very misleading since most plaque formations are not symmetrical. Therefore, the ultrasonographic shape and size of the plaque formation can vary according to the angle of insonation. This problem can be solved by ensuring that multiple approaches are used when imaging the vessel. In a transverse view, for example, the total circumference of the vessel and plaque formation can be more accurately seen.

Longitudinal View

- The vessels are followed from the clavicle to the mandible with anterior, oblique, lateral, and posterior projections to identify and evaluate plaque formations. The examiner should avoid pressing down too hard with the transducer as stimulation of the carotid sinus can decrease heart rate and alter heart rhythm.
- When activated, spectral analysis of the common carotid, internal carotid, and external carotid arteries is performed automatically by the scanner's computerized FFT algorithm. The examiner places and maintains a small sample volume in mid-stream or within the flow jet in order to obtain the most diagnostic information. In addition, the gain settings must be properly controlled: too much gain can be misinterpreted as turbulent flow, too little gain may result in an image that is difficult to evaluate.
- Comparison of flow characteristics from one side to the other, as well as from proximal to distal segments of the ipsilateral carotid system, is essential. Comparing the level of the bifurcation is also important, as increased flow in a collateralized external carotid artery has been misinterpreted as a patent internal carotid artery in cases of ipsilateral ICA occlusion: the ECA is mistaken for the ICA, and a branch of the ECA is thought to be the ECA.
- With a clear view of the common carotid artery, the probe is slowly angled more posterolaterally to identify the vertebral artery. As demonstrated in figure 22-1A, the artery will have vertical shadows running through it from the transverse processes of the vertebrae, giving it the appearance of a series of H's. Although flow direction (antegrade or retrograde) is routinely documented, the origin of the vertebral artery may also be evaluated for stenosis. Figure 22-1B illustrates the low-resistance flow patterns evident in the vertebral artery.

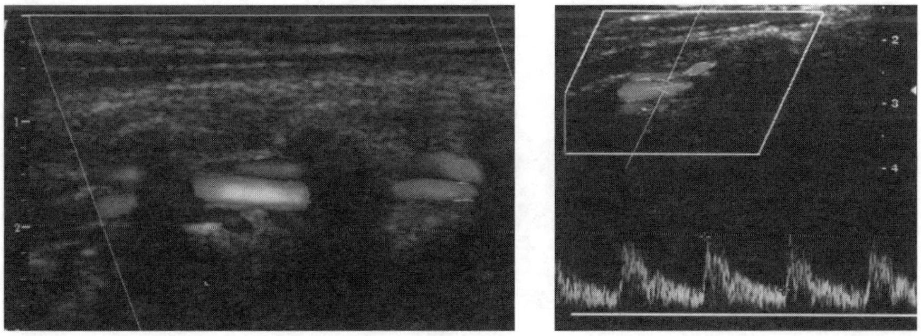

A B

Figure 22-1. B-mode image of the vertebral artery (**A**) and Doppler spectra evidencing the low-resistance flow patterns typical of this artery (**B**). See color plate 14.

Figure 22-2.
Sagittal view of normal common, internal, and external carotid arteries.

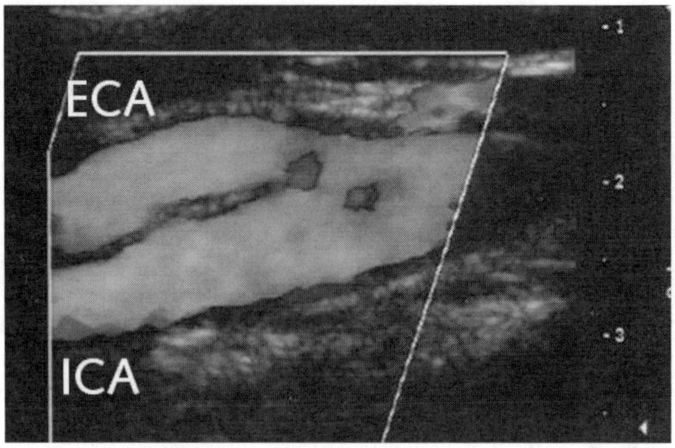

Transverse View

- The vessels are followed from the clavicle to the mandible to identify and evaluate plaque formations.
- The percentage stenosis calculations, i.e., diameter reduction, are documented.

6 Interpretation:

B-mode—Normal

- As demonstrated in figure 22-2, a solid color should be evident between the walls of the vessel, indicating the absence of atheromatous material (i.e., plaque), whose density differs from that of flowing blood.
- An anechoic line separating the endothelium from the medial layer of the arterial wall should be evident. This allows for measuring intima-media thickness (IMT), a tool currently regarded as a predictor of cardiovascular disease (figure 22-3). It can also be used to monitor the effects of risk-

Figure 22-3.
Sagittal view of the distal common carotid artery (CCA) and internal carotid artery (ICA) shows the anechoic flow channel. **A** The innermost bright line represents the intimal layer. The low-level echoes adjacent to the intima correspond to the medial layer. The outermost bright line signifies the adventitial layer. **B** One method of measuring the intima-media thickness (IMT) is to use calipers.

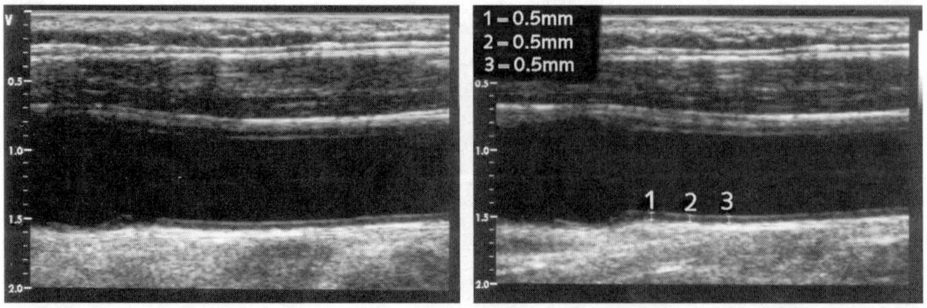

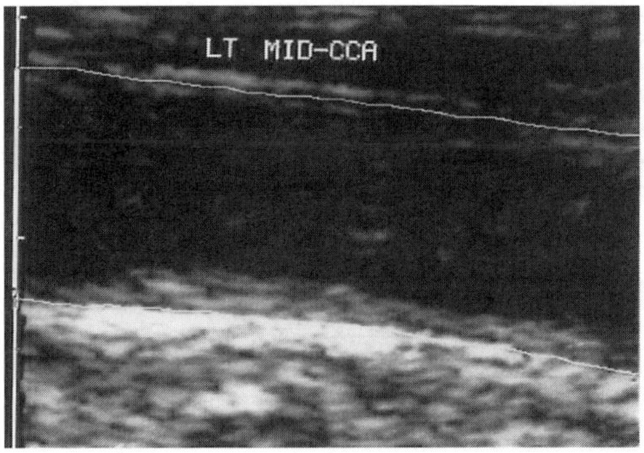

Figure 22-4. Homogeneous plaque formation on the posterior wall of the left common carotid artery.

factor modification and pharmacological agents such as lipid-lowering drugs. The technique involves obtaining multiple sagittal views of the common carotid and internal carotid arteries. Far-wall measurements of the distal common carotid artery and proximal internal carotid artery are most widely used. The measurements can be made manually with calipers or with software-assisted analysis programs. Although IMT normally varies with the age of the patient, an IMT $\leq$ 0.9 mm is generally considered to be within normal limits for people older than 60.

B-mode—Abnormal

- *Fatty streaks:* Low-level echoes of similar appearance (homogeneous) can be found in persons of all ages.
- *Fibrous (soft) plaque:* Low- to medium-level echoes of similar appearance (figure 22-4).
- *Complex plaque:* Low-, medium-, and high-level echoes (heterogeneous) indicating soft and dense areas (figure 22-5).

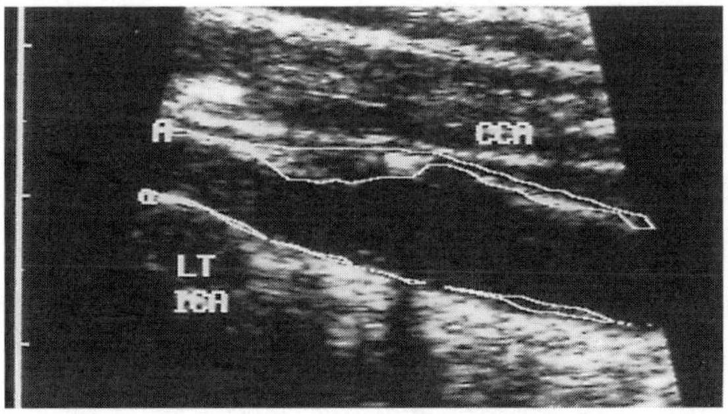

Figure 22-5. Heterogeneous plaque formation in the bulb area of the left internal carotid artery.

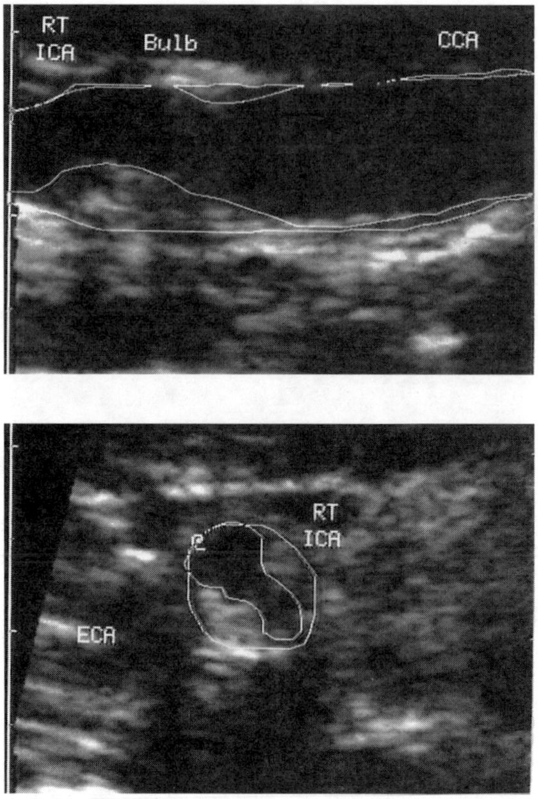

Figure 22-6. Plaque should be evident in multiple views. **A** Plaque formation in sagittal view. **B** Same formation in transverse view.

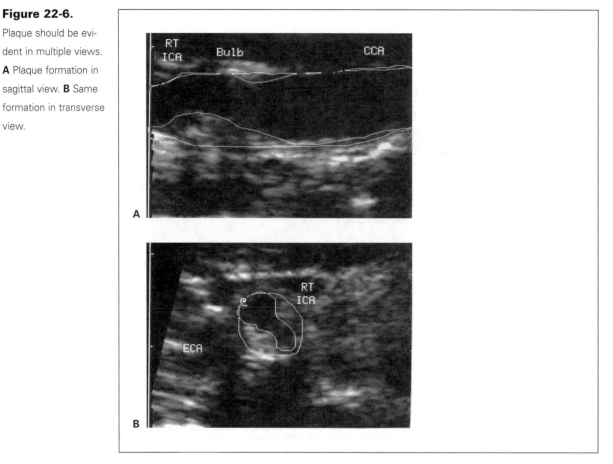

- *Calcification:* Very bright, highly reflective echoes. The acoustic shadow from the calcium deposit prevents a thorough evaluation of the vessel and may result in the calculation of an erroneous percentage stenosis.
- *Thrombosis:* Because fresh thrombus has the same echogenicity of flowing blood, careful interrogation with Doppler is necessary to ensure that there is no hemodynamically significant lesion.
- *Surface characteristics:* The surface characteristics of plaque may be described as smooth, slightly irregular, grossly irregular, or craterlike.
- *Stenosis:* Ideally, plaque should be visible from at least two of the longitudinal, or sagittal, projections and in the transverse view (figures 22-6A and B).
- *Occlusion:* Depending on the type of occlusive process, the vessel may be filled with highly echogenic material or be anechoic. Although Doppler findings are essential to the diagnosis of a probable occlusion, some B-mode characteristics are helpful: A vessel may be completely filled with echoes or moving in a horizontal, pistonlike motion.

Note: Since standard duplex scanning may not be able to detect a tiny residual lumen, i.e., a "string sign," it is important to utilize enhanced Doppler processing to display all flow patterns regardless of speed or direction of flow. Referred to as *Doppler energy* or *power Doppler*, the display of a solid color within the flow channel designates a residual lumen; the absence of color suggests an occlusion. However, caution is advised in diagnosing a carotid occlusion based on B-mode or color information alone.

Doppler—Normal

- The internal carotid artery (ICA) signal is slightly more high-pitched and continuous than the external carotid artery (ECA). Since the brain is a low-resistance vascular bed, the flow in the ICA is less pulsatile, with increased flow during diastole. As shown in figure 22-7A, the waveform of the ICA has a rapid upstroke; the downstroke has a high diastolic component. A dicrotic notch may not be clearly evident. The terms triphasic, biphasic, and monophasic are not applied to the blood flow patterns of the cerebral vasculature as they are to those of the peripheral vasculature.
- Since the ECA supplies blood to vascular beds that have a higher peripheral resistance, e.g., the face and the scalp, the ECA signal is more pulsatile than the ICA. It appears more similar to the signals from the peripheral vessels, i.e., common femoral artery. As shown in figure 22-7B, the ECA has a rapid upstroke and downstroke with a very low diastolic component. A dicrotic notch is clearly seen, and tapping of the superficial temporal artery (STA) causes oscillations in the waveform (figure 22-7C).

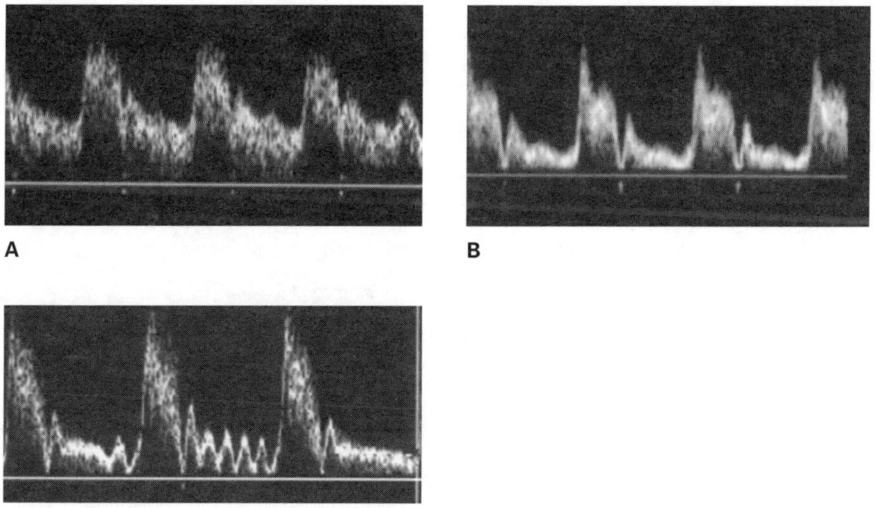

A B C

Figure 22-7.
A The waveform of the internal carotid artery has a rapid upstroke, a high diastolic component, and a dicrotic notch that may not be clearly evident. **B** The external carotid artery has a rapid upstroke and downstroke, a very low diastolic component, and a dicrotic notch that is clearly seen. **C** Manual tapping on the superficial temporal artery just anterior to the ear produces oscillations in the Doppler signal from the external carotid artery.

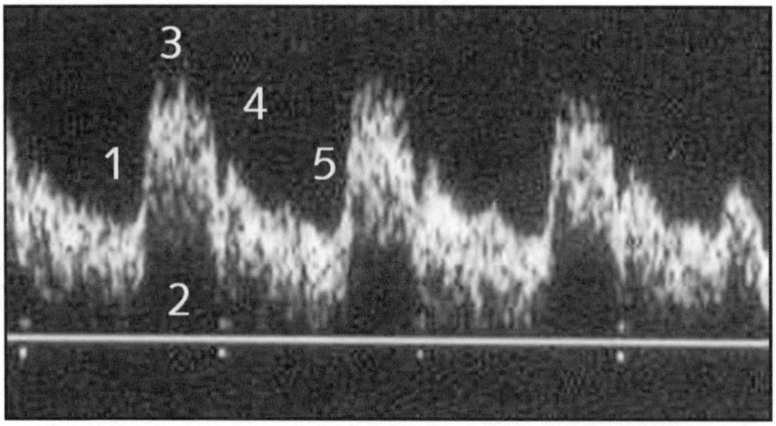

Figure 22-8. Spectral analysis of a pulsed Doppler waveform showing time on the x-axis and varied frequencies or velocities on the y-axis. Components of the waveform: (1) systolic upstroke, (2) frequency (spectral) window, (3) peak systole, (4) dicrotic notch, (5) end diastole.

- The common carotid artery (CCA) has flow characteristics of both the ICA and ECA.

- Because the sample volume of a pulsed Doppler can be more precisely placed in center stream, the signals will have a narrow band of frequencies in systole with a "blank" area, spectral window, under that narrow band. The presence of these qualities, as seen in figure 22-8, is consistent with laminar flow.

- A continuous wave Doppler, on the other hand, is not able to regulate its sample size or depth, and so a frequency window is not as apparent. Figure 22-9A shows the difference in CCA Doppler signals when a CW Doppler is used versus a pulsed Doppler. Figure 22-9B shows the ICA signals, with figure 22-9C showing the ECA signal.

Doppler—Abnormal

- A stenosis is characterized by a higher-pitched sound than normal, with a very high-pitched hissing or squealing type of signal evident at a significant stenosis. The waveform of a stenotic vessel has higher velocities that correspond with the accelerated flow through the narrowing (figure 22-10A). The very high-pitched hissing or squealing type of signal evident at a significant stenosis has higher than normal velocities in systole and diastole (figure 22-10B). Table 22-1 on page 236 describes the expected correlation among velocities, frequencies, and percentage stenosis.

- In spectral analysis, the band evident along the top of the waveform during systole may fill in the spectral window. This vertical thickening is considered *spectral broadening* and is consistent with multiple frequencies, i.e., turbulent flow.

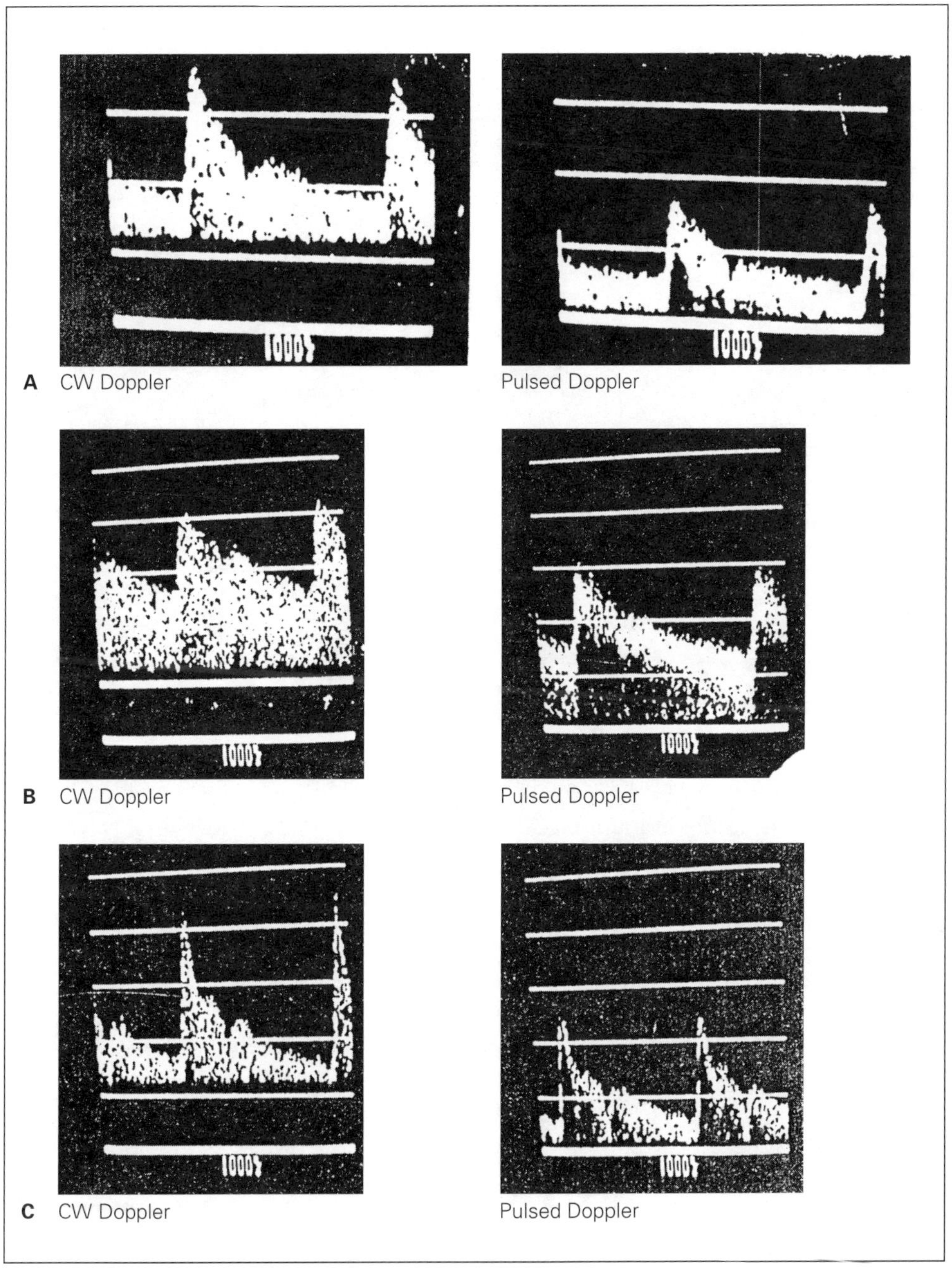

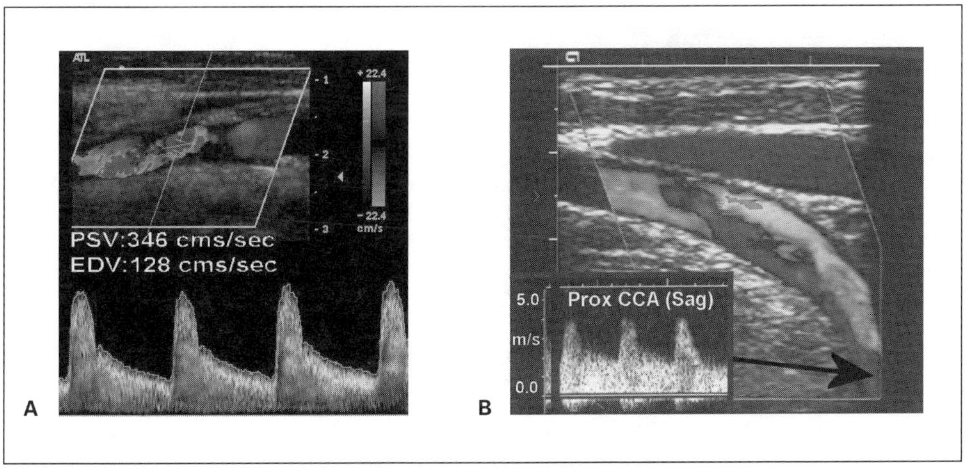

Figure 22-9.

A Normal spectral analysis of common carotid by continuous wave (CW) and pulsed Doppler. **B** Normal spectral analysis of internal carotid by CW and pulsed Doppler. **C** Normal spectral analysis of external carotid by CW and pulsed Doppler.

Figure 22-10.

A Sagittal view of a stenosis of internal carotid. There is significant focal reduction of blood flow filling within the vessel. Elevated peak systolic and end diastolic velocities confirm a hemodynamically significant lesion. Spectral broadening reflects flow turbulence consistent with significant flow alterations. A bruit related to this lesion may be evident in the cervical carotid area, although the bruit may disappear if or when the lesion becomes more severe, i.e., > 90%. **B** Longitudinal view consistent with flow alterations in the proximal common carotid, suggesting a more proximal lesion in the CCA. **Inset:** Very significant flow alterations detected just superior to the clavicle. Abnormal flow characteristics should prompt further investigation as to their cause.

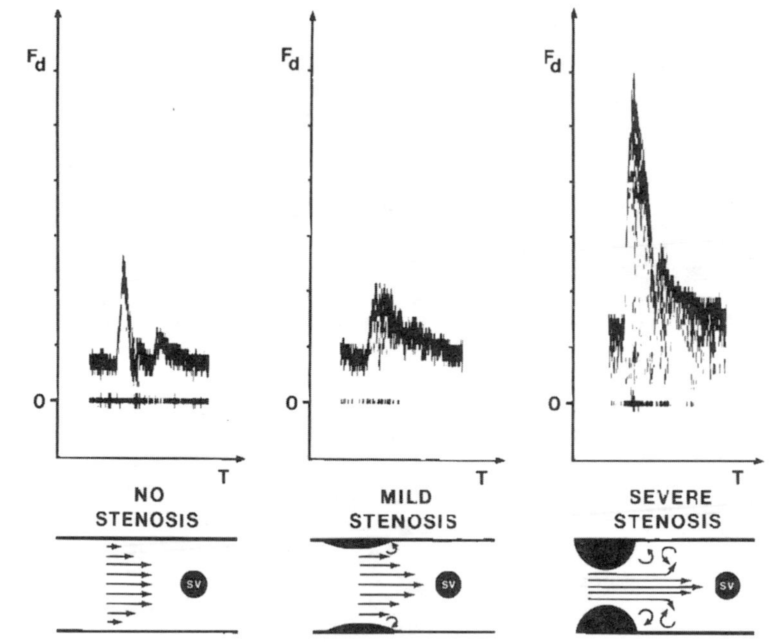

Figure 22-11. Laminar flow in a normal artery evidenced by a narrow band of frequencies during systole leaving a clean window beneath the envelope. Slight turbulence from a mild stenosis broadens the frequency band without changing the peak systolic velocity. High peak velocities and spectral broadening throughout the cardiac cycle characterize high-grade lesions. From Roederer GO, Langlois Y, Strandness DE: Comprehensive noninvasive evaluation of extracranial cerebrovascular disease. In Hershey FB, Barnes RW, Sumner DS [eds]: *Noninvasive Diagnosis of Vascular Disease.* Pasadena, Appleton Davies, 1984.

- As illustrated in figure 22-11, as the stenosis progresses, systolic and diastolic frequencies increase. The spectral band becomes very broad, resulting in the complete loss of the *window*.
- Distal to a stenosis, disturbed flow patterns are evident, i.e., dampened (decreased velocities), more rounded upstroke (increased acceleration time), low-resistance flow (increased flow in diastole), and poststenotic turbulence (spectral broadening observed, oftentimes with bidirectional flow).
- Although an absent signal may indicate occlusion, a tight stenosis cannot be ruled out. For example, the flow stream may be extremely difficult to detect due to slow flow resulting in no audible signal.

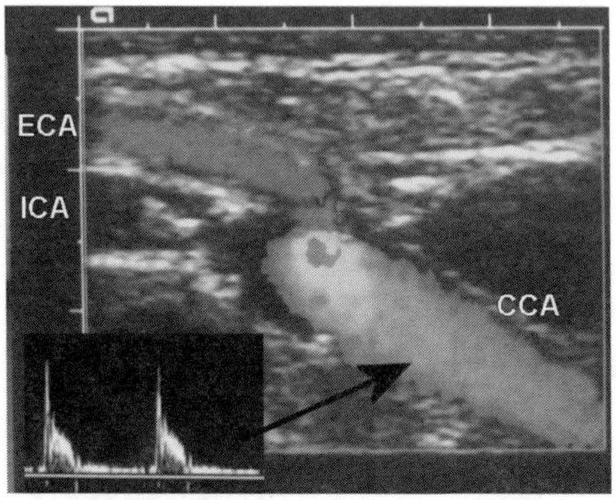

Figure 22-12. Longitudinal view consistent with an occlusion of the internal carotid artery (ICA). ICA occlusion is suspected when there is an absence of a Doppler arterial signal in the ICA, absence of color flow Doppler (or power Doppler) characteristics in that vessel, and, as illustrated in the inset, a very high-resistance signal in the ipsilateral common carotid artery.

- The loss of a diastolic component in the ipsilateral common carotid artery is usually consistent with an occlusion of the internal carotid artery (figure 22-12). Often in this situation, the contralateral CCA and ICA serve as collateral pathways with increased systolic and diastolic flow in those vessels.
- Although disease at the carotid siphon occurs infrequently compared to the extracranial system, it also must be considered when high-resistance flow patterns are evident in the extracranial ICA (figure 22-13).
- Poor cardiac output or stroke volume may result in bilaterally diminished CCA flow velocities. Diminished velocities unilaterally are suggestive of proximal disease, e.g., innominate or common carotid artery occlusive disease (figure 22-14).
- Increased cardiac output or stroke volume, as seen in some young healthy individuals, may result in elevated CCA flow velocities bilaterally.
- Antegrade flow in systole with sustained reversal of flow during diastole may indicate aortic regurgitation or insufficiency.

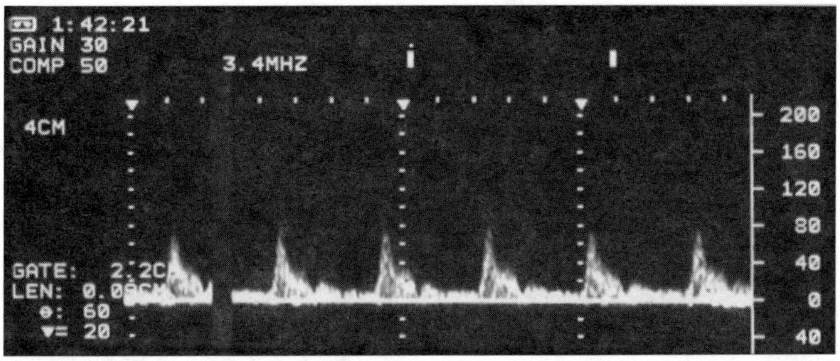

Figure 22-13. The Doppler signal obtained from the internal carotid artery reflects high-resistance rather than the expected low-resistance flow pattern. This is consistent with a distal obstruction, and in this case the blockage was at the carotid siphon.

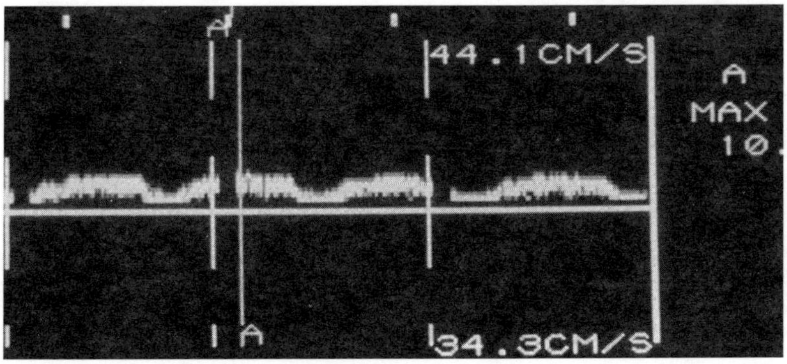

Figure 22-14. The low velocities and increased acceleration time, resulting in a rounded waveform of the common carotid artery, are consistent with proximal disease. If bilateral, consider poor cardiac output; if unilateral, consider disease in a proximal vessel.

Guidelines for Interpretation

● Interpretation guidelines for determining diameter stenosis based on peak systolic frequency (PSF) or peak systolic velocity (PSV) as well as end diastolic velocity (EDV) or end diastolic frequency (EDF) are available. Although the majority of vascular laboratories use velocity criteria, a few continue to use frequency information. While the literature supports various methods for interpreting the spectral findings, table 22-1 includes the basis of the most widely used guidelines developed by D.E. Strandness, MD, of the University of Washington using a 60 degree angle of insonation.

Table 22-1. Guidelines for determining percentage diameter stenosis.

% Stenosis	*PSF*	*EDF*	*PSV*	*EDV*
Normal*	< 4 kHz	NA	<125 cm/sec	NA
1–15%*	< 4 kHz	NA	<125 cm/sec	NA
16–49%*	< 4 kHz	NA	<125 cm/sec	NA
50–79%	> 4 kHz	< 4 kHz	>125 cm/sec	<140 cm/sec
80–99%	> 4 kHz	> 4 kHz	>125 cm/sec	>140 cm/sec
Occluded	Absent	Absent	Absent	Absent

*The degree of spectral broadening differs in these categories.

● Two important clinical trials evaluating the efficacy of carotid endarterectomy established therapeutic benefit for *asymptomatic* patients with greater than 60% diameter reduction of the internal carotid artery (the Asymptomatic Carotid Atherosclerosis Study, ACAS) and for *symptomatic* patients with greater than 70% stenosis of the internal carotid artery (the North American Symptomatic Carotid Endarterectomy Trial, NASCET).

ACAS**: PSV of ≥ 290 cm/sec and EDV of ≥ 80 cm/sec is consistent with ≥ 60% diameter reduction of internal carotid artery

NASCET***: ICA PSV / CCA PSV ratio of ≥ 4.0 is consistent with ≥ 70% diameter reduction of internal carotid artery

** Moneta G, Edwards JM, Papanicolaou G, et al: Screening for asymptomatic internal carotid artery stenosis: duplex criteria for discriminating 60%–99% stenosis. J Vasc Surg 21:989–994, 1995.

*** Moneta G, Edwards JM, Chitwood RW, et al: Correlation of North American Symptomatic Carotid Endarterectomy Trial (NASCET) angiographic definition of 70%–99% internal carotid artery stenosis with duplex scanning. J Vasc Surg 17:152–159, 1993.

The ICA/CCA ratio is calculated by dividing the highest peak systolic velocity of the ICA by the peak systolic velocity of the ipsilateral CCA at a standard distance from its bifurcation into the internal and external carotid arteries. Peak systolic velocities can vary in the CCA with faster flow normally detected proximally. Therefore, velocity measurements should be obtained at a consistent location, e.g., approximately 4 centimeters proximal to the bifurcation, making sure that this part of the CCA is not bulbous. If it is, move a bit more proximal to measure the peak systolic velocity of the CCA to ensure consistency.

- The Society of Radiologists in Ultrasound Consensus Conference included representatives of the various disciplines interpreting carotid duplex studies. It should be noted that the following criteria are suggestions from that group and not the results of scientific investigation. According to many physician "readers" (the physicians who interpret the ultrasound examinations), these criteria should not replace existing criteria validated by local, well-established vascular laboratories that have consistent protocols and quality assurance. The criteria in table 22-2 do, however, provide a suggested guide for those vascular laboratories that are getting started and have not yet had the opportunity to validate their own criteria with actual cases. New laboratories should establish their own consistent exam protocols as well as quality assurance (QA) policies and procedures, ensuring the formulation of validated diagnostic criteria. Others also have suggested that both new and well-established vascular laboratories incorporate the ACAS and NASCET criteria for ICA stenosis into their diagnostic paradigm. They concluded that the key to successful physician interpretation of carotid duplex exams is the use of consistent protocols and the application of validated diagnostic criteria. To that end, the following guidelines are suggested*:

  Doppler criteria cannot be used to predict a single-percent ICA stenosis.

  Doppler criteria cannot be used to predict ICA stenosis in 10% increments.

  Doppler criteria are not accurate for <50% stenosis.

  Doppler is not accurate for subcategorizing <50% stenosis.

* Grant EG, Benson CB, Moneta GL et al: Carotid artery stenosis: gray-scale and Doppler ultrasound diagnosis. Society of Radiologists in Ultrasound Consensus Conference. Radiology 229:340–346, 2003.

Table 22-2.
Some of the diagnostic criteria suggested by the Society of Radiologists in Ultrasound Consensus Conference.

	PSV	*EDV*	*ICA/CCA Ratio*
No plaque	<125	<40	<2.0
<50% DR	<125	<40	<2.0
50–69% DR	125–230	40–100	2.0–4.0
70–99% DR	>230	>100	>4.0

(DR = Diameter reduction)

- The Doppler evaluation is critical in diagnosing a probable occlusion. Findings may include:

  The absence of a Doppler arterial signal.

  The absence of a solid color within the flow channel with power Doppler or Doppler energy.

  Very low or absent diastolic component evident in the ipsilateral common carotid artery.

  With internal carotid artery occlusion, compensatory flow may be evident as increased flow in other vessels, e.g., the ECA, contralateral ICA, or either vertebral artery.

Interrelated, Corroborative Findings

Figure 22-15 is an example of an occlusion of the CCA. (See also color plate 15.) Although an unusual finding, it illustrates how the B-mode, Doppler, and color flow Doppler findings can coincide: no flow detected in the CCA, the vessel filled with heterogeneous echoes, and retrograde flow detected in the external carotid artery (demonstrating a collateral pathway through which flow to the ICA is maintained).

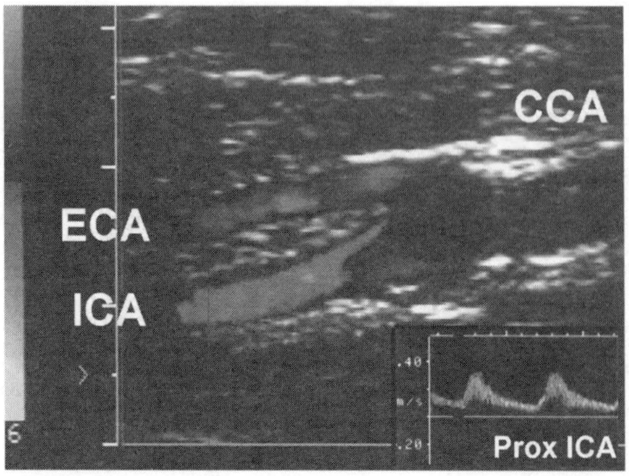

Figure 22-15.
Longitudinal view of an occluded common carotid artery. The retrograde blood flow detected in the external carotid artery represents the collateral pathway that is maintaining blood flow to the internal carotid artery (ICA). **Inset:** Flow in the ICA.

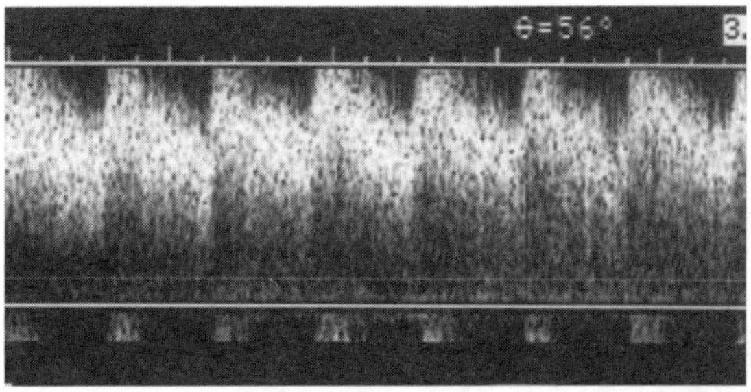

Figure 22-16. Aliasing is a misrepresentation of the Doppler signal that occurs because of limitations of the equipment being used. Velocities that exceed the PRF are wrapped around and displayed as coming up from the baseline.

Technical Issues and Doppler Artifacts

- *Aliasing*, illustrated in figure 22-16, is a misrepresentation of the Doppler signal that occurs because of the limitations of the equipment being used, i.e., a low pulse repetition frequency (PRF). Since the maximum frequency is one-half the PRF, flow information greater than one-half the PRF (Nyquist limit) cannot be displayed. True peak flow is not available, and the waveform has a flat, crew-cut appearance. Methods of increasing the PRF, and thereby increasing the Nyquist limit, include changing to a lower-frequency transducer, altering the angle of insonation to decrease the depth of the vessel, or using a continuous wave Doppler.
- *Crosstalk*, a form of mirror imaging, illustrated in figure 22-17, is an artifact produced by the presence of strong reflectors or an overly high gain setting. The Doppler shifts that are duplicated above and below the baseline appear in the absence of flow disturbances. Normal helical, or nonaxial, flow will also be displayed as some forward and reverse flow as the blood travels in a spiral motion (figure 22-18). The asymmetrical flow patterns demonstrate that the laminar flow is not traveling parallel to the walls of the vessel (axial), but rather in a corkscrew pattern (nonaxial).

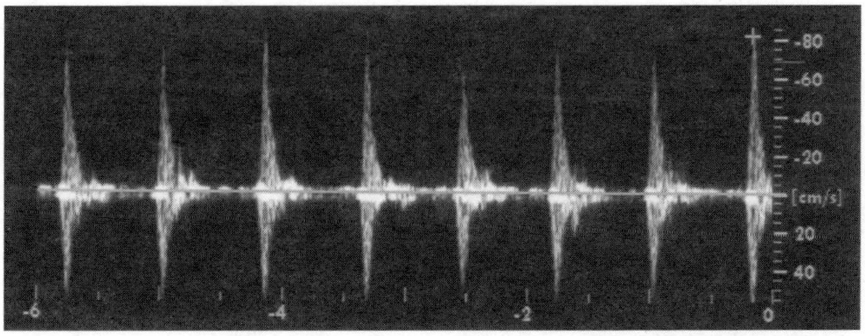

Figure 22-17. Doppler spectra evidencing mirror imaging artifact known as *crosstalk*.

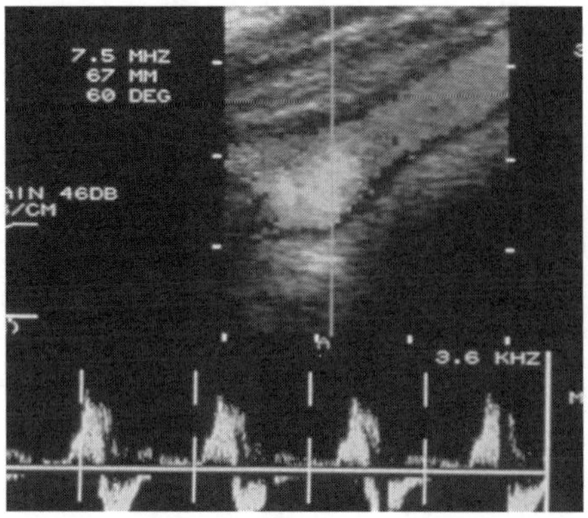

Figure 22-18. Doppler spectra depicting helical flow patterns.

7 Intraoperative monitoring:

- Capabilities:

  B-mode imaging can identify defects secondary to carotid endarterectomy, such as stricture of the suture line, intimal flaps, areas of platelet aggregation, or residual plaque.

  Doppler evaluates the hemodynamic significance of an irregularity.

- Technique:

  The transducer is placed inside a sterile sleeve/plastic bag after acoustic gel has been put into the sleeve.

  The wound is filled with sterile saline.

  The examination proceeds according to the previously described technique. Areas of flow disturbances identified with color flow imaging require thorough evaluation of the vessel wall(s) in gray scale. B-mode ultrasound can identify subtle defects that might be concealed by color flow imaging (figure 22-19).

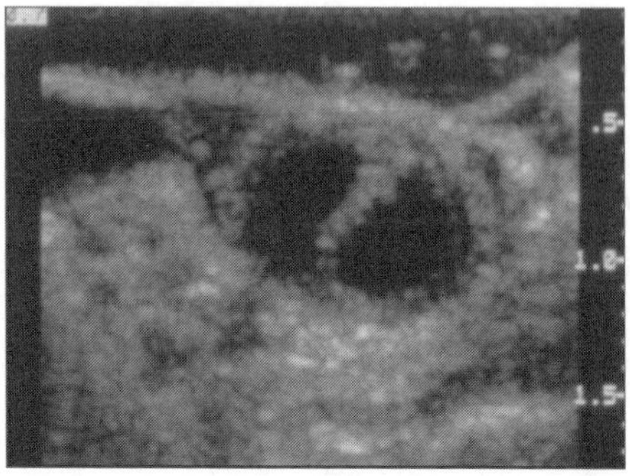

Figure 22-19. Transverse view of vessel with a wall defect that is more obvious with B-mode imaging than with color flow imaging.

CHAPTER 23

Transcranial Doppler (TCD)

1 Capabilities:

- Transcranial Doppler (TCD) is capable of detecting intracranial stenoses and occlusions.
- TCD can assess collateral circulation in known cases of severe carotid stenosis or occlusion.
- TCD can evaluate the onset, severity, and time course of vasoconstriction caused by subarachnoid hemorrhage.
- TCD is useful in evaluating intracranial arteriovenous malformations.
- TCD can be used to assess patients with suspected brain death.

2 Limitations:

- Recent eye surgery may eliminate the transorbital approach.
- Some patients have hyperostosis of the temporal bone, making adequate penetration of the ultrasound wave impossible.
- If performed without imaging, vessels may be inaccurately identified because of the lack of anatomic landmarks.

3 Patient positioning:

- The patient assumes a comfortable supine position.
- The patient should understand that they must lie quietly and avoid speaking during scanning.

4 Physical principles:

- The principles of pulsed Doppler are described on pages 224–226.
- A transcranial Doppler system usually uses a 2 MHz pulsed Doppler with spectrum analyzer.

Figure 23-1. Components of the intracranial circulation: (1) anterior communicating, (2) anterior cerebral, (3) ophthalmic, (4) internal carotid, (5) middle cerebral, (6) posterior communicating, (7) posterior cerebral, (8) basilar, (9) vertebral arteries. Courtesy of Keith Fujioka, BS, RVT, Pacific Vascular, Inc.

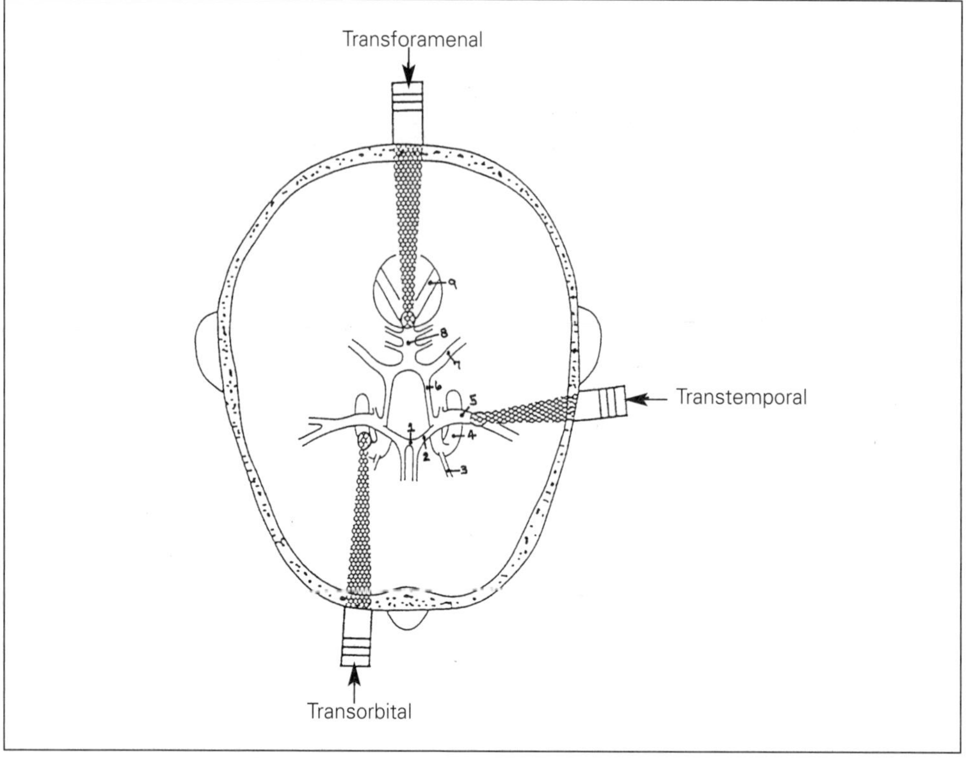

- The assumed angle of insonation is zero degrees.
- As illustrated in figure 23-1, three acoustic "windows" provide access to the intracranial vessels: transtemporal, transorbital, and transforamenal. The transtemporal approach allows for three views: anterior, middle, and posterior. The transforamenal approach may also be referred to as the *suboccipital approach*. Although a fourth window, submandibular, may be used to evaluate the intracranial internal carotid artery, it is not used as frequently as the others.
- Figures 23-2A, 23-2B, and 23-2C illustrate the three most common windows utilized in TCD evaluations.
- Accurate vessel identification requires appropriate sample volume size and depth and knowledge of the direction and velocity of blood flow, the relationship of the various flow patterns to one another, and common carotid artery compression and oscillation maneuvers.
- The standard method of quantifying velocity is time-averaged maximum of the mean velocity or more commonly, time-averaged mean velocity (TAMV). Peak systolic velocity measurements are not utilized in this application.

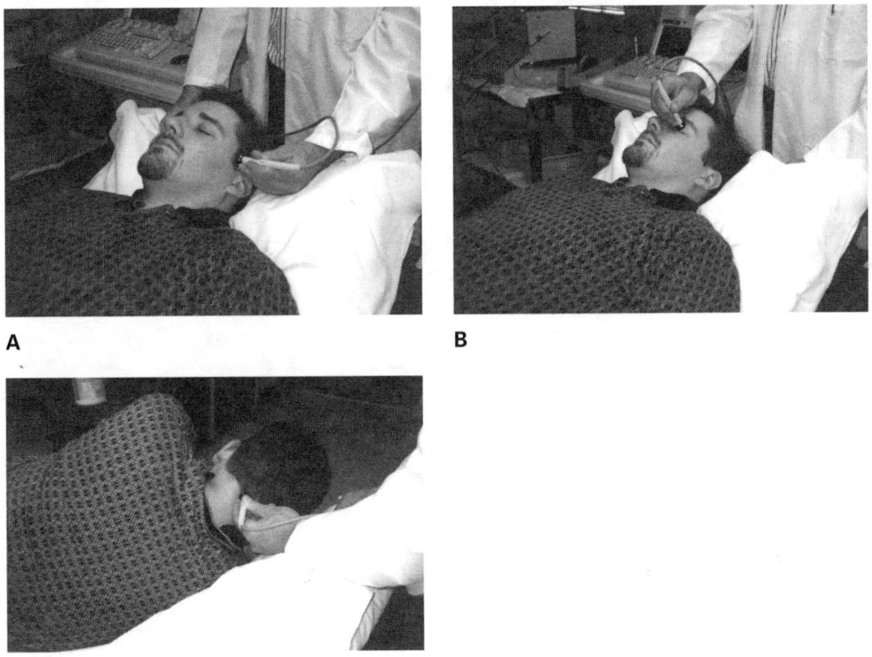

Figure 23-2. **A** Transcranial Doppler (TCD) examination from the transtemporal approach. **B** TCD examination from the transorbital approach. **C** TCD examination from the transoccipital (transforamenal) approach.

5 Techniques:

- A routine study begins with a unilateral transtemporal approach to identify the middle cerebral artery, anterior cerebral artery, posterior cerebral artery, and most distal internal carotid artery.
- The ipsilateral transorbital approach is used to evaluate the ophthalmic artery and carotid siphon.
- The preceding two steps are completed contralaterally.
- The foramen magnum approach is then used to evaluate the intracranial vertebral and basilar arteries.
- Table 23-1 on the following page provides some guidelines for the identification of intracranial vessels.
- Before common carotid artery compression or oscillation maneuvers are performed, the status of the carotid arteries must be evaluated with duplex ultrasonography. Oscillation maneuvers are contraindicated in cases of low-bifurcation, high-grade stenosis or occlusion of the internal carotid artery, and in the presence of complicated atheromatous plaque formation.

Table 23-1. Guidelines for identifying the intracranial vessels.

Vessel	*Window*	*Depth (mm)*	*Direction*	*Velocity (cm/sec)*	*Angle*
MCA	Transtemporal	30–60	Antegrade	55 +/– 12	Anterior & superior
Terminal ICA	Transtemporal	55–65	Bidirectional	55 +/– 12	Same as above
ACA	Transtemporal	60–80	Retrograde	50 +/– 11	Same as above
PCA	Transtemporal	60–70	Antegrade	39 +/– 10	Posterior
ICA	Transorbital	60–80	Parasellar: antegrade. Supra-clinoid: retrograde. Genu: both.	47 +/– 14	Varies
Ophthalmic	Transorbital	40–60	Antegrade	21 +/– 5	Medial
VA	Transforamenal	60–90	Retrograde	38 +/– 10	Right & left of midline
BA	Transforamenal	80–120	Retrograde	41 +/– 10	Midline

Courtesy of Colleen Douville, BA, RVT, Harborview Medical Center, Seattle, WA

Compression technique:

Palpate the common carotid artery.

Apply slow downward pressure.

Hold the compression for 2–4 cardiac cycles and then slowly release pressure.

Note any changes in blood flow during the compression maneuver.

Oscillation technique:

Palpate the common carotid artery.

Apply series of short, rapid, incomplete compressions.

Note any oscillatory patterns transmitted during the maneuver.

6 Interpretation:

- The interpretation criteria incorporate the flow characteristics evident on spectral analysis: direction, velocity, turbulent flow, pulsatility, systolic upstroke, and the hemispheric ratio of middle cerebral artery/internal carotid artery.

● Knowledge of the collateral pathways is essential in determining intracranial collateralization secondary to carotid artery disease:

Crossover collateralization occurs when antegrade flow is evident in the ipsilateral anterior cerebral artery. This abnormality can be attributed to flow from the contralateral anterior cerebral artery via the anterior communicating artery. Increased flow velocities (> 150%) in the contralateral anterior cerebral artery are also seen (compare the contralateral anterior cerebral artery with the contralateral middle cerebral artery). An additional finding that confirms crossover collateralization is ipsilateral middle cerebral artery flow velocities that diminish with contralateral compression of the common carotid artery and that respond positively to contralateral oscillation maneuvers of the common carotid artery.

External-to-internal collateralization is evident when there is retrograde flow in the ipsilateral ophthalmic artery. This abnormality can be attributed to flow from the external carotid artery branches and can be confirmed by noting reduction, obliteration, or reversal of ophthalmic arterial flow with compression of ipsilateral external carotid artery branches. Decreased pulsatility and increased velocity in the ipsilateral ophthalmic artery may also be evident. (Please refer to the discussion of periorbital Doppler in chapter 20 for more information.)

Posterior-to-anterior collateralization is evident when flow velocities in the ipsilateral posterior cerebral artery exceed those of the ipsilateral middle cerebral artery by, for example, > 125%. Increased flow velocities in the posterior cerebral artery with ipsilateral compression of the common carotid artery is an additional finding that confirms patency of the posterior communicating artery.

● It is essential that the examiner pay close attention to the variety of factors that may alter intracranial blood flow, such as age, sex, hematocrit, blood gases, and metabolic demand.

● The diagnosis of an occlusion is difficult because of the technical limitations of the study; it is most accurate in suspected occlusions of the internal carotid and middle cerebral arteries. The criteria are similar to those used in evaluating occlusion in other vessels: absence of the Doppler arterial signal, low diastolic component in the arterial segment just proximal to the suspected occlusion, and evidence of collateralization.

● Although vasospasm can occur in other cerebral vessels, the diagnosis of vasospasm is most accurate in the middle cerebral artery. Serial recordings of middle cerebral arterial flow patterns are necessary to document the increase in velocities associated with a vasospasm. Normal velocities are < 120 cm/sec with a hemispheric ratio < 3. The hemispheric ratio is calculated by dividing the time-averaged maximum velocity (TAMV) of the middle cerebral artery by the TAMV of the distal extracranial internal carotid artery.

- In an arteriovenous malformation, the arteries supplying the malformation usually have increased systolic and diastolic flow velocities with very low pulsatility indices. In addition, reduction of flow in adjacent arteries is usually evident.

- In addition to other methods to determine brain death, the TCD findings appear to correlate well with cerebral circulatory arrest. Changes in flow characteristics seem to follow the deterioration of cerebral perfusion as it progresses through six stages. Initially, the diastolic velocity decreases and the pulsatility index increases. This pattern progresses to the point at which end diastolic velocity reaches zero, followed by a reversal of diastolic flow. In the later stages, reverberatory flow develops, followed by a low-velocity spiked-systolic waveform. Absence of flow is the final step of cerebral circulatory arrest.

7 Intraoperative monitoring:

- Capabilities:

  TCD allows for the identification of flow abnormalities during many cerebrovascular and cardiovascular procedures, including carotid endarterectomy and cardiopulmonary bypass. Flow abnormalities may warrant changes in surgical technique.

- Technique:

  Continuous monitoring of middle cerebral arterial (MCA) blood flow requires the standard equipment along with a headset to stabilize the Doppler probe at the transtemporal window. A probe cover is not necessary as the transtemporal window is not in the surgical field; therefore sterility is not necessary.

- Interpretation:

  Significant decrease in middle cerebral artery flow velocities during the cross-clamping portion of carotid endarterectomy may signal the need for shunting. Auditory signals related to microemboli may lead the surgeon to alter operative technique.

CHAPTER 24

Other Conditions

Subclavian Steal

Subclavian steal (figure 24-1) is a condition in which blood destined for the brain is shunted away from the cerebral circulation because the subclavian or innominate artery has a high-grade stenosis or total occlusion proximal to the take-off of the vertebral artery. As illustrated in figure 24-2, blood flows retrograde down the vertebral artery in order to provide blood flow to the arm.

1 Patients are usually asymptomatic.

2 Subclavian steal is more frequent on the left side than the right.

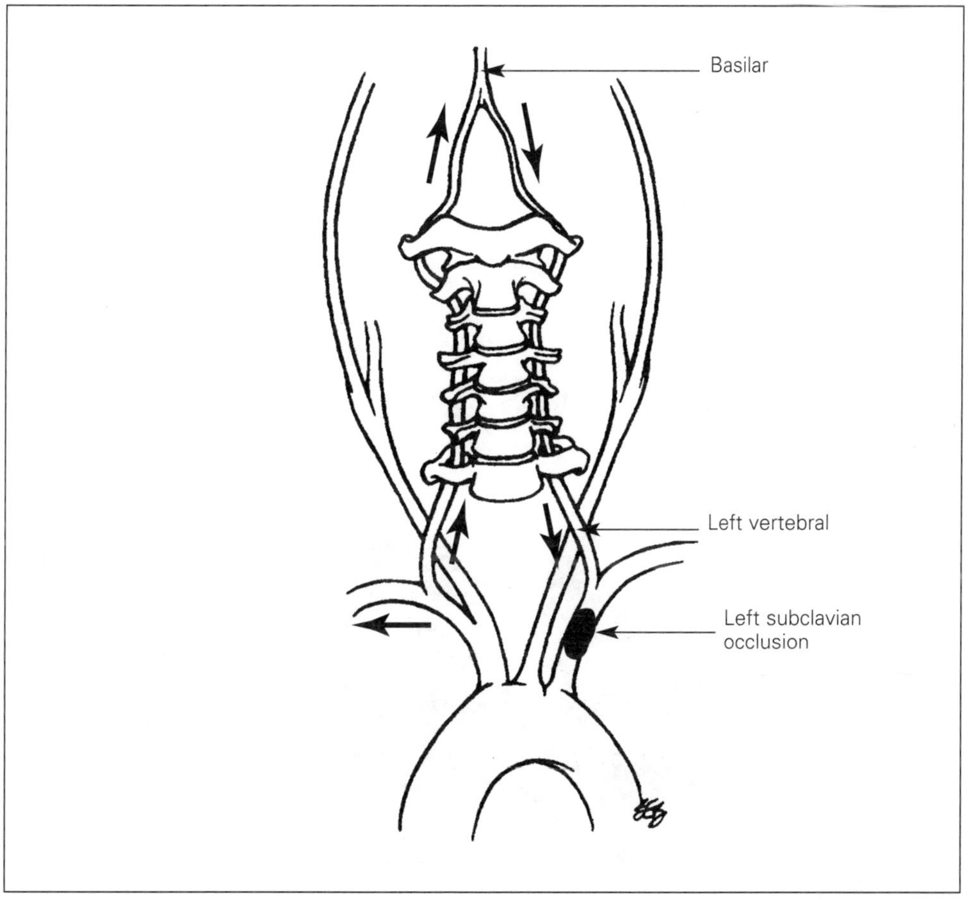

Figure 24-1. Example of subclavian steal. From Belanger AC: *Vascular Anatomy and Physiology: An Introductory Text.* Pasadena, Davies Publishing, 1999.

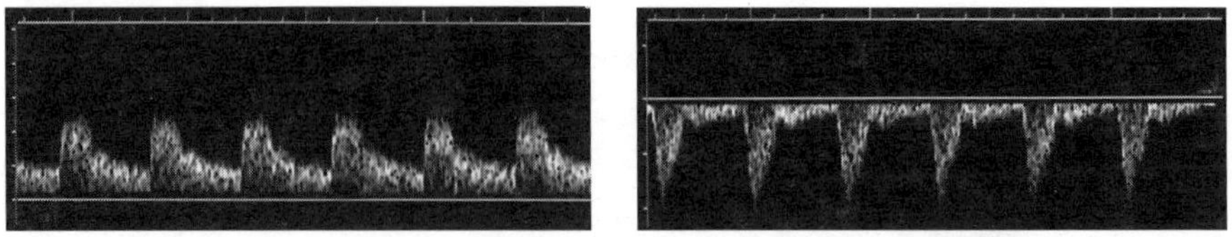

A B

Figure 24-2.
A Normal antegrade flow in the vertebral artery. **B** Abnormal retrograde flow in the vertebral artery.

3. There is a brachial blood pressure difference of 15–20 mmHg or greater from one side to the other.
4. Pulses in the affected arm are decreased, but arm claudication is rare.
5. Flow resistance in the vertebral artery increases because it is feeding a higher-resistance bed.
6. Surgical treatment may include a bypass graft or endarterectomy.
7. With subclavian stenosis rather than occlusion, the character of blood flow in the ipsilateral vertebral artery may be altered (figure 24-3).

Temporal Arteritis

Inflammation of the distal segment of the superficial temporal artery or its frontal and/or parietal branches.

1. Patients are symptomatic, complaining of headaches, tenderness on palpation of the distal superficial temporal artery or its branches, and possibly ipsilateral blindness.
2. A duplex study can assist in the diagnosis of this disease process but requires an experienced vascular technologist/sonographer with a record of high accuracy as confirmed with biopsy findings.

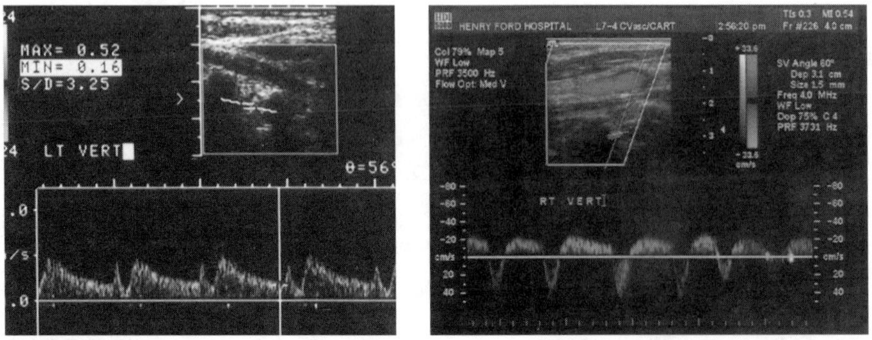

A B

Figure 24-3.
A The early systolic deceleration is consistent with a proximal stenosis of the subclavian artery. **B** With progression of the disease process, flow in the ipsilateral vertebral artery may exhibit bidirectional (to-and-fro) flow and progress further to complete flow reversal.

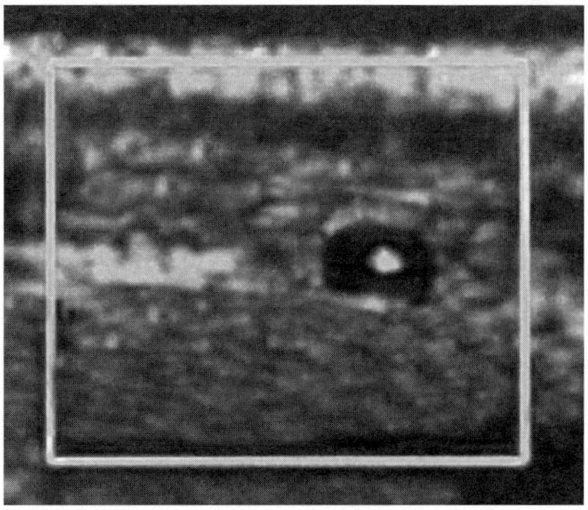

Figure 24-4. This magnified transverse view of the superficial temporal artery shows the anechoic halo around the vessel wall caused by edema. Courtesy of Marsha Neumyer, BS, RVT, FSVU.

3 B-mode findings may include visualization of a narrowed segment of the vessel wall and an anechoic halo around the vessel wall secondary to edema (figure 24-4).

4 Spectral analysis may document accelerated flow at the area of narrowing.

5 Medical treatment includes steroid therapy (e.g., prednisone).

6 When B-mode findings and Doppler flow characteristics are abnormal and consistent with temporal arteritis, the patient may not have to undergo a biopsy of the temporal artery. However, abnormal duplex findings may not be evident in the patient who has started steroid therapy.

Invasive Tests and Therapeutic Intervention

CHAPTER 25

See chapter 17 for more discussion of invasive studies of the arterial system, which are reviewed briefly here:

Arteriography

For quick reference, see plates 8–13 on pages 421–424. Plates 8 and 9 show normal and abnormal carotid systems, respectively. In plate 10, angiography of the aortic arch and extracranial cerebrovasculature demonstrates multiple disease in a patient admitted because of a CVA. Filling defects in the angiogram in plate 11 reveal irregular plaque in the carotid bulb and internal carotid artery, while plates 12 and 13 demonstrate fibromuscular dysplasia of the distal internal carotid artery and the so-called "string sign" of nearly total occlusion.

1 Type:

- Intraarterial injection
- Digital subtraction angiography, arterial

2 Technique:

- Intraarterial injection is usually performed according to the Seldinger technique in which a very thin catheter is inserted into an artery, fed into the arterial tree, and positioned for the injection of radiopaque dye to obtain a radiographic image of that vessel and the distal arterial system.
- The most commonly used arteries are the common femoral, with the axillary or brachial cannulated if necessary.

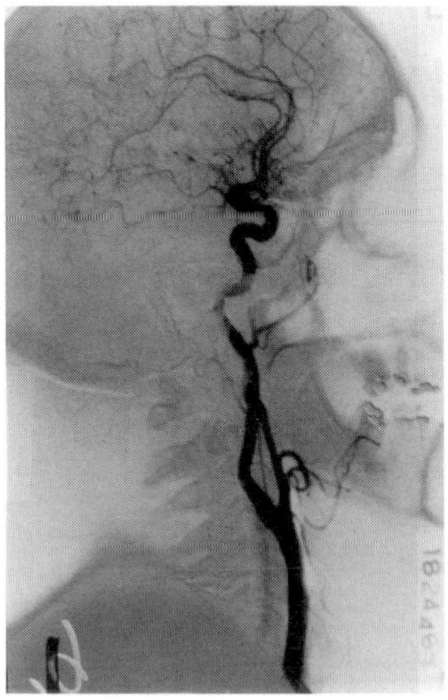

Figure 25-1. Normal carotid angiogram in which contrast medium completely fills the vessels.

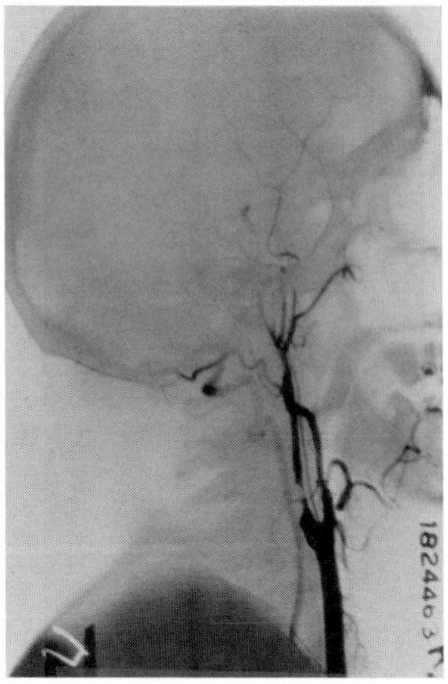

Figure 25-2. Abnormal carotid angiogram with filling defect in the internal carotid artery.

- Digital subtraction angiography (DSA) uses real-time digital video processing to detect small amounts of the contrast medium that has been injected into an artery.
- A rapid film changer technique is used to expose the films sequentially as the contrast agent moves through the vessel.
- Following removal of the catheter, pressure is applied to the puncture site and the patient is positioned supine for 2 to 6 hours. A closure device may be used and/or a variety of methods to maintain pressure on the insertion site to avoid hemorrhage and/or pseudoaneurysm formation.

3 Interpretation:

- Normal anatomy should appear on the films as the contrast medium completely fills the vessel (figure 25-1).
- Any deviation from normal—such as a *filling defect*—is evidence of an arterial abnormality (figure 25-2).

In figure 25-3, angiography of the aortic arch and extracranial cerebrovasculature demonstrates multiple disease in a patient admitted because of a

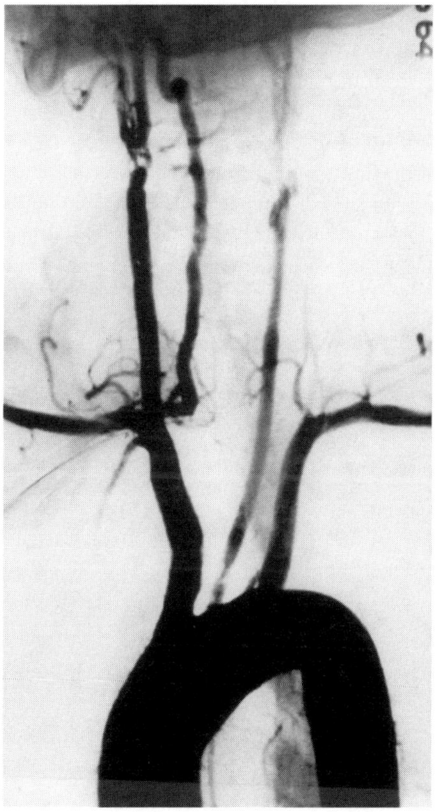

Figure 25-3. Arch and carotid angiogram showing multiple disease in a patient admitted with acute CVA. **Right bifurcation vessels:** High-grade stenoses suggested by filling defects. **Left proximal common carotid artery:** High-grade stenosis with poor vessel filling throughout the remainder of the carotid system but especially above the bifurcation. In this case the proximal disease increases the time it takes for the contrast agent to move distally within the vessel. For better visualization, delayed films would be required. If the patient had undergone a carotid duplex scan, Doppler flow in the CCA distal to the stenosis would be of poor quality (reduced strength, rounded peak), but resistance would not change. On the other hand, if the flow were approaching a significant stenosis, such as on the right, resistance (pulsatility) would increase. **Left vertebral artery:** No flow noted, indicating occlusion or—another possibility—a congenitally absent left vertebral artery.

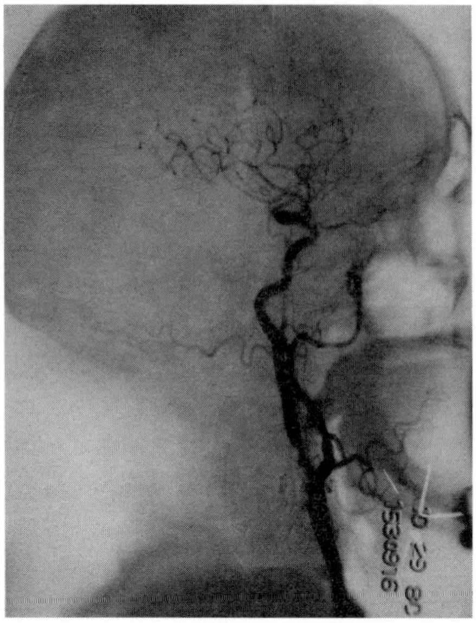

Figure 25-4. Angiogram from patient with asymptomatic carotid bruit. In this image the absence of contrast agent (filling defect) reveals the irregularity of the plaque in the bulb and internal carotid artery.

CVA. Filling defects in the angiogram in figure 25-4 reveal irregular plaque in the carotid bulb and internal carotid artery, while figure 25-5 demonstrates fibromuscular dysplasia of the distal internal carotid artery. The so-called "string sign" of nearly total occlusion is evident in figure 25-6.

- Percentage stenosis calculation—diameter reduction: To calculate the percentage stenosis on the basis of vessel diameter, subtract from 1 the diameter of the residual lumen (d) divided by the diameter of the true lumen (D) and then multiply by 100, as follows:

$$[1 - (d/D)] \times 100$$

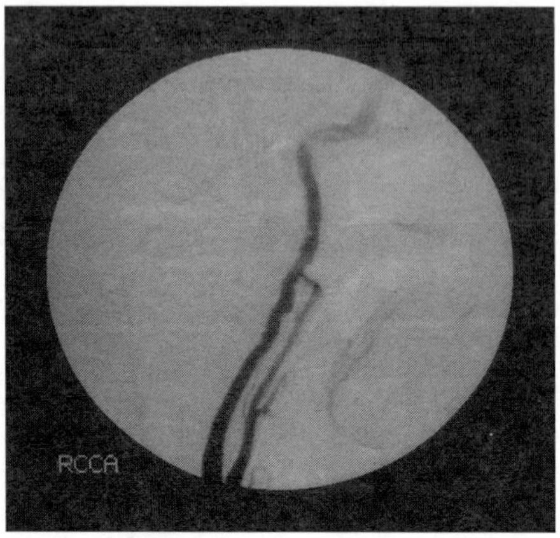

Figure 25-5. Angiogram showing internal carotid artery with fibromuscular dysplasia (FMD) of the distal portion. While most carotid disease is atherosclerotic and occurs at the bifurcation, FMD is usually located quite distal. On angiography, the vessel with FMD typically appears to be beaded, a characteristic referred to as a "string of beads." On duplex scanning, Doppler signals would show spectral broadening and increased velocities that vary with the location of the Doppler sample.

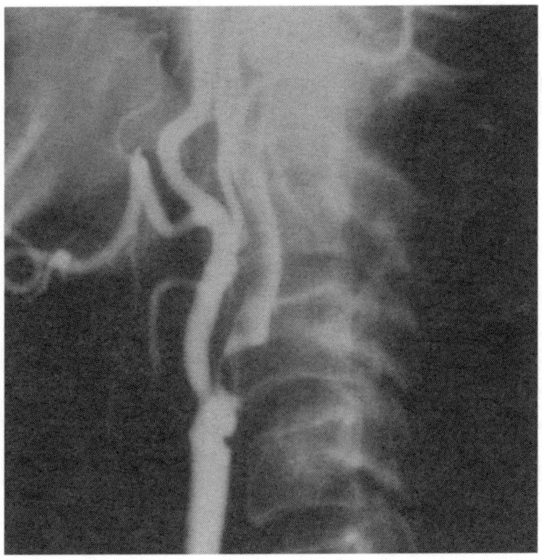

Figure 25-6. Angiogram depicting a "string sign" in the carotid bulb. Doppler peak flow velocities are reduced because of the low flow state of this preocclusive lesion. Auscultation of the neck would probably fail to elicit a bruit because the disease is so severe.

As figure 25-7A demonstrates, a sagittal view of the vessel being evaluated is necessary to make this calculation.

- Percentage stenosis calculation—area reduction: To calculate the percentage stenosis on the basis of the vessel's cross-sectional area, and assuming that the lesion is symmetrical (figure 25-7B), subtract from 1 the square of the residual diameter (d) divided by the square of the true diameter (D) and multiply by 100. In cases of symmetrical lesions, a 50% reduction in diameter is equal to a 75% reduction in cross-sectional area, which is considered to be hemodynamically significant.

$$[1 - (d^2/D^2)] \times 100$$

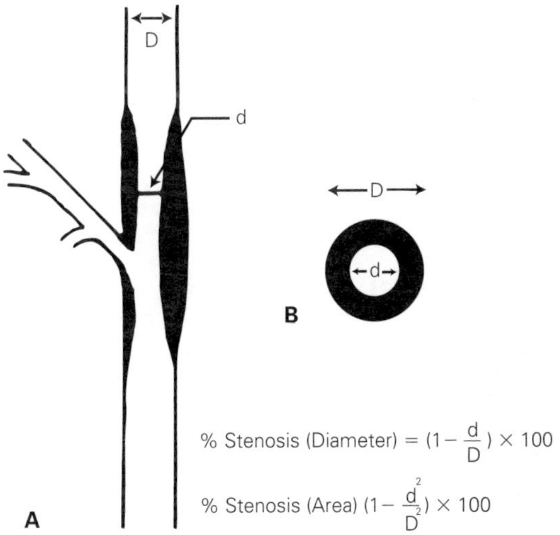

Figure 25-7. A hemodynamically significant stenosis reduces the diameter of an artery by 50% (**A**) and its cross-sectional area by 75% (**B**).

- Calculating percentage stenosis—Examples:

  Figure 25-7 illustrates a symmetrical lesion where D = 6 mm and d = 2 mm. Calculate the diameter reduction as follows:

  Equation used: $[1 - (d/D)] \times 100$

  Step one: Solve for d/D: $[1 - (.33)] \times 100$

  Step two: Subtract .33 from 1: $[.67] \times 100$

  Step three: Multiply by 100: 67%

  Calculate the area reduction of that same lesion as follows:

  Equation used: $[1 - (d^2/D^2)] \times 100$

  Step one: Solve for d^2/D^2): $[1 - (.11)] \times 100$

  Step two: Subtract .11 from 1: $[.89] \times 100$

  Step three: Multiply by 100: 89%

4 Limitations:

- Information regarding the hemodynamic status of the vessel under examination is not available.
- A two-dimensional view is standard.
- The calculation of percentage stenosis is based on the most stenotic portion of the vessel as compared to a "normal," more distal portion of the vessel.
- A patient may be at some risk for complications secondary to the invasive technique and/or the contrast medium used.

5 Complications:

- Hematoma at the puncture site
- Pseudoaneurysm at the puncture site
- Arterial occlusion
- Neurologic problems
- Allergic reaction

Magnetic Resonance Imaging (MRI)

- Uses radio frequency energy and a strong magnetic field to produce multiplane anatomic images.
- May show evidence of a cerebral infarction earlier than CT.

Magnetic Resonance Angiography (MRA)

- Provides high-quality images of blood flow without the need for x-rays.
- Extremely sensitive to the presence of stenosis, but tends to overestimate the disease process.
- Requires the use of a contrast agent.

Computed Tomography (CT)

- Provides high-quality images of anatomy.
- Employs ionizing radiation to obtain cross-sectional images .
- In cerebrovascular disease, used to evaluate the presence/absence of cerebral infarctions, tumors, masses, or anatomic variations.
- May include use of intravenous (IV) contrast.

Computed Tomography Angiography (CTA)

- Used to evaluate the status of the extracranial and intracranial vessels, i.e., patency, stenosis, occlusion, hemorrhage, arteriovenous malformations, and intracranial aneurysms.
- Requires use of intravenous (IV) contrast.

Therapeutic Intervention—Medical Therapy

1 Control of risk factors:

- Antihypertensive drugs decrease shear forces on the endothelial cells.
- Weight control and a low-cholesterol diet may enhance normal endothelial cell metabolism.
- The cessation of smoking eliminates nicotine's potentially harmful effects on the endothelium.

2 Pharmacologic therapy for occlusive disease:

- Aspirin is an antiplatelet drug that decreases platelet aggregation, which in turn decreases thrombotic activity.
- Other antiplatelet or antithrombotic medications may be prescribed.

Therapeutic Intervention—Surgical and Endovascular Therapy

1 Treatment options for stenosis:

A stenosis in the common or internal carotid artery may be treated by endarterectomy or placement of a stent. The decision to operate varies with the extent and severity of disease, the patient's symptoms, and the medical condition of the patient.

- Endarterectomy:

  An endarterectomy is the surgical removal of intraluminal atherosclerotic material. Postoperative surveillance with carotid duplex relies on the same interpretative guidelines used for diagnosis of a stenosis (pages 236–238). As seen in figure 25-8, the geometry of the vessel may change after carotid endarterectomy with a patch graft. Although the atherosclerotic process can result in restenosis, narrowing of the vessel wall within 6 months to 2 years following endarterectomy is usually attributable to neointimal hyperplasia. In addition to accelerated flow, restenosis is characterized by a thickened vessel wall with hypoechoic material at the operative site. Unlike an atherosclerotic stenosis, the narrowing from neointimal hyperplasia may regress over time (see color plate 13). Figure 25-9 demonstrates the rare finding of a pseudoaneurysm of a carotid endarterectomy vein patch graft site.

- Stent:

  As previously described on page 183, a stent is a metallic structure that may be covered with a synthetic material, e.g., Gore-Tex, and acts

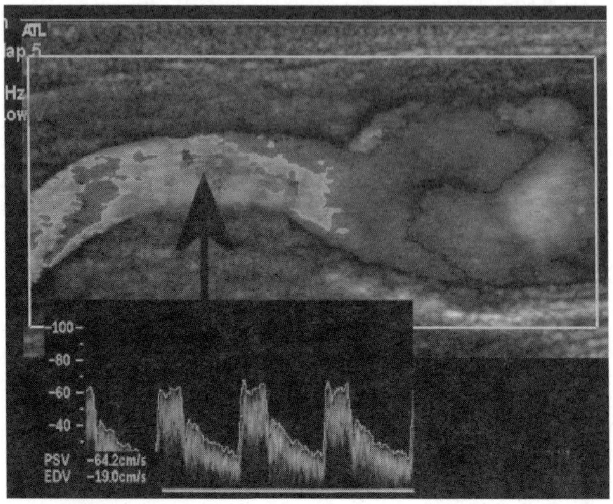

Figure 25-8. Sagittal view of an internal carotid artery (ICA) after carotid endarterectomy with vein patch grafting. Carotid endarterectomy is the most frequent surgical treatment for hemodynamically significant stenosis of the ICA. A vein patch is often used to prevent reduction of an already small arterial lumen when the surgeon sutures the vessel walls together. Dilatation of the wall is characteristically seen.

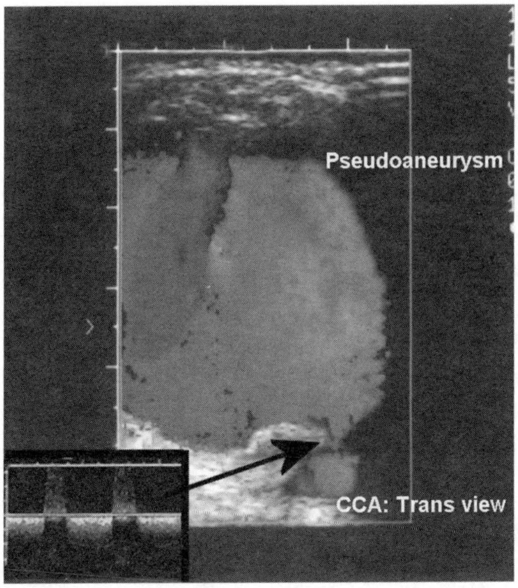

Figure 25-9. Longitudinal view of the ICA consistent with a pseudoaneurysm of the carotid endarterectomy vein patch graft site. Although rare, a pseudoaneurysm at the surgical site is a possible complication that must be considered. The aneurysmal sac and to-and-fro flow in the communication tract or neck (inset) is characteristic of a pseudoaneurysm.

as a type of scaffold. It is designed to maintain the intraluminal structure and patency of an artery that has been dilated using the percutaneous transluminal angioplasty (PTA) technique. There is increasing use of stents under certain conditions of carotid artery stenosis, and the technique continues to undergo extensive clinical investigation.

Postoperative surveillance of stented arteries with carotid duplex does not rely on the same interpretative guidelines used for diagnosis of a stenosis because some flow acceleration appears to be a normal finding in a stented vessel. Although interpretation criteria are being developed, two studies can provide some guidelines for poststent evaluations:

1. Stanziale SF, Boules TN, Selzer F, et al: Determining in-stent stenosis of carotid arteries by duplex ultrasound criteria. J Endovasc Ther 12:346–353, 2005. The results showed that PSV of ≥225 cm/s and ICA/CCA ratio of ≥2.5 was consistent with a ≥50–69% stenosis. A PSV of ≥350 cm/s and ICA/CCA ratio of ≥4.75 was consistent with a ≥70% stenosis.
2. Yung-Wei C, White C, Woods C, et al: Ultrasound velocity criteria for carotid in-stent restenosis. Catheterization Cardiovasc Interventions 69:349–354, 2006. The results showed PSV of ≥240 cm/s and ICA/CCA ratio of ≥2.45 was consistent with a ≥50–69% stenosis. A PSV of ≥450 cm/s and ICA/CCA ratio of ≥4.30 was consistent with a ≥70% stenosis.

The use of these criteria requires internal validation.

In addition to velocity criteria, B-mode findings would include intrastent (i.e., within the stent) changes such as visible restenosis, the development of neointimal hyperplasia, intimal dissection, vessel thrombosis, or stent deformations (e.g., incomplete apposition, stent-induced kinking of ICA).

2 Treatment options for occlusion:

- Pathology:

  Occlusion of a cervical carotid artery usually does not require surgical intervention. Blood is not flowing through the vessel, and therefore the risk of flow abnormalities or embolization no longer exists. In addition, it is impossible to surgically access the entire occluded vessel. However, a focal occlusion of the proximal internal carotid artery (ICA) may occur, i.e., the occluding plaque formation can be seen in its entirety below the mandible. Suspected mechanisms for maintaining patency of the distal, cervical ICA include a persistent hypoglosseal artery or collateralization through the vasa vasorum. In this rare situation, the entire occlusion is clearly visible and accessible to surgical intervention.

- Bypass graft:

  A bypass graft provides an alternate circulatory pathway so that blood can travel around an occlusion. Although bypass grafts are rarely performed for internal carotid arterial disease, a bypass graft may be performed for an occlusion of other vessels, such as the proximal common carotid, the subclavian, or the innominate arteries.

PART III Venous Evaluation

GROSS ANATOMY

VENOUS HEMODYNAMICS

PATIENT HISTORY, MECHANISMS OF DISEASE, AND PHYSICAL EXAMINATION

IMPEDANCE PLETHYSMOGRAPHY

STRAIN GAUGE PLETHYSMOGRAPHY

PHOTOPLETHYSMOGRAPHY

AIR PLETHYSMOGRAPHY

CONTINUOUS WAVE DOPPLER

DUPLEX SCANNING AND COLOR FLOW IMAGING

INVASIVE TESTS AND THERAPEUTIC INTERVENTION

CHAPTER 26

Gross Anatomy of the Central and Peripheral Venous Systems

The anatomic relationship of the veins to the heart is the same as for the arteries: closer to the heart is proximal; farther away from the heart is distal. For example, veins located closer to the heart, e.g., femoral, are considered more proximal to veins located at the ankle which would be considered more distal. Although this may express a different perspective, it does provide consistency in terms of anatomy.

Lower Extremity Veins

1 Deep veins:

- The *deep digital veins* form the metatarsal veins.
- The *metatarsal veins* form the *deep venous arches*.
- The paired *peroneal veins* empty the lateral region of the leg and carry blood cephalad into the *tibioperoneal trunk*. The paired deep veins of the calf follow the arteries and are called *venae comitantes* (corresponding veins).

- The paired *posterior tibial veins* empty the back of the leg and carry blood into the tibioperoneal trunk.
- The paired *anterior tibial veins* empty the front of the leg. The anterior tibial and tibioperoneal trunk veins join just below the knee to form the *popliteal vein.*
- Large muscular veins empty the soleal muscles into the posterior tibial and peroneal veins, and the gastrocnemius muscle into the popliteal vein.
- The popliteal vein becomes the *femoral vein* at the *adductor canal* or *Hunter's canal.* The confluence of the femoral vein and deep femoral/profunda femoris vein forms the *common femoral vein.*
- The common femoral vein becomes the *external iliac vein* just above the inguinal ligament.
- The external iliac vein unites with the *internal iliac vein* to become the *common iliac vein.* The left common iliac vein passes beneath the right common iliac artery to empty into the inferior vena cava. This pressure point is recognized by some to account for the increased incidence of left lower extremity deep venous thrombosis (May-Thurner syndrome).
- The confluence of the common iliac veins form the *inferior vena cava* at the level of the 5th lumbar vertebra and continues to the right atrium of the heart.
- Figure 26-1 illustrates sagittal and cross-sectional views of the deep veins most commonly evaluated by ultrasonography.

2 Superficial veins:

- The digital veins form larger venous channels of the foot that eventually form the great saphenous vein.
- As illustrated in figure 26-2 on page 266, the *great saphenous vein* is the longest vein in the body, originating on the dorsum of the foot and traveling medially to the saphenofemoral junction in the groin just below the inguinal ligament.
- The *small saphenous vein* ascends the back of leg to join the popliteal vein (figure 26-2).
- Note on terminology: The *great* and *small saphenous veins* are the currently acceptable terms for what formerly were called the *greater* and *lesser saphenous veins.*

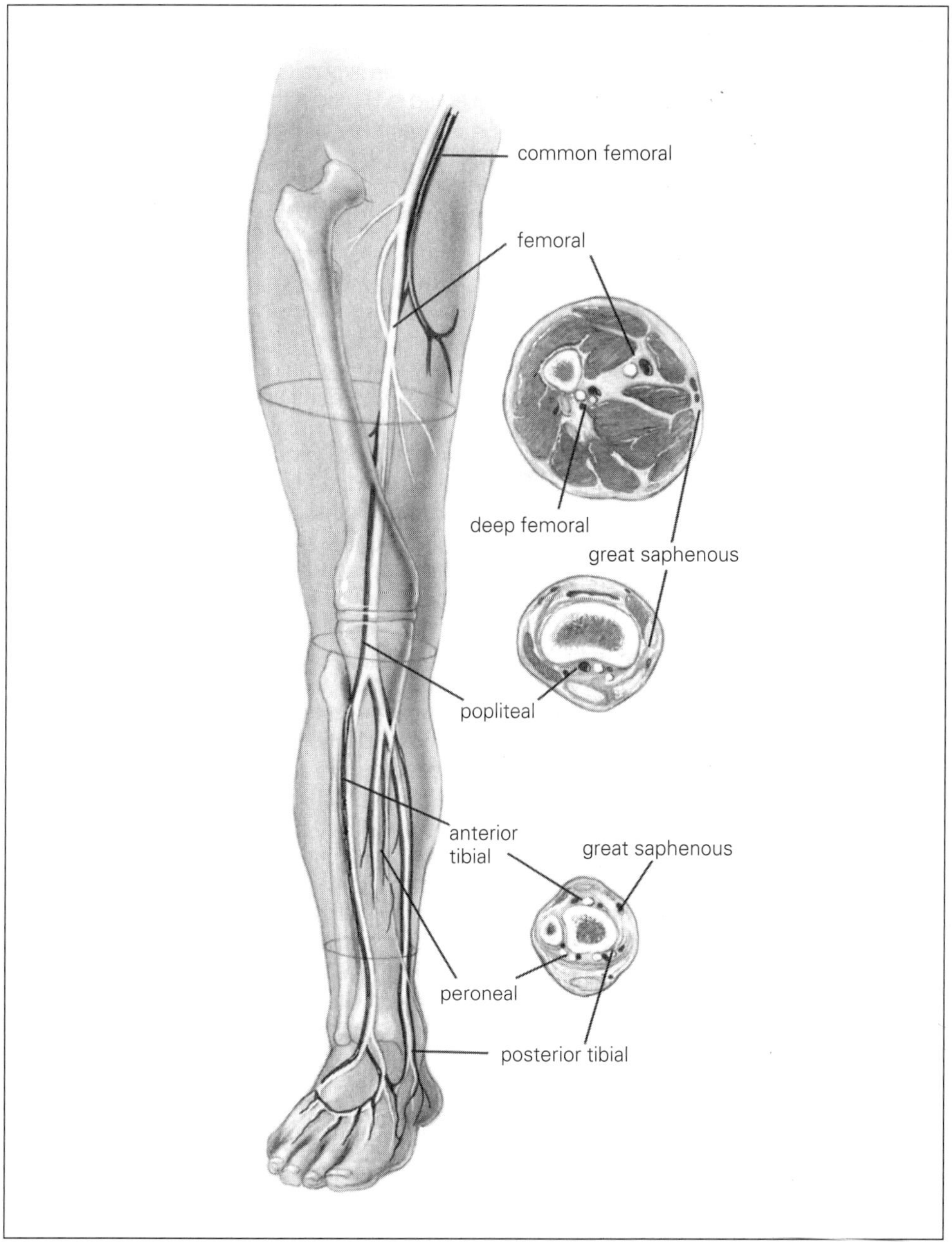

Figure 26-1. Deep venous system of the lower extremity. From Salles-Cunha SX, Andros G: *Atlas of Duplex Ultrasonography: Essential Images of the Vascular System.* Pasadena, Appleton Davies, 1988.

3 Perforating veins:

- The *perforating veins* (commonly called *perforators*) form communications between the superficial and the deep systems.
- Venous drainage of the lower leg occurs as blood travels from the superficial veins through the perforators into the deep veins (figure 26-3).
- Each perforating vein has at least one valve to maintain this unidirectional flow.

Figure 26-2. Main veins of the lower extremity. From Ridgway DP: *Introduction to Vascular Scanning: A Guide for the Complete Beginner,* 3rd edition. Pasadena, Davies Publishing, 2004.

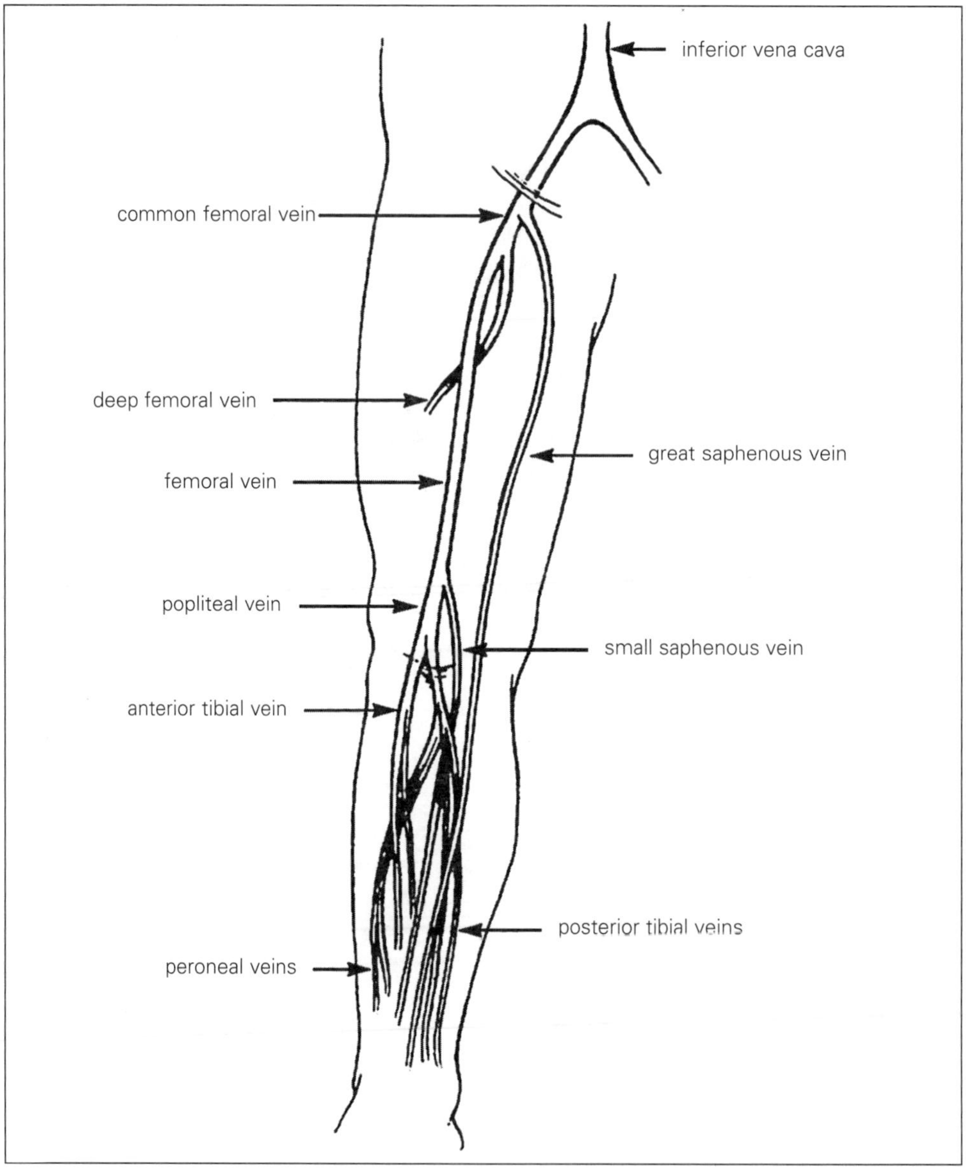

Figure 26-3. Perforating veins carry blood from the superficial system into the deep system.

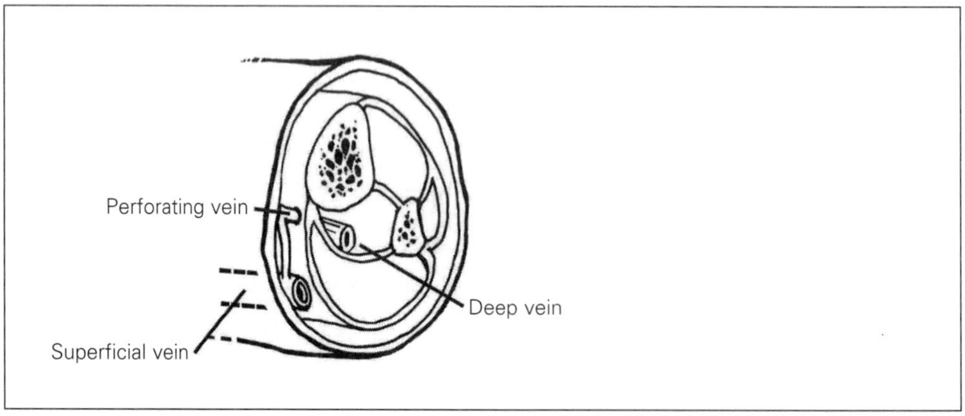

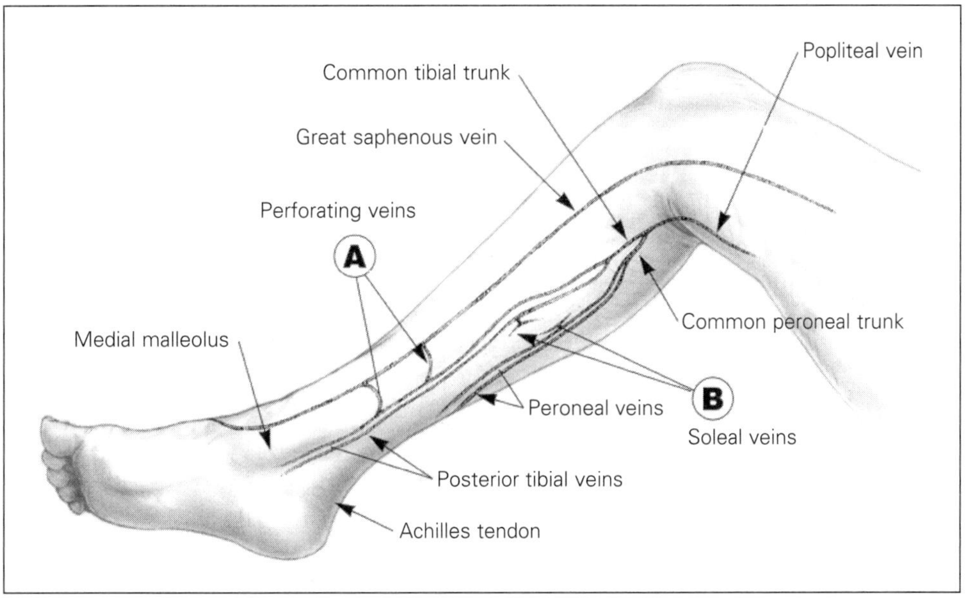

Figure 26-4. The soleal veins can be identified as they connect into the posterior tibial or peroneal veins. The perforating veins also can be identified as they connect the superficial veins with the deep veins. Reprinted with permission from Talbot SR, Oliver MA: *Techniques of Venous Imaging.* Pasadena, CA, Davies Publishing, 1992.

- The posterior tibial veins have two important perforators near the medial malleolus ("A" in figure 26-4).
- The posterior communicating branch of the great saphenous vein in the medial lower calf is connected to a third perforator.
- The *posterior arch vein* (figure 26-5) is important because it represents a superficial connection of the three ankle perforating veins—an anatomic

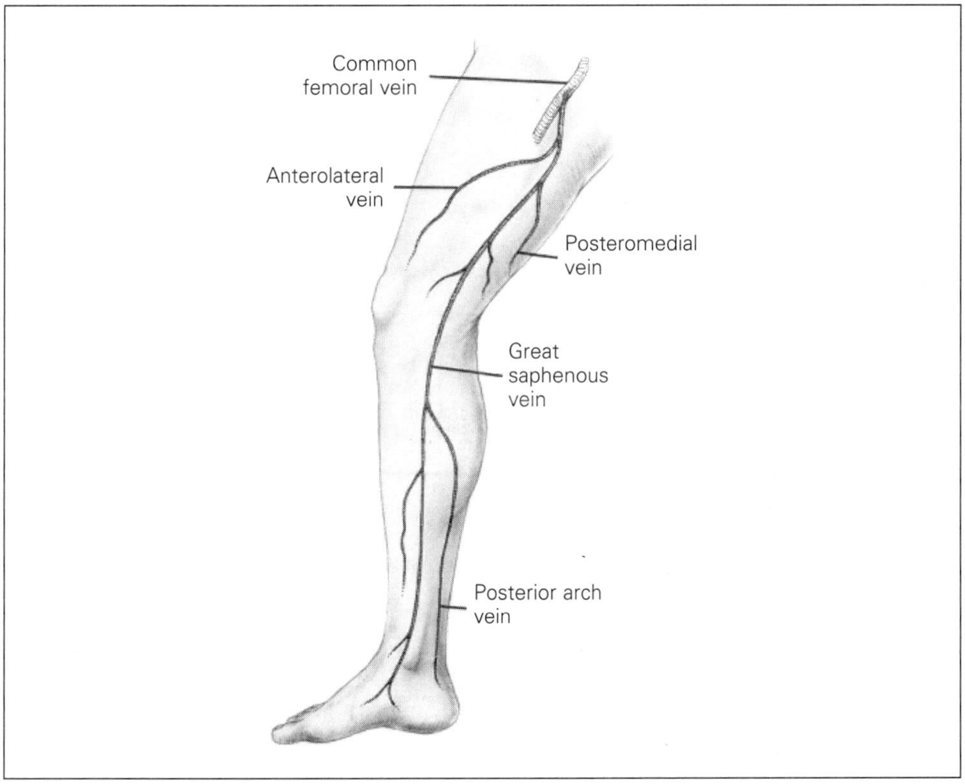

Figure 26-5. Great saphenous vein and major branches. Reprinted with permission from Talbot SR, Oliver MA: *Techniques of Venous Imaging.* Pasadena, CA, Davies Publishing, 1992.

Figure 26-6. The great saphenous vein with some of its major perforators as well as the posterior arch vein that joins it. Cockett I, II, and III perforators are evident in the distal portion of the posterior arch vein.

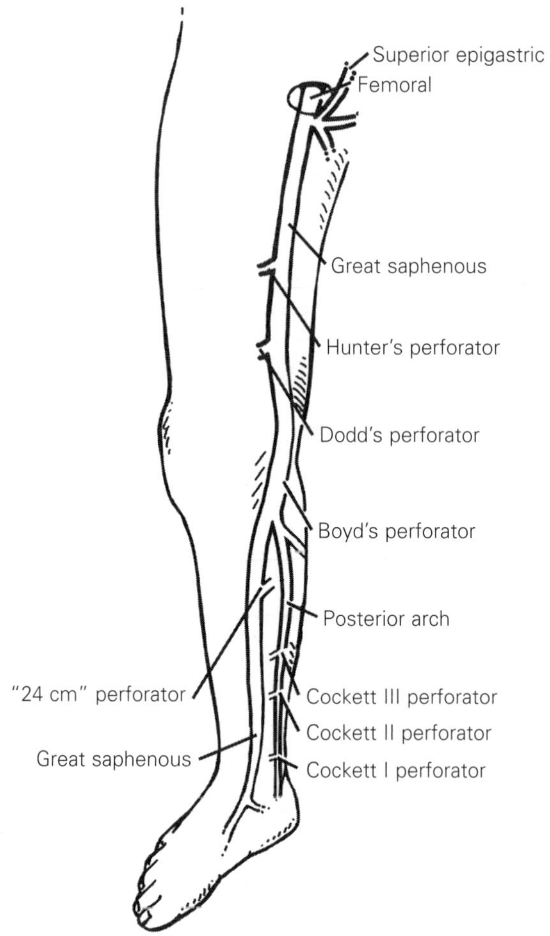

fact of major importance in the development of a venous stasis ulcer. Figure 26-6 shows the anatomic location of Cockett I, II, and III perforators.

- The small saphenous vein of the posterior calf has an important lateral perforating branch.

4 Venous sinuses:

- In the brain, *venous sinuses* are spaces between the dura mater and periosteum that receive venous return. They eventually terminate in the *internal jugular veins.*

- In the lower extremities, the venous sinuses are dilated channels located in the calf muscles that allow venous blood to accumulate and then to drain into the posterior tibial and peroneal veins. These large sinusoid saccular muscular veins of the soleal and gastrocnemius muscles act as a major part of the calf-muscle pump ("B" in figure 26-4).

Upper Extremity Veins

1 The *deep digital veins* form the *palmar arches* of the hand and empty into the radial and ulnar veins.

2 The paired *radial veins* travel along the lateral aspect of the forearm. (The paired deep veins of the arm and forearm follow the arteries and are called *venae comitantes* [corresponding veins].)

3 The paired *ulnar veins* travel along the medial aspect of the forearm.

4 The radial and ulnar veins form the *brachial veins*, usually near the elbow or *antecubital fossa*, but possibly anywhere in the upper arm.

5 The paired brachial veins become the *axillary vein*, usually near the armpit or axilla. Normally, another landmark for the brachials becoming the axillary is where the brachial and basilic veins join.

6 The axillary vein becomes the *subclavian vein* when it is joined by the cephalic vein, usually near the lateral border of the first rib.

7 The subclavian vein joins the internal jugular to form the *brachiocephalic* or *innominate vein*.

8 The right and left innominate veins form the *superior vena cava*, which carries blood to the right atrium (figure 26-7).

9 Figure 26-8 illustrates the sagittal view of the veins of the upper extremity commonly evaluated by ultrasonography.

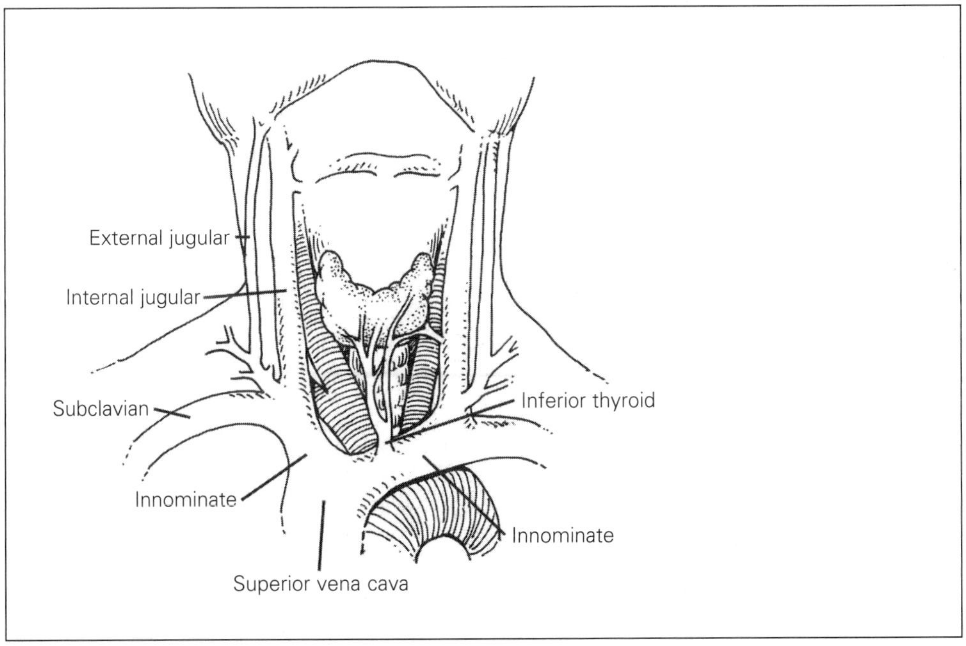

Figure 26-7. The right and left innominate veins, formed by the confluence of the subclavian vein and internal jugular vein, join together to form the superior vena cava.

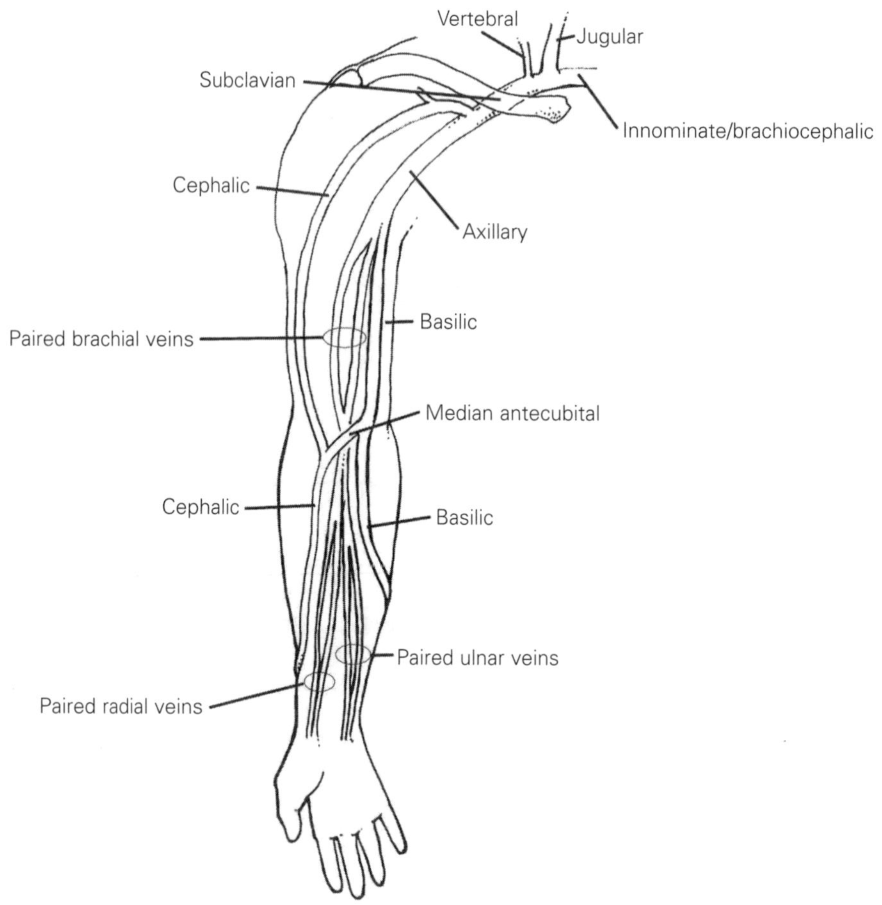

Figure 26-8. Sagittal view of the veins of the upper extremity commonly evaluated by ultrasonography.

The Superficial Veins

1. The digital veins form the *cephalic vein* on the lateral aspect of the forearm and arm with the *basilic vein* on the medial aspect.
2. The cephalic vein travels laterally up the arm and, at its confluence with the *axillary vein,* becomes the subclavian vein.
3. The basilic vein ascends medially up the arm, becoming the axillary vein where it joins the brachial vein.

The Central Veins

1. Blood in the head and upper extremities eventually empties into the *innominate veins,* which then form the *superior vena cava* that in turn empties into the right atrium.
2. The lower half of the body eventually empties into the common iliac veins, which join to form the *inferior vena cava,* which empties into the right atrium.

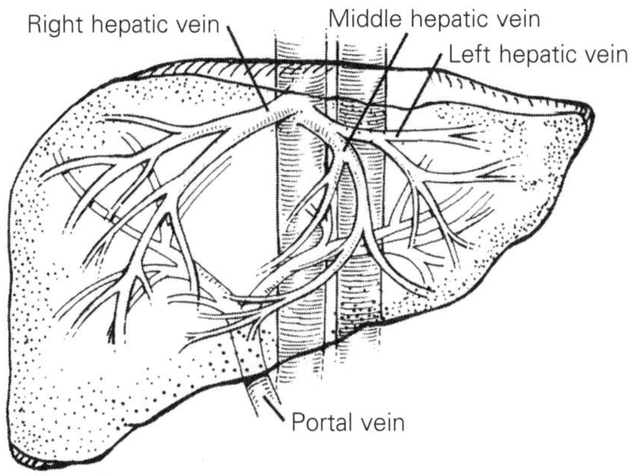

Figure 26-9. The portal vein carries blood into the liver and hepatic veins return the blood to the inferior vena cava. This is unique in that blood passes through two sets of veins as it travels back to the heart.

The Abdominal Veins

1 The portal system drains blood from the abdominal part of the digestive tract, i.e., the pancreas, spleen, gall bladder, and mesentery, and carries it into the *portal vein*. The portal vein, formed by the superior mesenteric and the splenic veins, carries blood *into* the sinusoids of the liver. Flow into liver is called hepatopetal. Although the hepatic artery does carry blood to the liver, the portal vein is the predominant source of blood flow (about 80%) to that organ. The hepatic veins carry blood *from* the liver into the inferior vena cava. Flow away from the liver is called hepatofugal. Unlike other parts of the venous system, here the blood passes through two sets of veins as it travels back to the heart (figure 26-9).

2 The right and left renal veins drain blood from the kidneys and carry it directly into the inferior vena cava. The left renal vein crosses the aorta anteriorly and is usually longer than the right. Both right and left renal veins are positioned in front of their respective corresponding renal arteries.

The Structural and Functional Anatomy of Veins

1 The veins transport blood toward the heart as it carries away the waste products of cellular activity.

2 The veins are collapsible tubes that expand in response to increased *transmural pressure*, also referred to as *distention pressure*. They are not completely passive structures but have some element of reactivity, which may be referred to as *venomotor tone*. The size of the vein can change when smooth muscle cells contract in response to stimulation of the sympathetic nervous system, as with temperature changes, exercise, stress, or trauma.

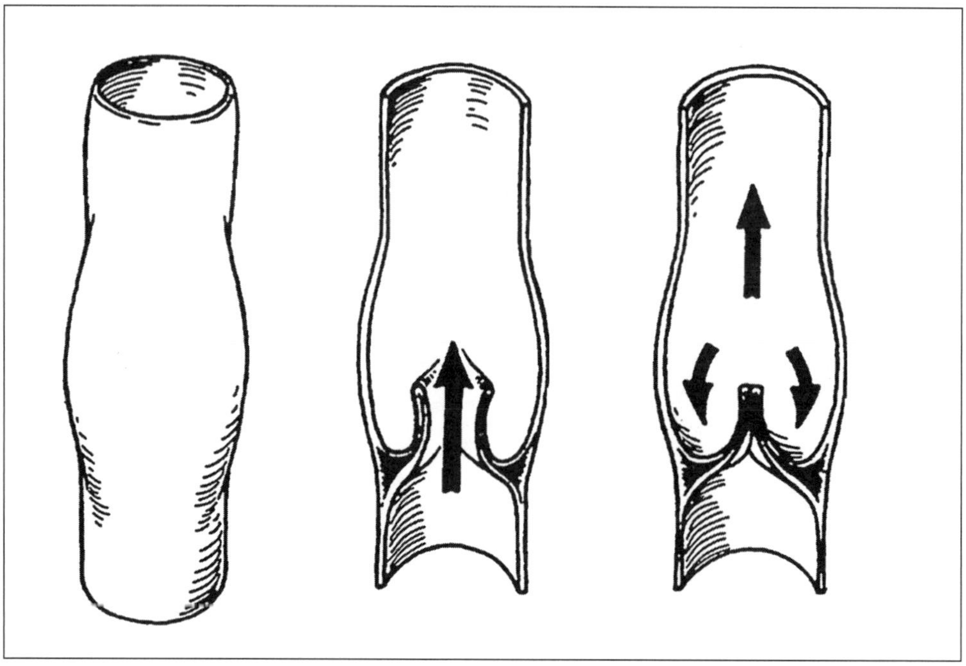

Figure 26-10. Competent venous valves maintain unidirectional flow within the venous system. From Belanger AC: *Vascular Anatomy and Physiology: An Introductory Text.* Pasadena, Davies Publishing, 1999.

3 The inner layer of a vein—*tunica intima*—is a single layer of endothelial cells.

4 The middle layer of a vein—*tunica media*—is a thicker layer of smooth muscles and collagenous fiber.

5 The outer layer of a vein—*tunica adventitia*—is a thin fibrous layer surrounding elastic tissue. This layer contains the *vasa vasorum,* tiny vessels that distribute blood to and from the walls of the larger veins.

6 As illustrated in figure 26-10, the bicuspid valves of the venous system are structural elaborations of the intimal layer that keep the blood flowing in only one direction by snapping shut during retrograde blood flow. Although valves are evident in the upper extremity veins, venous disease of the upper extremity is uncommon. The valves of the lower extremities are more susceptible to disease secondary to the effects of venous thrombosis, increased venous pressure from gravity, increased intra-abdominal pressure, and/or venous obstruction. Dilatation at the site of venous valves along with the retrograde flow behind the venous valve may be normally represented as reversed flow with color duplex studies, e.g., red color at the valve site as compared to the blue color depicted in the patent vein.

7 Typical location and number of valves:

- Veins with valves:

  Great saphenous vein: approximately 12 valves, most below the knee

  Small saphenous vein: 6–12 valves

  Perforators: *each* contains 1 valve and sometimes as many as 3

  Infrapopliteal (deep) veins: 7–12 valves *each*

  Popliteal and femoral veins: 1–3 valves *each*

  External iliac vein: contains valves approximately 25% of the time

  Common femoral vein: 1 valve

  Internal jugular vein: 1 valve

  Axillary vein: 1 valve

  Cephalic vein: variable in distal segment

  Basilic vein: variable in distal segment

- Veins without valves:

  Soleal sinuses

  External iliac vein: contains valves approximately 25% of the time

  Internal iliac veins

  Common iliac veins

  Inferior vena cava

  Subclavian veins

  Innominate veins

  Superior vena cava

CHAPTER 27 Venous Hemodynamics

Venous Resistance

1 Unlike arteries, veins are highly compliant, expanding into a circular cross-sectional shape as intraluminal pressure increases and collapsing into elliptical and then "dumbbell" configurations as intraluminal pressure decreases to the point at which it is exceeded by the pressure of surrounding tissue.

2 This difference between the pressures within *(intraluminal)* and outside *(interstitial)* of the veins is called the *transmural pressure*, and it is the transmural (across the wall) pressure that determines the cross-sectional shape of the veins: The higher the intraluminal pressure, the higher the transmural pressure; and the higher the transmural pressure, the larger and more circular the venous cross section.

3 Because the veins are seldom completely full of blood as the arteries are, their flattened shape offers a great deal of flow resistance. On the other hand, because the fully distended cross-sectional area of a vein is about three to four times that of the corresponding artery, the veins can carry more blood without an increase in pressure. If the veins were not so compliant, increased flow would increase pressure, decreasing the pressure gradient to the heart and reducing flow. See figure 27-1.

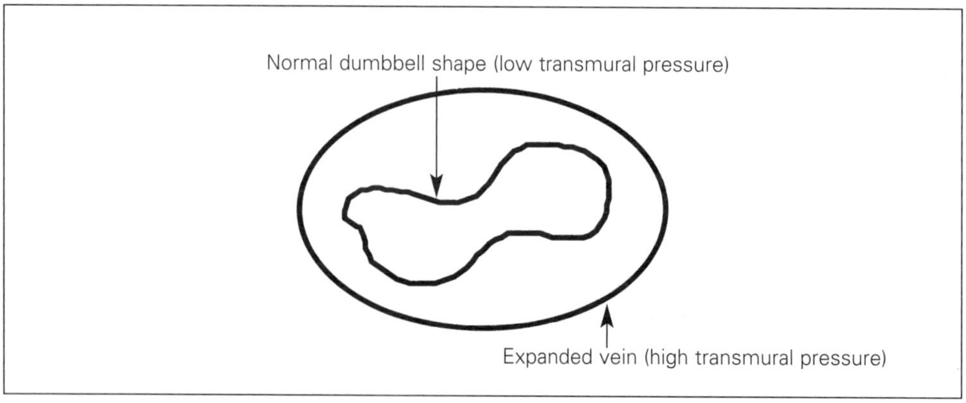

Figure 27-1. The normal flattened shape of the vein offers a great deal of flow resistance. But vein walls are highly compliant, expanding into a more circular shape as blood flow increases.

Hydrostatic Pressure

1 Hydrostatic pressure is equivalent to the weight of a column of blood extending from the heart to the level where the pressure is being measured. The formula, HP = pgh, describes hydrostatic pressure (HP) as being directly related to the product of specific gravity of blood (p), acceleration due to gravity (g), and distance from the heart (h).

2 Hydrostatic pressure in the supine individual is negligible, e.g., around 0 mmHg. When the individual stands, the hydrostatic pressure measured at the ankles typically rises to approximately 100 mmHg, depending on the person's height.

- Since hydrostatic pressure is related to gravity and distance from the heart, the pressure will actually decrease if the extremity is raised above the level of the heart. (See figure 27-2.)

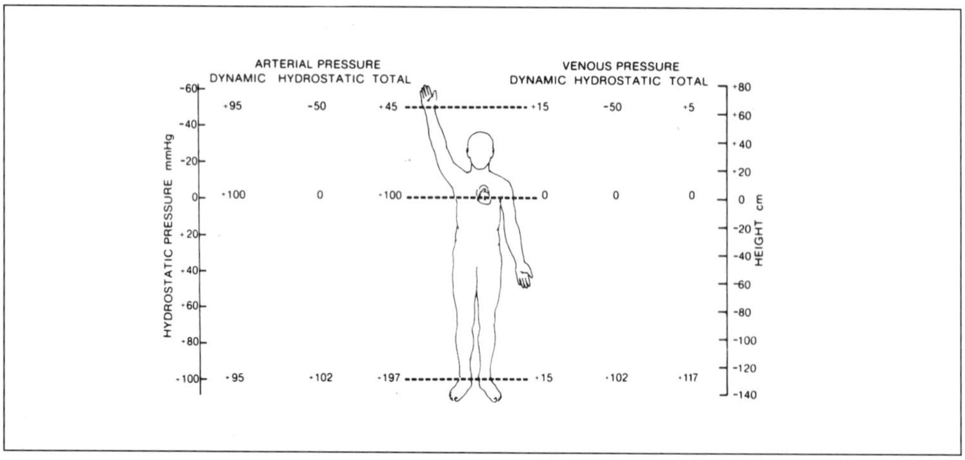

Figure 27-2. Effect of hydrostatic pressure on venous and arterial pressures. The reference point for zero pressure is at the right atrium. If the subject were supine, total intravascular pressure would closely approximate the dynamic pressure. Reprinted with permission from Sumner DS: The hemodynamics and pathophysiology of venous disease. In *Vascular Surgery.* Edited by RB Rutherford. Philadelphia, WB Saunders, 1977, pp 147–163.

Pressure/Volume Relationships

1 The shape of veins is determined by the transmural pressure, e.g., the difference between the pressure inside the vein (the intraluminal pressure) and the pressure of the surrounding tissue (the interstitial pressure).

2 Even small increases in intraluminal pressure cause the vein to expand and change from its collapsed dumbbell-shaped configuration to a circular shape. But much greater pressure changes are required to expand the vein further since the walls of an expanded, circular vein are no longer very compliant.

Effects of Muscle Pump Mechanism

1 The contraction of leg muscles squeezes the veins and propels the blood toward the heart. If the venous valves are competent, the blood will travel in only one direction. This normal pattern of venous flow decreases venous pressure and pooling and increases venous return to the heart and cardiac output.

2 With incompetent valves, though, the opposite occurs: venous pressure and venous pooling increase, decreasing venous return to the heart and cardiac output.

During Muscle Contraction

3 Although at rest the veins act as reservoirs for blood collection, during activity they propel blood toward the heart.

4 The power source for this propulsion is the contraction of the leg muscles, especially the calf muscles, and may be referred to as the "venous heart."

5 In a normal venous system the contraction of the calf muscles forces blood cephalad toward the heart. When the calf muscle relaxes, blood moves from the superficial veins into the deep venous system (figure 27-3).

6 Venous valves maintain the unidirectional movement of the blood.

7 In cases of valvular incompetence, control of blood flow is deficient, the blood traveling both antegrade and retrograde.

During Muscle Relaxation

8 Upon relaxation of a calf or leg muscle, a potential "space" with very low to absent pressure is created in the deep venous system, and the blood flows from the superficial veins into the deep system via the perforators.

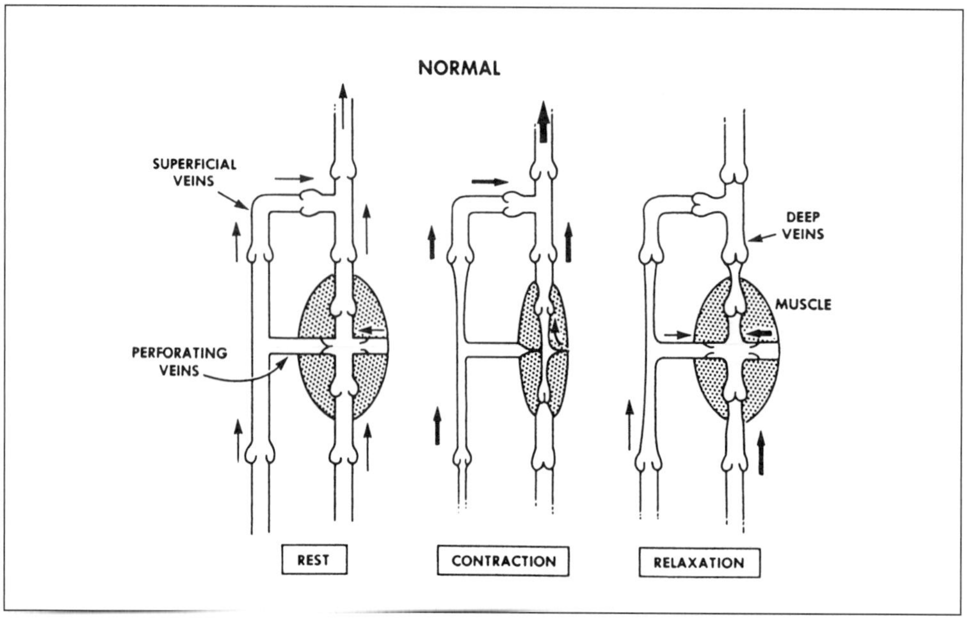

Figure 27-3. Dynamics of the muscle pump mechanism in a normal limb. Reprinted with permission from Sumner DS: Venous dynamics–varicosities. Clin Obstet Gynecol 24:743–760, 1981.

This movement occurs because of the pressure gradient established by the normal muscle pump mechanism.

9 This physiologic process reduces peripheral venous pressure.

During Respiration

10 Respiration greatly affects venous flow. *Inspiration* decreases intrathoracic pressure but increases intraabdominal pressure as the diaphragm descends.

11 The decrease in intrathoracic pressure decreases the pressure gradient between the veins in the arms, head, and neck and those in the thoracic cavity, increasing the inflow of blood from the arm and head veins. The increase in intraabdominal pressure decreases the pressure gradient between the veins in the legs and those of the abdomen, reducing the outflow of blood from the peripheral veins of the lower extremity (figure 27-4).

12 The opposite occurs during *expiration,* which increases venous flow from the lower extremities, while halting flow from the upper extremities.

13 Flow in the portal vein of an adult is minimally phasic. There is almost no variation with respiration.

During the Valsalva Maneuver

14 As the patient performs a *Valsalva maneuver* by taking in a deep breath and then bearing down (as if having a bowel movement), intrathoracic and intraabdominal pressures increase significantly. All venous return is

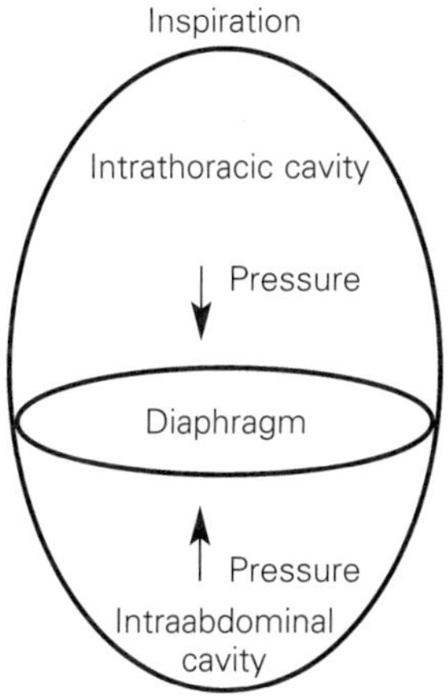

Figure 27-4. The increase in intra-abdominal pressure causes a cessation of blood flow from the lower extremities into the abdomen.

halted. The resultant decrease in the volume of blood returning to the heart will cause cessation of the spontaneous common femoral venous signal. *Augmentation* of the venous signal should be evident as the patient releases the deep breath and stops bearing down. These are normal responses to the Valsalva maneuver.

15 If the venous signal is augmented as the patient bears down, flow reversal from incompetent valves is likely. This is an abnormal response to the Valsalva maneuver.

16 The Valsalva maneuver should not be performed in patients with severe coronary artery disease, acute myocardial infarction, or moderate to severe hypovolemia.

Patient History, Mechanisms of Disease, and Physical Examination

CHAPTER 28

Again, it is very important to obtain a complete history and to confirm appropriate indications for testing prior to performance of the study.

Patient History: Signs and Symptoms

Acute

1 In the past, the clinical diagnosis of acute deep venous thrombosis (DVT) based on the patient's medical history and physical exam was considered notoriously unreliable, i.e., ≤ 50%, with some of the most common clinical findings of:

- Swelling
- Pain
- Redness or erythema
- Warmth

2 Currently it is recognized that the accuracy of the clinical diagnosis of DVT in the outpatient population can be significantly improved depending on the pretest probability of disease. Although several methods for defining that probability have been validated, table 28-1 contains one of the widely used outpatient scoring systems developed by Wells et al.

Table 28-1. Clinical scoring system for estimating pretest probability of DVT.*

Clinical Criteria	***Points***
Active cancer within 6 months	1
Paralysis, paresis, or recent plaster immobilization of lower extremities	1
Recently bedridden for > 3 days or major surgery within 4 weeks	1
Localized tenderness along the distribution of the deep veins	1
Entire leg swollen	1
Calf swelling > 3 cm (compared to contralateral leg)	1
Pitting edema	1
Collateral superficial veins	1
Alternative diagnosis as likely or more likely than DVT	−2

Patients with ≤ 0 points are classified as having a low pretest probability of DVT, those with 1 or 2 points a moderate probability, and those with ≥ 3 points a high probability.

* Wells PS, Anderson DR, Bormanis J, et al: Value of assessment of pretest probability of deep-vein thrombosis in clinical management. Lancet 350:1795–1798, 1997.

3 Differential diagnosis in patients with a clinically suspected deep venous thrombosis includes:

- Muscle strain
- Direct injury to the leg
- Muscle tear
- Baker's cyst
- Cellulitis
- Lymphangitis
- Heart failure
- Extrinsic compression
- Complications of chronic venous insufficiency

Chronic

4 Patients with chronic venous disease may have:

- Swelling
- Heaviness
- Discoloration or ulcerations
- Varicosities

Risk Factors and Contributing Diseases

The development of venous thrombosis is based on three factors known as *Virchow's triad:*

1 Trauma to the vessel (endothelial damage)

- Intrinsic trauma, e.g., damage to the vessel wall from a catheter or intravenous drugs.
- Extrinsic trauma can occur at any site, e.g., as the result of an accident.

2 Venous stasis. *Examples:*

- Bedrest or immobility
- Myocardial infarction
- Congestive heart failure
- Hypotension
- Chronic obstructive pulmonary disease
- Obesity
- Pregnancy
- Previous deep venous thrombosis
- Extrinsic compression
- Paraplegia
- Surgery:

  Patient is immobile because of anesthesia.

  Calf muscle pump not activated.

  Mechanical ventilation may alter pressure gradients.

  Extrinsic compression from positioning and/or the procedure itself.

  Injury to endothelium causes alteration in its anti-thrombogenic properties.

3 Hypercoagulability. *Examples:*

- Pregnancy
- Cancer
- Oral contraceptives and/or hormone replacement therapy
- Inherited states, e.g., factor V Leiden, protein C or protein S deficiencies, antithrombin III deficiency.

Mechanisms of Disease

Thrombosis

1 Intraluminal thrombi, predominately composed of red blood cells trapped within a fibrin web, frequently originate at the cusps of the venous valves or in the soleal sinuses because of stagnation. Thrombi can also originate from other sources:

- Paget-Schroetter syndrome, also known as *stress* or *effort thrombosis*, usually involves thrombosis of the subclavian or axillary vein secondary to intense, repetitive activity. This venous component of thoracic outlet syndrome is frequently associated with repetitive motion of the upper extremity as seen with heavy lifting or strenuous throwing of a baseball or football.
- May-Thurner syndrome is compression of the left common iliac vein by the right common iliac artery as the artery crosses over it. In some cases the artery causes enough compression to thicken the vein wall as well as alter flow to the point of thrombosis.
- Superior vena cava (SVC) syndrome is an obstruction of the SVC, usually by a malignant lesion, although thrombosis from catheter placement or procedures can also occur. Dyspnea (difficulty breathing) is the most common symptom, but facial and extremity swelling may also be evident.

2 As the clot propagates, flow becomes restricted, venous pressure increases (venous hypertension), the vein walls stretch, and the venous valves can be damaged. Valve sites may dilate, which decreases ability of walls to coapt, allowing retrograde blood flow. All of these factors may lead to chronic venous insufficiency.

3 The great clinical danger in cases of deep venous thrombosis is that the entire thrombus or a part of it may break loose, travel into the pulmonary circulation, and cause pulmonary embolism, an oftentimes life-threatening development. Prompt diagnosis and treatment of a deep venous thrombosis is therefore a priority.

Valvular Incompetence

4 Valves no longer maintain unidirectional movement of the blood, and the calf muscle pump no longer forces blood cephalad toward the heart or from the superficial system into the deep.

5 Incompetent venous valves allow blood to travel both antegrade and retrograde, increasing pressure and creating venous hypertension, which can precipitate edema and varicosities.

6 In addition, fluid, red cells, and fibrinogen may leak into the surrounding tissue as a result of the increased venous pressure. The breakdown of red blood cells creates the hemosiderin deposits that cause the brawny discoloration of the skin. The breakdown of other substances can also prevent the tissue from getting proper oxygen and nutrients, resulting in ulceration. The combination of chronic swelling, brawny discoloration, and ulcer formation in the distal calf-to-ankle area, or *gaiter zone*, is referred to as the *post-thrombotic* or *postphlebitic syndrome* if related to a previous deep venous thrombosis (DVT).

Congenital Venous Disease

7 Patients may also have congenital venous disease such as:

- Avalvular (valveless) vein(s)
- Incompetent valves
- Arteriovenous malformation(s) (AVM)
- A variety of syndromes, including, for example, Klippel-Trenaunay, which includes hypoplastic or absent deep veins (i.e., absent iliacs with varicosities of the superficial system resulting in an enlarged limb)

Portal Hypertension

8 The elevated pressure in the portal vein is often the result of an obstruction of blood flow. Although it is usually related to some form of advanced chronic liver disease such as cirrhosis, it can also be caused by severe congestive heart failure or a proximal venous occlusion of, for example, the hepatic vein or inferior vena cava. The elevated pressure results in a reverse pressure gradient that causes portal venous flow away from the liver (hepatofugal flow).

Physical Examination

1 Varicose veins:

- *Primary varicose veins* are dilated, tortuous veins that may be hereditary, i.e., the result of a congenital absence of one or more valves. The increased venous pressure that causes primary varicose veins is unrelated to obstruction of the deep venous system.
- *Secondary varicose veins,* on the other hand, are caused by obstructive conditions of the deep venous system, such as previous deep venous thrombosis, pregnancy, or obesity.

2 Skin changes:

- *Edema* becomes evident as fluid accumulates in the tissues.

- Redness (*erythema*) may be caused by an inflammatory process and/or cellulitis.

- A brownish (*brawny*) discoloration represents chronic venous insufficiency. Increased venous pressure causes fluid, red blood cells, and fibrinogen to leak into the surrounding tissue. The breakdown of the red blood cells creates a hemosiderin deposit that gives off a brownish coloration. Brawny discoloration is usually seen in the area of the lower leg to ankle, known as the gaiter zone, secondary to incompetent perforators that connect the posterior arch vein to the deep system.

- Lipodermatosclerosis is thickening and hardening of the skin and can be found in patients with chronic venous insufficiency. The high venous pressure allowing red blood cells and fluid to leak into the tissue can lead to chronic inflammation. When the skin and the fat under the skin are inflamed for years, the tissue becomes hard. As shown in figure 28-1 (see also color plate 16), over time the tissue becomes depressed, changing the contour of the ankle area in a manner referred to as a "bottle-neck deficiency."

- Whiteness (*pallor*) may be seen in phlegmasia alba dolens. This limb-threatening condition results from arterial spasms that occur secondary to extensive, acute iliofemoral vein thrombosis. The limb is very swollen, pale, and painful.

- A bluish discoloration (*cyanosis*) may be seen in phlegmasia cerulea dolens, another limb-threatening complication of an acute iliofemoral vein

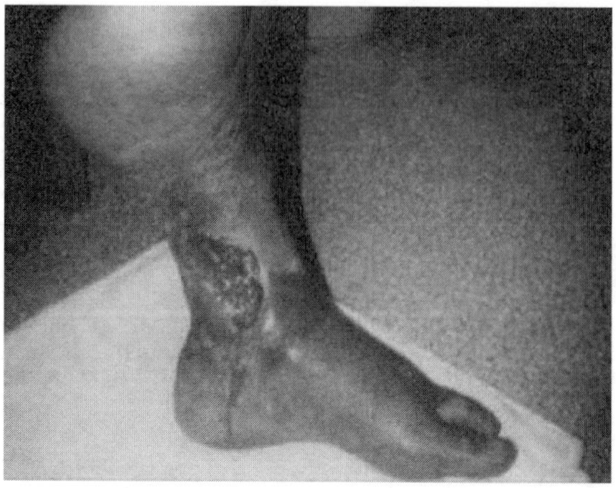

Figure 28-1. An example of lipodermatosclerosis in a patient with chronic venous insufficiency. The tissue becomes hard and over time can change the contour of the ankle area resulting in a "bottle-neck deficiency."

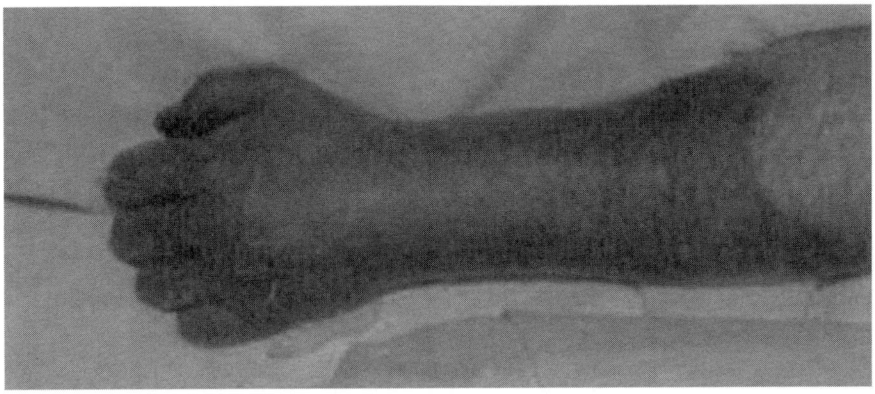

Figure 28-2. An example of phlegmasia cerulea dolens, a potentially limb-threatening complication of acute deep venous thrombosis. An unusual finding in the lower extremity; upper extremity involvement, as seen here, is rare and did result in amputation of the affected limb. See color plate 17.

thrombosis. The severely reduced venous outflow causes a marked reduction in arterial inflow. Tissue hypoxia can develop, leading to venous gangrene. The limb is very swollen, dark blue, and painful (figure 28-2).

3 Ulceration:

- Tissue breakdown, caused by a lack of oxygen and nutrients, usually occurs near the medial malleolus where Cockett's I, II, and III perforators are located.
- Incompetence of one or all of the three perforators that carry blood from the posterior arch vein into the deep venous system plays a major role in the development of venous stasis ulcers. Since these perforators are located near the medial malleolus, ulcerations are most frequently seen at that site.
- Although it may be difficult to discriminate venous stasis ulcers from ulcers secondary to arterial insufficiency, table 28-2 provides some general guidelines.

Table 28-2. Differentiating venous and arterial ulcers.

Characteristics	***Venous Ulcers***	***Arterial Ulcers***
Location	Near medial and lateral malleolus	Tibial area, toes, bony prominences
Pain	Mild to severe	Severe
Appearance	Shallow, irregular shape	Deep, regular shape
Bleeding	Venous ooze	Little
Other findings	Stasis changes: Brawny discoloration, lipodermatosclerosis, varicosities often present	Trophic changes: Shiny skin, loss of hair, thickened toenails

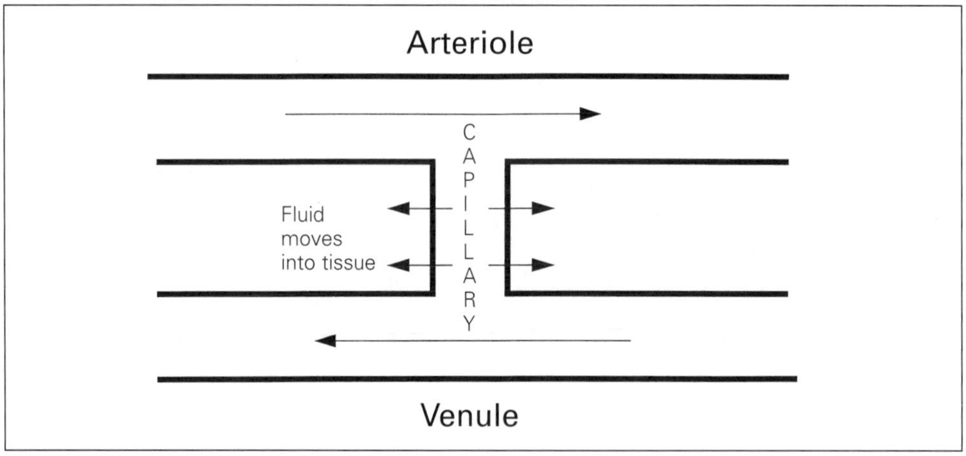

Figure 28-3. Increased capillary pressure increases movement of fluid into the interstitial spaces throughout the microcirculation.

4 Edema:

- *Edema* is a condition in which the body tissue contains excessive fluid. It is one of the most consistent signs of elevated peripheral venous pressure.

- Although the causes of edema are various, edema from venous disease occurs because an obstructive process such as deep venous thrombosis increases capillary pressure. This increase in pressure not only prevents the normal reabsorption of fluid that has moved into the interstitial spaces at the level of the arterioles, but also increases the movement of fluid into the interstitial spaces throughout the microcirculation (figure 28-3).

- *Pitting edema* is an accumulation of fluid in the subcutaneous tissue. When manual pressure is applied on the tissue, some fluid is displaced, causing a depression in the skin surface. Pitting edema may be secondary to fluid retention, an electrolyte imbalance, renal dysfunction, congestive heart failure or other causes of elevated venous pressure.

- In *nonpitting edema* the tissue is so engorged with fluid that it cannot be displaced with manual pressure. One cause of *lymphedema*, a nonpitting edema, is the result of an obstruction in the lymphatic system. Normally, the lymphatic system drains excess fluid from tissues. Fluid accumulates when lymph nodes and/or lymph vessels are removed or damaged, as is frequently seen after many types of cancer surgery.

Impedance Plethysmography

CHAPTER 29

1 Capabilities:

Impedance plethysmography (IPG) can detect thrombi in the iliac, femoral, and popliteal veins.

2 Limitations:

- False positive results may be caused by extrinsic compression (from tight clothing, tumors, ascites, pregnancy, obesity, improper positioning, pain, or anxiety and other factors).

  Improper patient positioning can cause mechanical compression of the veins.

  Pain or anxiety may cause muscle contraction that in turn causes venous compression.

  Pregnancy or obesity can cause compression of the vena cava.

  Peripheral arterial occlusive disease can decrease venous filling.

  Chronic obstructive pulmonary disease can elevate central venous pressure.

  Improper application of the electrodes may result in inaccurate readings.

  Venous capacitance often improves when the test is repeated, probably secondary to changes in compliance and/or a response to mild reactive hyperemia.

- False negative studies may result from chronic venous occlusion with development of large collaterals.
- IPG does not detect isolated calf thrombus.

3 Patient positioning:

- The patient should be supine in a hospital bed with the legs externally rotated.
- The hips and knees should be flexed.

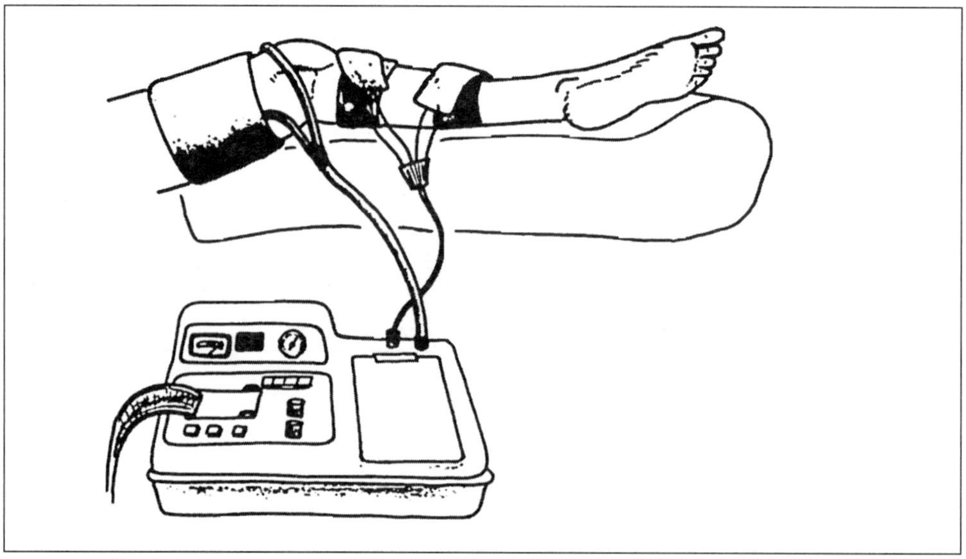

Figure 29-1. Lower limbs are positioned above the level of the heart to measure venous filling and to facilitate venous emptying. The knee is slightly flexed with the leg externally rotated. From Hershey FB, Barnes RW, Sumner DS (eds): *Noninvasive Diagnosis of Vascular Disease.* Pasadena, Appleton Davies, 1984.

- As shown in figure 29-1, the extremities are elevated so that the heels are above the calves and the calves are above the level of the heart.

4 Physical principles:

- Plethysmography is any technique that measures volume changes. In the peripheral vasculature, these volume changes are caused by changes in blood volume.

- Impedance is the hindrance to the passage of an alternating electrical current. In plethysmography, changes in resistance are the major cause of changes in impedance.

- Ohm's Law provides the basis of IPG: It expresses the relationship between voltage, current, and resistance in a circuit. The total current (I) is equal to the voltage (V) divided by the resistance (R):

$$I = V/R \textit{ or } R = V/I \textit{ or } V = R \times I$$

- An analogy to Ohm's Law is the garden hose. The water pressure in the hose is similar to voltage. Resistance occurs in the hose because of the friction of the water molecules against the inner wall of the hose and because of the adjustable nozzle at the end of the hose. If resistance is increased by closing down the nozzle, flow (or current) in the hose will decrease. If the pressure (or voltage) is increased and resistance is kept the same, flow (or current) will increase. Current (I), then, is like the volume of water flowing in the hose. Voltage (V) is similar to the pressure within the hose. Resistance (R) is comparable to the effects of friction, geometry, and directionality within the hose, including the effects of closing or opening the nozzle. These concepts are very similar to Poiseuille's equation, which describes the relationship among pressure, volume flow, and resistance (see the discussion of fluid dynamics in chapter 2):

Q (volume flow) = P (pressure)/R (resistance)

or

R = P/Q *or* P = R × Q

- Ohm's Law, then, is analogous to Poiseuille's equation in that current (I) is like volume flow (Q), voltage (V) is similar to pressure (P), and forms of resistance (R) are evident in both electrical and fluid circuits.

- Blood volume in the calf increases as the inflated thigh cuff prevents venous outflow. The change in the size of the limb is irrelevant; it's the change in blood volume that is important.

- Since blood is a good conductor of electricity, as blood volume increases in the limb, resistance to the flow of electrons through the limb decreases. If blood volume decreases in the limb, resistance increases. Changes in electrical impedance are the result of changes in blood volume, and a harmless current can be used to evaluate these alterations.

  When calculating for changes in total current, i.e., I = V/R, an increase in R (resistance) results in a decrease in I (current); a decrease in R (resistance) results in an increase in I (current).

  When calculating for changes in voltage caused by changes in resistance, i.e., V = I × R, an increase in R (resistance) results in an increase in V (voltage); a decrease in R (resistance) results in a decrease in V (voltage).

- Using DC *(direct current)* coupling, a record of the volume changes can be made on a strip-chart recorder (plethysmography). Increased volume/decreased resistance is displayed as a positive deflection and decreased volume/increased resistance as a negative deflection.

- *DC coupling* is an electrical voltage that is either positive or negative with current flowing in only one direction. These characteristics make DC coupling very useful in evaluating the slower-flow states of the venous system, including the blood content of the skin. Any battery, e.g., car or flashlight batteries, operates in DC mode. AC *(alternating current)* coupling, on the other hand, is an electrical voltage that reverses its polarity (positive or negative voltages) 60 times a second. Used in arterial studies, AC coupling requires more intense changes to produce a measurable signal. In the United States, standard household outlets deliver 120 volts of AC current.

Figure 29-2. Two-wire method of impedance plethysmography.

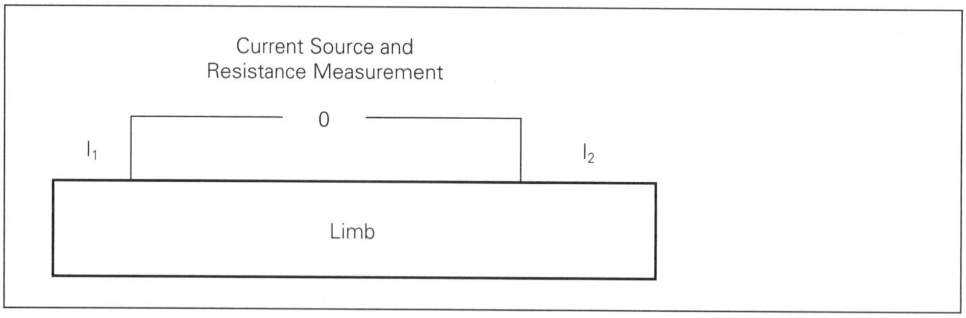

- One of two circuits can be used to measure resistance:

  The two-wire method passes a current from I_1 to I_2 and measures changes in resistance between them (figure 29-2). But since the connection of the electrode to the skin introduces an unpredictable amount of resistance itself and cannot be calibrated, the two-wire method is rarely used.

  The four-wire method, on the other hand, has the current passed from I_1 to I_2, but resistance is measured by using V_1 and V_2 (figure 29-3). By measuring the decrease in voltage between V_1 and V_2, changes in resistance are more accurately determined.

5 Technique:

- A thin bead of electrically conductive gel or paste is applied to the aluminum strips of the IPG electrodes if individually packaged; pre-gel pads are not used.

- The electrodes are carefully placed on the proximal and distal portions of the calf following its natural contour.

- The occluding pneumatic cuffs are placed on the thighs and inflated to 50–60 mmHg. Because venous outflow is obstructed, the veins fill to their maximum capacity. The cuffs are then rapidly deflated to record the outflow patterns.

Figure 29-3. Four-wire method of impedance plethysmography.

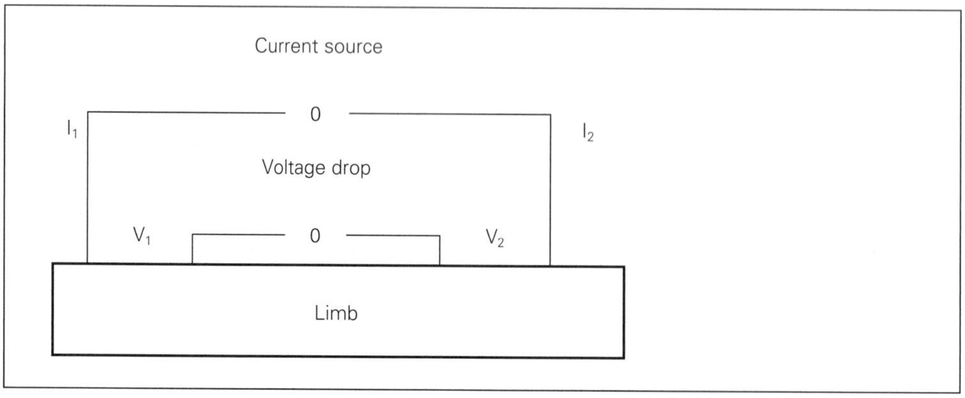

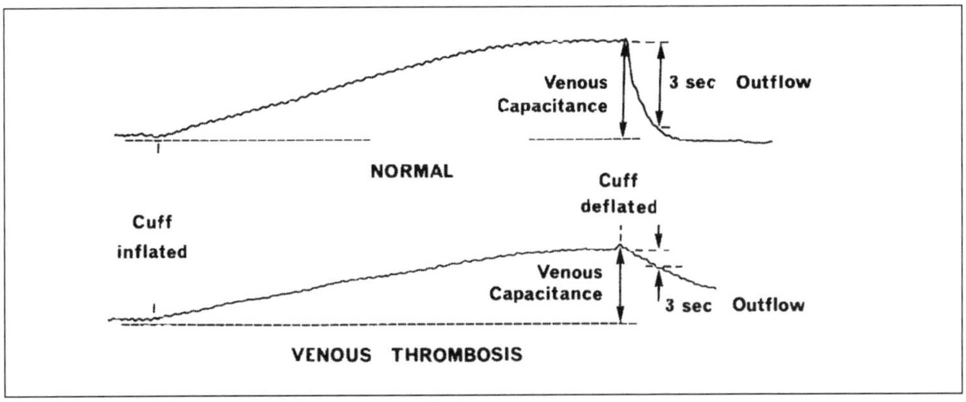

Figure 29-4.
Top The maximum filling capability of the veins (venous capacitance) is represented by the maximum rise on the tracing. Outflow refers to the amount of venous emptying that occurs during the 3-second period following thigh-cuff deflation. **Bottom** Venous thrombosis results in reduced venous capacitance and venous outflow. From Hershey FB, Barnes RW, Sumner DS (eds): *Noninvasive Diagnosis of Vascular Disease*. Pasadena, Appleton Davies, 1984.

- Calibration or standardization is completed by pressing the STD button, which removes a specific amount of air from the unit. The amplitudes of the tracings are adjusted to ensure that the sizes of the tracings are equal. In this way a significant diminution in the tracing can be interpreted as a significant reduction in blood volume. Calibration is critical to the usefulness of any plethysmographic technique because differences in the size of the tracing have diagnostic value.

- Abnormal results following the initial run require that the test be repeated. If the results remain abnormal, subsequent repetitions, with adjustments to cuff placement and patient position, are required before the study can be considered to be abnormal.

6 Interpretation—Quantitative:

- *Capacitance* is the maximum filling capability of the veins as venous outflow is momentarily halted by an occluding cuff. As illustrated in figure 29-4, venous capacitance is represented by the maximum rise of the tracing (as compared to the initial baseline) and is referred to as *maximum venous capacitance* (MVC).

- *Outflow* is the amount of venous emptying that occurs after deflation of the occluding thigh cuff. The tracing should fall to the baseline within 3 seconds of thigh cuff deflation (figure 29-4, top). The measured "fall" at 3 seconds is calculated and plotted on a scoring grid and is called *maximum venous outflow* (MVO).

- In cases of venous thrombosis there is reduced venous capacitance and venous outflow (figure 29-4, bottom).

- The absence or presence of venous obstruction is determined by where the outflow and capacitance values fall on the scoring grid. As

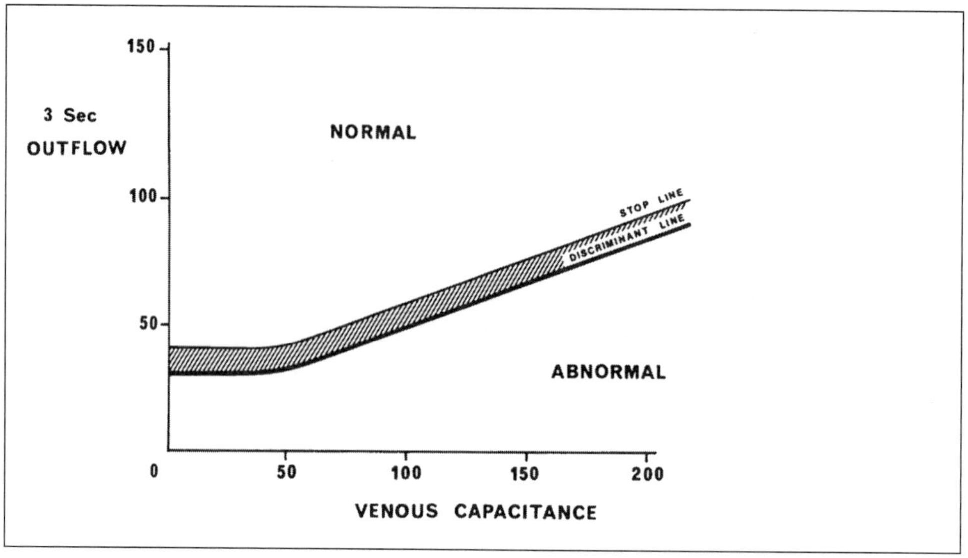

Figure 29-5. Scoring grid for IPG. The X-axis represents venous capacitance and the Y-axis venous outflow. The values plotted above the discriminating line are within normal limits, while those below the line are considered abnormal. From Hershey FB, Barnes RW, Sumner DS (eds): *Noninvasive Diagnosis of Vascular Disease.* Pasadena, Appleton Davies, 1984.

figure 29-5 illustrates, the Y-axis of the grid represents venous outflow with venous capacitance represented on the X-axis. Values plotted above the discriminating line are within normal limits, while those below the line are considered abnormal.

- Because venous capacitance is related to the initial volume of blood in the calf, several repeat runs are required to ensure accurate diagnosis of deep venous thrombosis when the first values fall below the discriminating line on the scoring grid.

Strain Gauge Plethysmography (SPG)

CHAPTER 30

1 Capabilities:

Strain gauge plethysmography (SPG) is used to detect venous obstruction in the large veins above the knee, e.g., femoral.

2 Limitations:

- The extreme sensitivity of SPG limits its usefulness in some clinical applications.
- Extrinsic compression can result from tight clothing, tumors, ascites, improper positioning, pain, anxiety, pregnancy, or obesity.

  Improper patient positioning can cause mechanical compression of the veins.

  Pain or anxiety can result in muscle contraction resulting in venous compression.

  Pregnancy or obesity can cause compression on the vena cava.
- Peripheral arterial occlusive disease can decrease venous filling.
- Chronic obstructive pulmonary disease (COPD) can elevate central venous pressure.

3 Patient positioning:

- The patient should be supine in a hospital bed.
- The legs should be externally rotated with the hips and knees flexed.
- The extremities should be elevated so that the heels are above the calves, and the calves are above the level of the heart (figure 30-1).

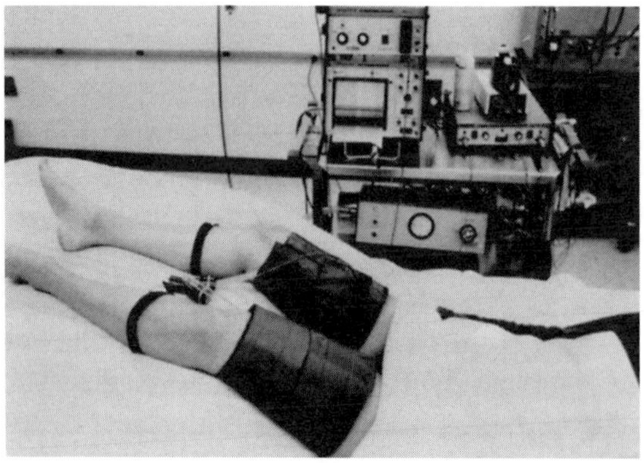

Figure 30-1. Proper patient positioning for SPG facilitates venous outflow. The legs should be elevated above the level of the heart. From Hershey FB, Barnes RW, Sumner DS (eds): *Noninvasive Diagnosis of Vascular Disease.* Pasadena, Appleton Davies, 1984.

4 Physical principles:

- Plethysmography is a technique that measures volume changes.
- Strain gauge plethysmography utilizes a mercury-in-Silastic strain gauge that indirectly senses changes in blood volume by measuring the circumference of the limb. A silicone rubber tube filled with mercury is wrapped around the limb with sufficient stretch to make good contact. Copper electrodes are at both ends of the gauge, and voltage is applied across the gauge.
- The limb expands as the occluding thigh cuff is inflated and venous outflow is reduced. As limb circumference increases, the tube stretches and the resistance across the gauge increases. Increased resistance reduces voltage across the gauge, which is reflected by a change in the tracing.
- Volumetric changes are recorded on a strip-chart recorder, with increased limb volume/increased resistance across the gauge displayed as a positive deflection and decreased volume/decreased resistance across the gauge displayed as a negative deflection. Compared to impedance plethysmography, where the size of the limb is irrelevant and where blood volume within the calf veins cause a change in resistance, strain gauge plethysmography depends on an overall increase in the size of the limb. In other words, impedance plethysmography measures changes in resistance through the calf while strain gauge plethysmography measures changes in the circumference of the calf.
- As the inflated thigh cuff prevents venous outflow, blood in the calf increases and allows maximum venous capacitance to be determined.

- With the quick release of the occluding cuff, strain gauge plethysmography can monitor volumetric changes associated with venous outflow.

5 Technique:

- A pneumatic cuff is placed around the thigh and the mercury-in-Silastic gauge is wrapped around the widest part of the calf.
- The length of the unstretched gauge should be about 90% of the circumference of the limb.
- Calibration or standardization is completed. As noted above, calibration is critical to the usefulness of any plethysmographic technique since differences in the size of the tracing have diagnostic value.
- Inflate the thigh cuff to approximately 50 mmHg. As the calf begins to expand in response to the obstruction of venous outflow, the gauge is stretched and the voltage changes. The voltage change is interpreted as a positive deflection on the recorder and considered to represent maximum venous capacitance.
- As the volume increase stabilizes, usually within 45 seconds, the occluding cuff is quickly deflated. The rate at which the calf empties is also recorded and is considered maximum venous outflow.

6 Interpretation—Quantitative:

- As figure 30-2 (top) demonstrates, venous capacitance is represented by the maximum rise of the tracing (as compared to the initial baseline).
- Outflow is the amount of venous emptying that occurs after deflation of the occluding thigh cuff. The tracing should fall to the baseline within 3 seconds of thigh cuff deflation. The measured "fall" at 3 seconds is calculated and plotted on a scoring grid.
- With venous thrombosis, both venous capacitance and venous outflow are reduced (figure 30-2, bottom).

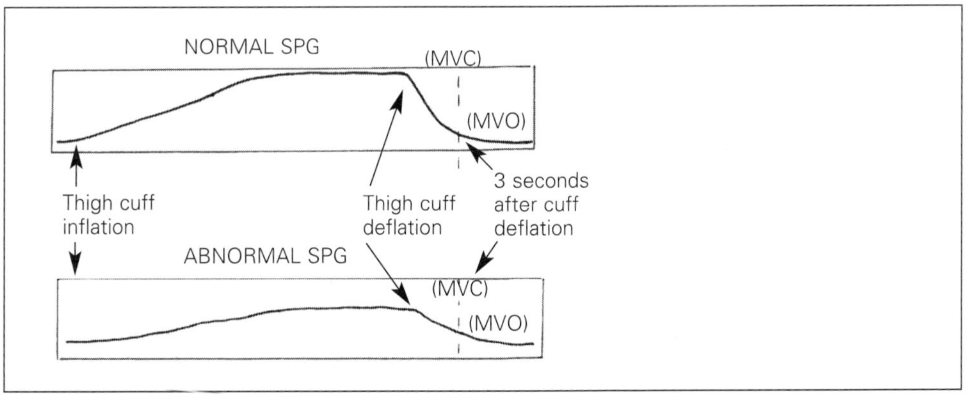

Figure 30-2. Normal (top) and abnormal SPG tracings.

- As with impedance plethysmography, the absence or presence of venous obstruction is determined by where the outflow and capacitance values fall on the scoring grid (see figure 29-5 on page 294). The values plotted above the discriminating line are within normal limits, while those below the line are considered abnormal.

- Because venous capacitance is related to the initial volume of blood in the calf, several repeat runs are required to ensure accurate diagnosis of deep venous thrombosis when the first values fall below the discriminating line on the scoring grid.

Photoplethysmography (PPG)

CHAPTER 31

1 Capabilities:

Photoplethysmography, which documents capillary blood volume, evaluates the presence and severity of venous insufficiency.

2 Limitations:

- Photoplethysmography is contraindicated in a patient with acute deep venous thrombosis.
- Improper placement of the PPG sensor (over a varicose vein, for example) can result in inaccurate information.
- Thickening of the skin may prevent adequate penetration of the infrared light.
- The PPG sensor must be placed on intact skin.

3 Patient positioning:

The patient is seated with legs dangling, i.e., non–weight bearing.

4 Physical principles:

- Although photoplethysmography is not true plethysmography because it does not measure volume changes, it can determine changes in the blood content of the skin (the microcirculation), which reflects intravenous flow. In addition, tiny arterial pulsations are usually evident superimposed on the tracing.
- The PPG photocell, which consists of a light-emitting diode and a photosensor, transmits light into the subcutaneous tissues that is reflected back to the photosensor. The light is not absorbed.
- Since blood attenuates light in proportion to its content in the tissue, the difference between the transmitted and reflected signal can be amplified and converted into a waveform.
- DC (direct current) coupling is used. DC coupling is an electrical voltage that is either positive or negative with the current flowing in only one direction. Because it detects slower changes in blood content, it is used

for venous studies. Any battery, e.g., car or flashlight battery, operates in DC mode. AC (alternating current) coupling, on the other hand, is an electrical voltage that reverses polarity (positive or negative) 60 times a second. The current flows in both directions and requires more intense changes to produce a measurable signal. For this reason it is used for arterial studies. In the United States, standard household outlets deliver 120 volts of AC current.

- Although photoplethysmography cannot be calibrated volumetrically as some other types of plethysmography (e.g., air plethysmography), it is important to maintain the same "size" or "gain" setting throughout the

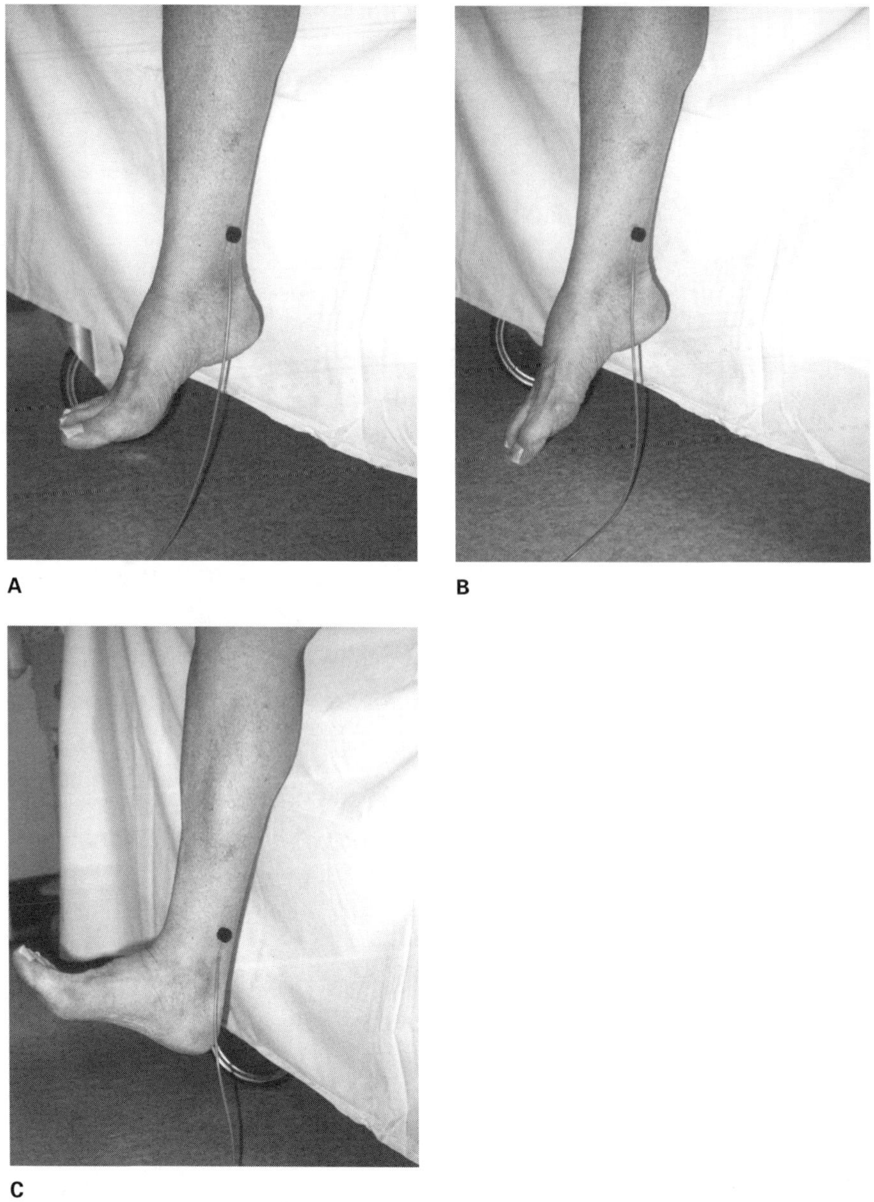

Figure 31-1. Venous refill (or "reactive") time (VRT) to help determine competency of the valves. The leg is non–weight bearing. The PPG sensor is placed above the medial malleolus on intact skin. **A** The initial foot position is relaxed, as in this example. **B** Plantar flexion. **C** Dorsiflexion. The changes in foot position activate the muscle-pumping action of the calf, which in turn empties the venous blood from the lower leg. The patient, who is told and shown how to move his foot, performs five sets of the plantar flexion/dorsiflexion maneuvers.

entire study to ensure that a significant difference in the tracing can be reliably interpreted as a significant difference in blood volume.

5 Technique:

- The PPG sensor is applied to the patient's lower leg, approximately 5–10 cm above the medial malleolus. Care must be taken so that the sensor is not over a varicosity.

- The patient is instructed to complete a series of exaggerated dorsiflexions to empty the veins (figure 31-1).

- If the patient cannot perform adequate dorsiflexions (e.g., the PPG tracing may not display the expected stepwise, negative deflection), the examiner can manually compress the calf. If manual compressions are necessary, however, they must be completed on both extremities to ensure consistency (figure 31-2).

- The PPG tracing continues to be recorded after the dorsiflexions to determine and document the venous refill (or "venous reactive") time (VRT).

- If the VRT is greater than 20 seconds, the study is considered to be normal. If the VRT is less than 20 seconds, a tourniquet is applied just above the knee to eliminate the influence of the superficial venous system. If a blood pressure cuff is used instead of a tourniquet, the examiner inflates it to 50 mmHg. The dorsiflexions are repeated to determine the venous refill time. Please refer to the manufacturer's interpretation guidelines for the laboratory's particular type of equipment since "normal" may vary from 18–25 seconds.

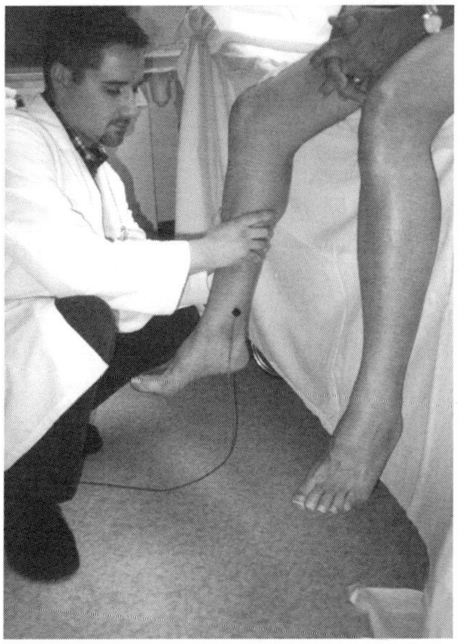

Figure 31-2.
If a patient is unable to flex the foot well enough to provide a good calf-pumping exercise, the technologist can manually compress the calf, being careful to show the patient what will occur. Manual compressions of the calf are performed approximately five times.

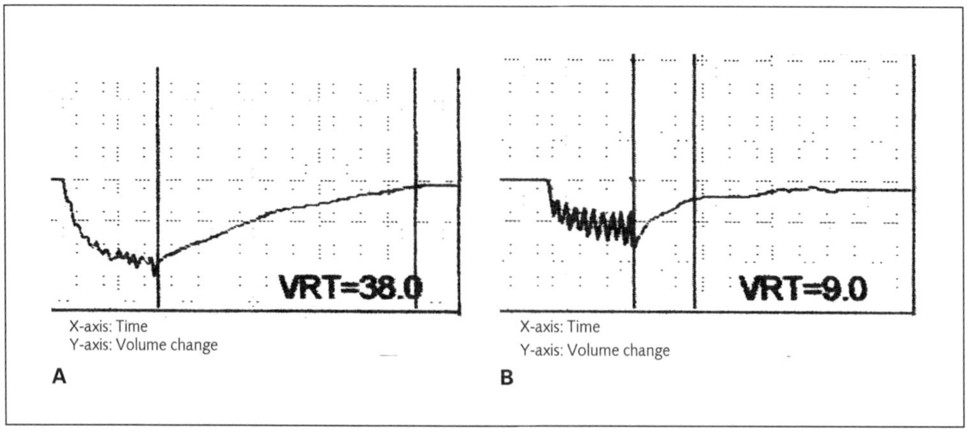

Figure 31-3. A normal PPG (**A**) documents slower venous filling, whereas the abnormal tracing (**B**) shows faster filling secondary to incompetent valves which produce retrograde flow.

- If the VRT is now more than 20 seconds, the examiner may reposition the tourniquet just below the knee and then instruct the patient to repeat the dorsiflexions.

6 Interpretation—Quantitative:

- A VRT of ≥ 20 seconds without a tourniquet is within normal limits. As illustrated in figure 31-3, venous filling occurs more slowly if the blood is traveling via the normal artery-arteriole-capillary-venule-vein route. Very fast refilling indicates that the vessel is filling via retrograde venous flow.
- A VRT of < 20 seconds without the tourniquet that normalizes to > 20 seconds with the tourniquet above the knee is consistent with reflux in the great saphenous vein.
- A VRT of < 20 seconds without the tourniquet that normalizes to > 20 seconds only when a tourniquet is applied below the knee is consistent with reflux in the small saphenous vein, although incompetence in the small saphenous vein is rarely observed.
- A VRT of < 20 seconds with and without tourniquet application is consistent with insufficiency of the deep system.
- Troubleshooting:

  Artifact from patient movement may result in a technically unsatisfactory study.

  Absent deflections or grossly irregular tracings require confirmation that the equipment is operating in DC rather than AC mode and/or in the PPG setting rather than the Doppler, VPR, or other setting, depending on the equipment.

  Deflections that are off the scale or barely discernible require adjustments to the amplifier gain/sensitivity/size setting.

CHAPTER 32

Air Plethysmography

1 Capabilities:

- Air plethysmography (APG) can determine the presence or absence of venous insufficiency
- APG can quantify venous reflux.

2 Limitations:

- Air plethysmography cannot be performed if the patient is not able to maintain positions or perform exercise.
- The presence of a cast, traction, or heavy nonremovable bandages prevents the study from being completed.
- Air plethysmography cannot be used to diagnose incompetent perforators or isolated incompetent distal veins.

3 Patient positioning:

The study is started with the patient supine. The patient then assumes a variety of positions (described below) and performs exercises to activate the calf-muscle pump.

4 Physical principles:

- Air plethysmography is true plethysmography because it measures volume changes in the extremity.
- The pneumatic cuff wrapped around the limb is connected to a pressure transducer that detects volume changes. These changes are amplified and converted to analog display.
- The technique documents volume changes secondary to position changes and exercise.

5 Technique:

- With the patient supine, a large cuff is applied to the lower leg (below knee to ankle) and inflated to about 6 mmHg. The test leg is somewhat elevated with the foot positioned on a hard cushion measuring 12–15 inches in height. Manual calibration is completed.
- The leg is passively elevated to empty the venous system, and the zero venous volume is documented.
- The patient stands quickly, bearing weight on the contralateral (nontest) leg. The increased venous volume is documented on a strip chart recorder as the test leg fills up with venous blood due to gravitational pressure (now increased because the patient is standing).
- The patient then stands with weight distributed equally over both feet and performs 1 tip-toe exercise to activate the calf muscle pump. A reduction in venous volume and venous pressure should normally be evident.
- 10 tip-toe maneuvers are then completed.
- The resultant decrease in calf venous volume is documented and used to calculate the ejection volume (EV) and the venous filling time (VFT).
- The patient quickly resumes the supine position and the leg being tested is elevated to empty the veins.
- If the findings are abnormal, the study is repeated after a tourniquet is applied to eliminate the influence of the superficial system.

6 Interpretation—Quantitative:

- The venous filling index (VFI) documents the rate of venous refilling. This is based on refilling of the test leg after the patient quickly stands from the supine position. (While the patient is still supine, the technologist first elevates the test leg to empty the veins.) The VFI is calculated using venous volume (VV) and venous filling time (VFT):

$$VFI = 90\%VV / VFT\ 90$$

A normal VFI is ≤ 2.0, minor-to-moderate reflux is > 2.0–10.0, and severe reflux is > 10.0.

- The ejection fraction (EF) measures the calf muscle pump function after one toe-up exercise by quantifying how much venous blood is left in the leg after one contraction of the calf muscle pump. The EF is calculated using the ejection volume (EV) and functional venous volume (VV):

$$EF = EV / VV \times 100$$

An ejection fraction greater than 60% is considered normal.

● The residual volume fraction (RVF) is calculated as the percentage of venous volume remaining after 10 toe-up movements and is equivalent to the ambulatory venous pressure in mmHg.

$$RVF = RV / VV$$

A residual volume fraction of less than 35% is considered normal.

● Figure 32-1 illustrates the changes in venous volume related to various positioning and exercise.

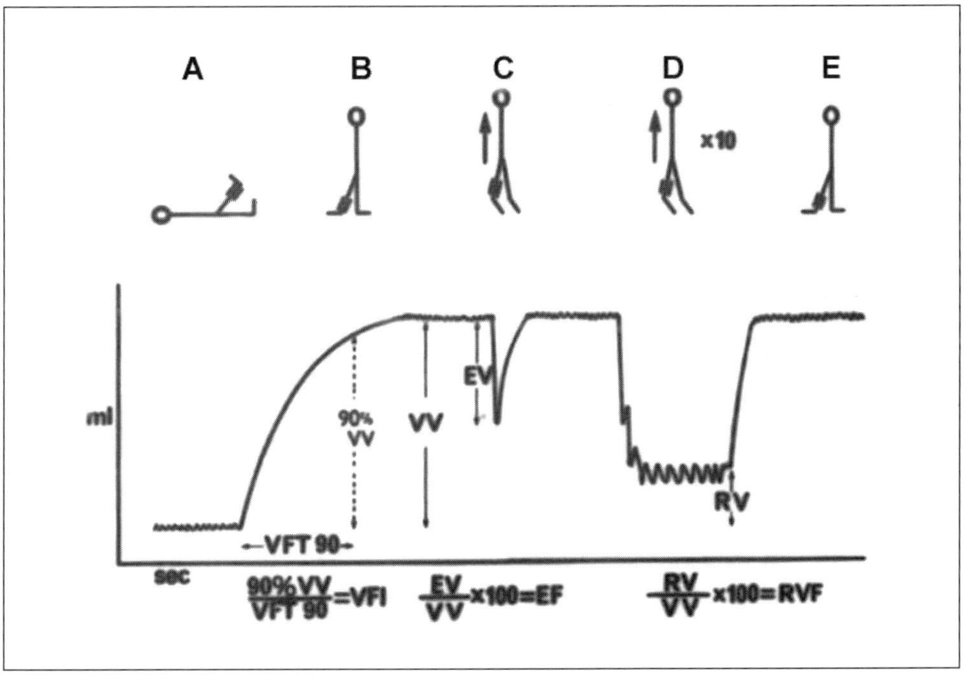

Figure 32-1. Air plethysmography measures volume changes. As illustrated, expected alterations in venous volume should occur as the patient changes his or her position and completes simple exercises. After position E, the patient quickly resumes a supine position, and the test leg is elevated to empty the veins. These alterations in volume are used to interpret the presence of venous insufficiency.

Continuous Wave Doppler

CHAPTER 33

1 Capabilities:

- A venous Doppler study can evaluate the deep venous system for obstruction.
- Venous incompetence can also be evaluated.

2 Limitations:

- It may be difficult to differentiate whether abnormal flow patterns are the result of deep venous obstruction or extrinsic compression.
- The paired deep veins in the calf make the diagnosis of isolated calf-vein thrombosis extremely difficult.
- Sources of a *false-positive* study (noninvasive test is positive for disease, but it is wrong) include:

  Extrinsic compression, e.g., tight clothing, tumors, ascites, pregnancy, obesity, improper patient positioning, or pain causing muscle contraction.

  Peripheral arterial disease (PAD), which can cause decreased venous filling.

  Chronic obstructive pulmonary disease (COPD), which can result in elevated central venous pressure, thereby altering pressure gradients and reducing venous flow patterns.

  Improper Doppler angle or probe pressure, which can obliterate venous flow and/or result in accelerated flow that may be misinterpreted as flow in a diseased vessel.

- Sources of a *false-negative* study (noninvasive test is negative for disease, but it is wrong) include:

  Partial thrombosis

  Collateral development

  Presence of bifid system (multiple deep veins)

- The continuous wave Doppler study requires a very experienced technologist because the information obtained is more subjective than that obtained by other modalities, such as venous duplex ultrasonography.

3 Patient positioning:

- The patient should be supine, with the body slightly shifted to the side being evaluated. The leg should be slightly externally rotated with the hip and knee flexed slightly.

- The extremities should be lower than the level of the heart to facilitate venous filling (e.g., with the patient in a reversed Trendelenburg position of approximately 30 degrees).

- If abnormal venous signals are obtained, repositioning and reevaluation are essential before a reliable conclusion can be reached, e.g., an extreme side-lying position may diminish extrinsic compression on IVC from pregnancy, ascites, or tumor formation.

4 Physical principles:

- Continuous wave (CW) Doppler uses two piezoelectric crystals, one to continuously emit ultrasound and the other to continuously receive the reflected waves. This results in a fixed sample size and no range resolution or ability to place the sample volume at a specific depth. Continuous wave ultrasound also does not provide the ability to create an anatomic image.

- To maximize the quality of the Doppler venous signal, a 5 MHz probe should be held at approximately 45–60 degrees to the skin surface.

- The deep veins are found adjacent to the corresponding artery. Therefore, correct vessel identification depends on hearing the accompanying arterial signal.

- Although venous flow is synchronous with respiration, pulsatile lower extremity venous flow may be evident in patients with fluid overload (e.g., overhydration or congestive heart failure). See also page 312.

- The examination is based on the evaluation of auditory venous signals in the resting limb and in the limb undergoing proximal and distal compression maneuvers.

5 Technique:

- The examiner begins with the asymptomatic side, placing the probe at the inguinal ligament, identifying the pulsatile common femoral artery, and then moving the probe medially in order to insonate the common femoral vein.

- The signal should be spontaneous, indicating a patent vessel, phasic with respirations, and augmented with distal compression and proximal release. The Valsalva maneuver is an effective method of applying proximal compression to the common femoral vein: the patient takes in a deep breath, holds it, bears down as if having a bowel movement, and then releases it. **Please note:** This maneuver should not be performed if the patient has a history of arrhythmias because it could precipitate cardiac irregularities. If the patient is unable to perform the Valsalva maneuver, the technologist can place his or her hand just above the inguinal ligament and press down to apply a manual proximal compression.

- The examiner then evaluates the common femoral vein on the symptomatic side for the same flow patterns and responses to compression maneuvers.

- The femoral, popliteal, and posterior tibial veins are evaluated as previously described.

6 Interpretation—Qualitative:

- *Spontaneity:* The venous signal must be clearly heard at all sites with the exception of the tibial veins and great saphenous vein of the lower extremity and the radial and ulnar veins of the upper extremity, which normally have no spontaneous Doppler signals. In addition, the patient who is cold will have veins that are more collapsed, reducing venous flow. Signals that are evident only following distal compression are abnormal and consistent with venous obstruction.

- *Patency:* Spontaneous flow or flow following distal augmentation in the tibial, radial, and ulnar veins provides documentation that the vessel is patent, i.e. open. Figure 33-1 illustrates one of the steps to document patency of the posterior tibial vein.

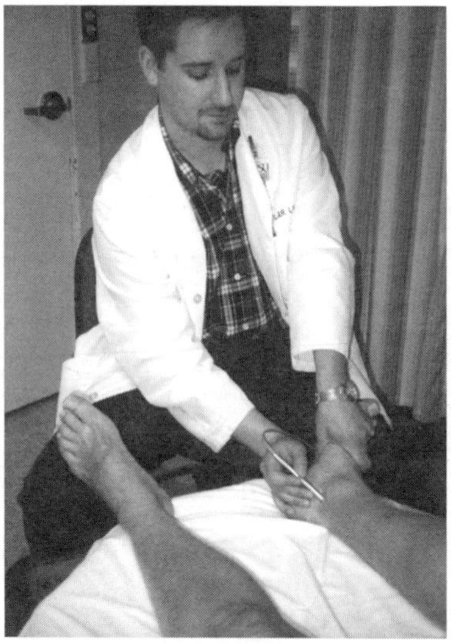

Figure 33-1. In this example, distal compression of the foot augments the venous signals. Upon release of distal compression, the venous Doppler flow signals should not show any further augmentation or directional change.

- *Respiratory patterns/phasicity:* As demonstrated in figure 33-2, the lower extremity venous signal is normally phasic with respiration, increasing with expiration and decreasing with inspiration. Flow patterns of the upper extremity are the opposite: decreasing with expiration and increasing with inspiration. When the patient is positioned so that the head is lower than the heart, i.e., slight Trendelenburg position, venous flow of the lower extremity is not related as much to respiration but rather is more cardiac-dependent. Continuous flow patterns of veins of the upper or lower extremity are usually consistent with proximal venous obstruction, but may be normally evident in patients with shallow respirations (figure 33-3).

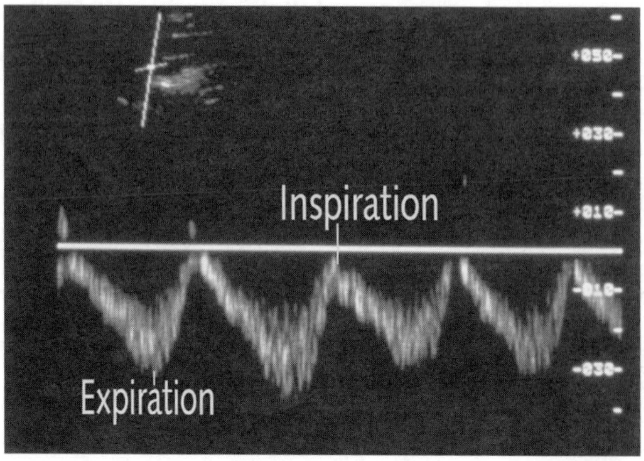

Figure 33-2. Normal phasicity of a lower extremity venous signal.

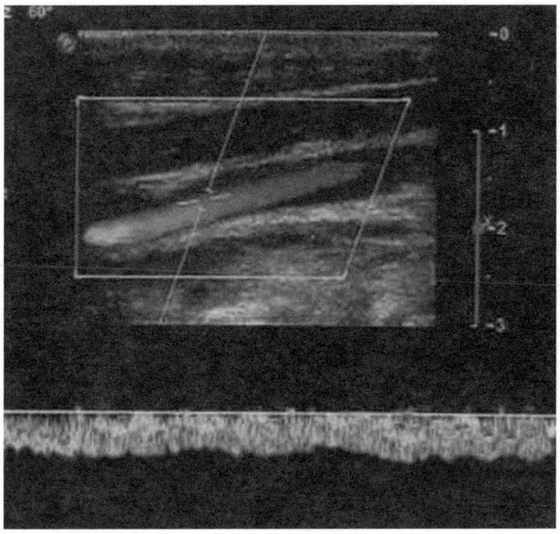

Figure 33-3.
In this example, the continuous flow patterns are consistent with a proximal venous obstruction.

- *Augmentation with distal compression and proximal release:* As shown in figure 33-1, manual compression is applied distal to the transducer. Since distal compression should augment (increase) the venous signal as illustrated in figure 33-4, the absence of augmentation during distal compression is consistent with obstruction.

- *Results on proximal compression:* The venous signal should not change. With maximum proximal compression, venous flow should be halted. The effect of proximal compression on venous flow is similar to that of the Valsalva maneuver (see pages 278–279). Augmentation during proximal compression is indicative of valvular incompetence, signifying *venous reflux* (i.e., retrograde blood flow).

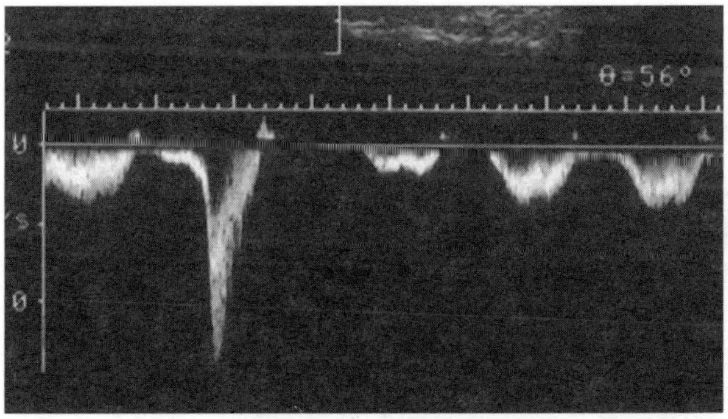

Figure 33-4.
Normal augmentation with distal compression.

Figure 33-5. Longitudinal view of the subclavian vein. Flow patterns reflect phasicity as well as pulsatility due to the vein's close proximity to the heart.

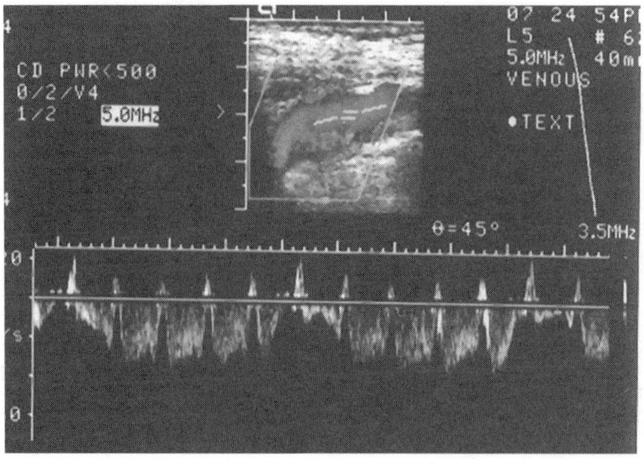

- *Valsalva maneuver:* Normally, venous flow is augmented *following* this maneuver; decreased augmentation suggests obstruction. Augmentation *during* the maneuver signifies reflux (retrograde flow) and is consistent with valvular incompetence.

- *Extrinsic compression:* Pressure on the vessels from surrounding tissues and/or structures (e.g., tumors, pregnancy, ascites, etc.) can alter normal flow patterns.

- *Venous pulsatility:* Pulsatile venous flow is commonly—and normally—heard in the subclavian vein because of its close proximity to the heart (figure 33-5). In contrast, pulsatile venous flow of the lower extremities is not normal. Rather, it is evident in cases of fluid overload (e.g., from overhydration) or increased venous pressure (e.g., from congestive heart failure) (figure 33-6). Pulsatile flow may also be the result of tricuspid regurgitation/insufficiency.

Figure 33-6. Pulsatile venous flow in the peripheral veins is suggestive of fluid overload, most commonly from congestive heart failure.

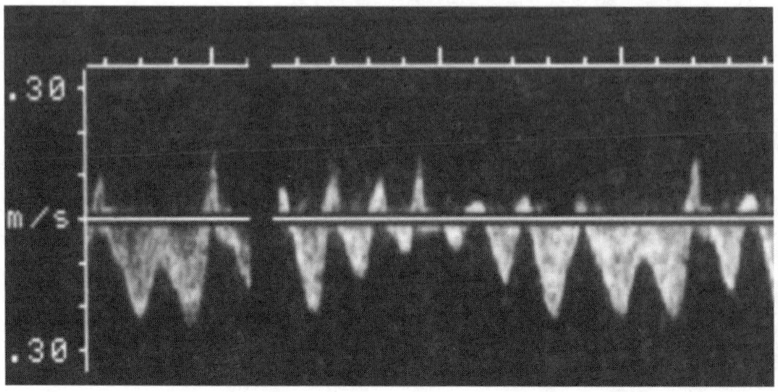

CHAPTER 34

Duplex Scanning and Color Flow Imaging

1 Capabilities:

Peripheral Veins

- Venous imaging can rule out or identify venous thrombosis in the veins of the arms or legs.
- It can differentiate an acute from a subacute process.
- It can be used to evaluate nonocclusive thrombosis.
- It is more accurate than other noninvasive techniques in diagnosing calf lesions.
- It is better able to distinguish between extrinsic compression and intrinsic obstruction.
- It can document the presence of nonvascular soft tissue masses (e.g., Baker's cysts).
- It can diagnose venous incompetence by augmentation of the Doppler signal with proximal compression or by directly imaging incompetent valves. Color flow imaging reveals visible flow reversal during proximal compression.
- It can reveal recanalized channels or evidence of collateralization.

Abdominal and Pelvic Veins

- Venous imaging can help to document the presence of elevated systemic venous pressure by visualizing the abdominal veins.
- It can rule out or reveal venous thrombosis.
- It can evaluate the patency of inferior vena cava interruption devices.
- It is sometimes capable of assessing portocaval shunts.
- It can help to evaluate some liver diseases by assessing the portal venous system.

2 Limitations:

- *Peripheral veins of the lower extremity:* It may be difficult to thoroughly evaluate all of the infrapopliteal veins secondary to edema, scarring, recent surgery, or obesity. (Changing to a lower frequency transducer may help to improve visualization in the obese patient.) In addition, visualization may be difficult because of vessel size, depth, and course.
- *Peripheral veins of the upper extremity:* It may be difficult to thoroughly evaluate the subclavian and brachiocephalic/innominate veins because of the bony structures of the chest.
- *Abdominal and pelvic veins:* It may be difficult to thoroughly evaluate all of the abdominal veins because of the depth of vessels and the presence of bowel gas.
- Sources of false-positive studies (noninvasive test is positive for disease, but it is wrong) include:

  Extrinsic compression from a tumor, ascites, or pregnancy.

  Peripheral arterial disease (PAD), which results in decreased venous filling leading to decreased venous outflow.

  Chronic obstructive pulmonary disease (COPD), which results in increased central venous pressure.

  Improper Doppler angle.

  Increased probe pressure, which can alter venous flow velocities and decrease phasicity and/or spontaneity.

3 Patient positioning:

- *Peripheral veins of the lower extremity:* To facilitate venous filling, the patient is placed in a reversed Trendelenburg position and turned toward the symptomatic side with the hip externally rotated and the hip and knee flexed slightly.
- *Peripheral veins of the upper extremity:* The patient is in a supine or low Fowler's position.
- *Abdominal and pelvic veins:* The patient may be placed in one or more of the following positions to optimize visualization:

  Supine with head slightly elevated

  Left and/or right lateral decubitus with head of bed elevated slightly

  Reversed Trendelenburg position

4 Physical principles:

In addition to the previously described physical principles of Doppler and B-mode imaging (see pages 224–226), the following principles and issues are of particular concern to venous imaging studies:

- The vein must be properly identified by appropriate landmark(s), such as an accompanying artery or adjacent bone.
- The peripheral vein must be clearly visualized in transverse view to ensure complete compressibility.
- Velocity signals should be obtained in sagittal view to maximize the Doppler shift.
- If color flow imaging is used, adjustments recommended by the manufacturer must be made to maximize color filling and flow patterns. Such adjustments may include modifying the color scale to detect the slower velocities of venous flow, changing wall filters, and increasing color gains.
- A combination of transverse and sagittal views are necessary to ensure the accuracy of diagnostic information.

5 Technique:

Peripheral Veins of the Lower Extremity—Deep Venous Thrombosis (DVT)

- The examiner begins imaging at the inguinal ligament in the groin by identifying the common femoral vein and artery (figure 34-1A). The common femoral vein should be visibly free of thrombus, that is, the vessel should be *anechoic*. Scanning should begin in the transverse orientation to assess the complete compressibility of the vein walls. *Coaptation*, a word derived from Latin meaning "to fit together," is another term for compressibility. With color flow imaging, sagittal scanning may be performed first to observe the characteristics of blood flow, followed by compression maneuvers in transverse view to rule out or confirm intraluminal thrombus.
- With the probe positioned for a sagittal view, the Doppler signals are evaluated for spontaneity, phasicity, and augmentation with distal compression and augmentation with proximal release.
- The saphenofemoral junction—where the great saphenous joins the common femoral vein (figure 34-1B)—must be carefully evaluated because thrombosis in the superficial system at or near the deep system may require more aggressive treatment than isolated thrombosis in the superficial system alone.

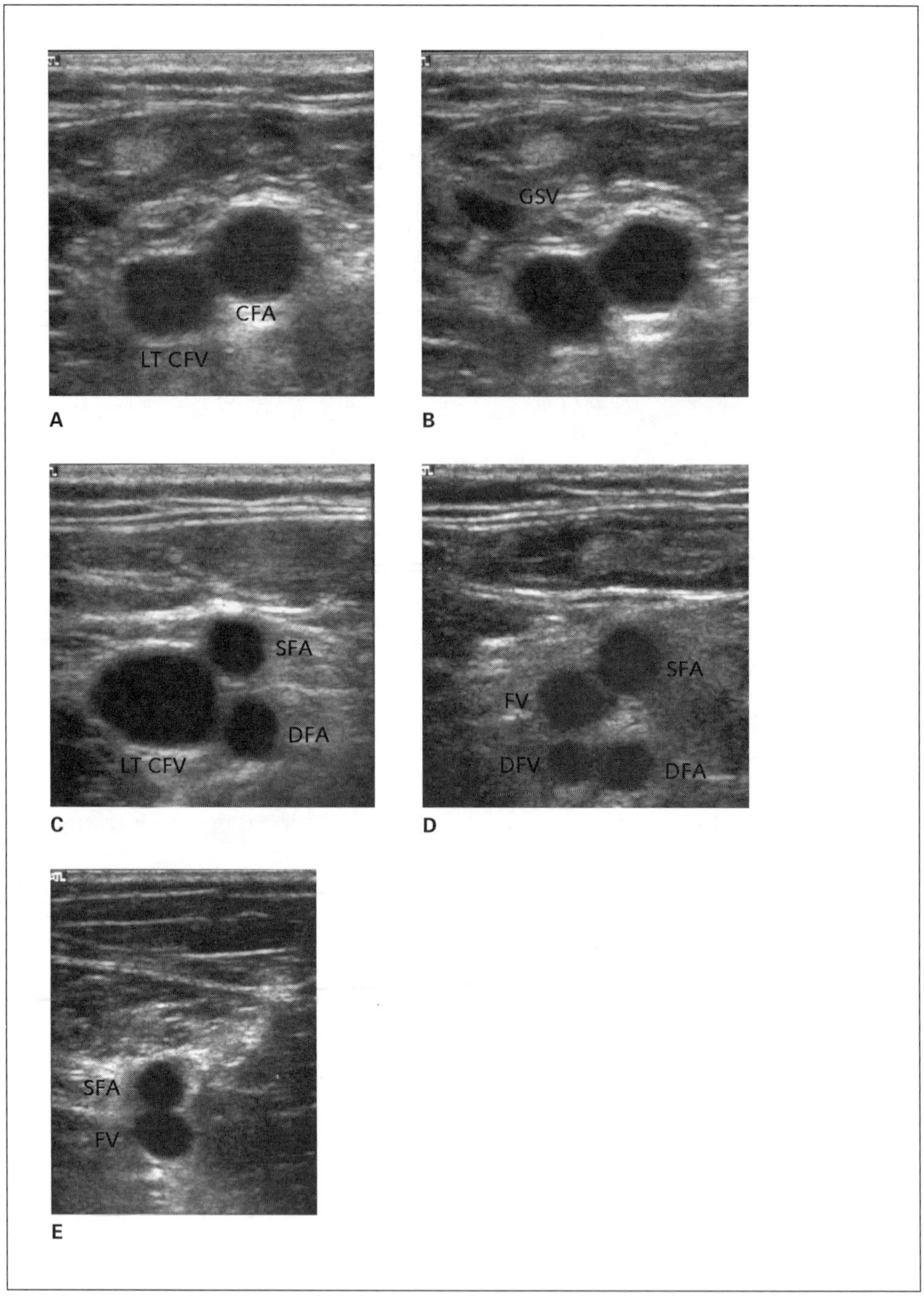

Figure 34-1. Transverse view of the common femoral vein (CFV) through the saphenofemoral junction into the distal femoral vein. **A** Common femoral vein and artery at the groin level. **B** Near saphenofemoral junction. GSV = great saphenous vein. **C** Common femoral vein and bifurcation of common femoral artery into superficial femoral (SFA) and deep femoral (DFA) arteries. **D** Femoral and deep femoral vessels. **E** Femoral vein (FV) and superficial femoral artery in distal thigh.

- As illustrated in figures 34-1C and D, the femoral vein is evident as the probe is slowly moved down the medial aspect of the thigh. Vessel wall compressibility must be evaluated in the transverse view every few centimeters. Doppler venous signals are obtained in the sagittal view.

- In the distal third of the thigh, the femoral vein dives deep (figure 34-1E). As it passes through the adductor hiatus, imaging may become

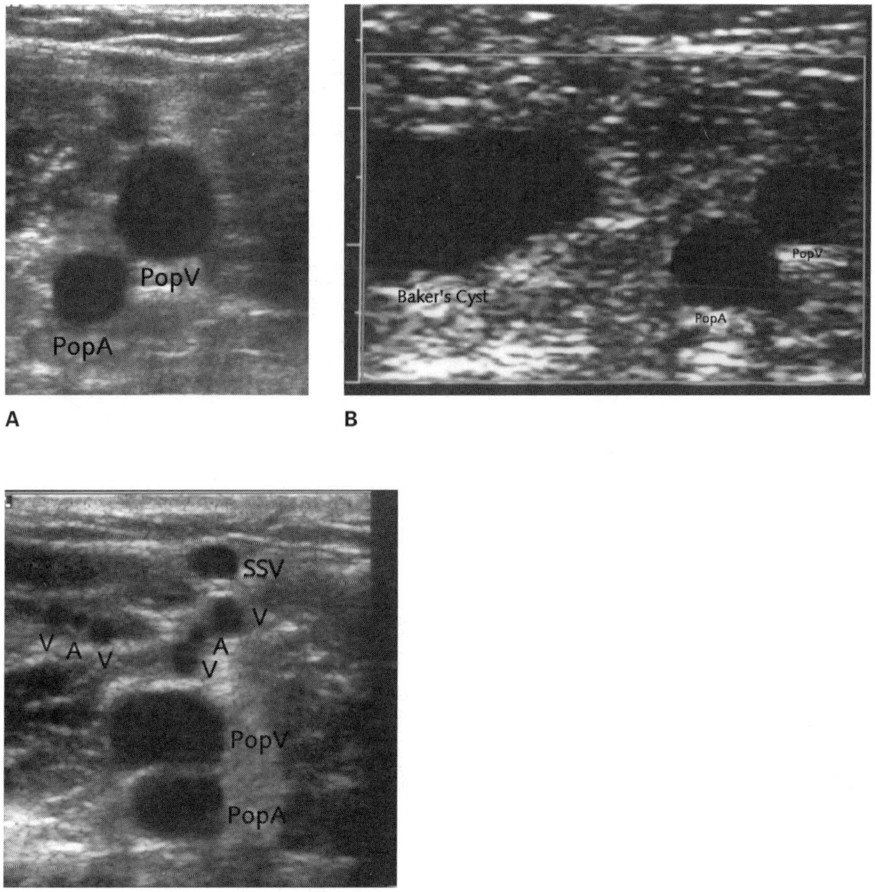

Figure 34-2. Transverse view of the popliteal fossa. **A** Popliteal vein (PopV) and artery (PopA). **B** Baker's cyst in medial popliteal fossa. **C** Small saphenous vein (SSV) and the paired gastrocnemius veins (V) with their accompanying artery (A).

more difficult. Although the anterior approach may still be useful, moving the probe to the posterior aspect of the distal thigh near the popliteal fossa may improve visualization.

- As illustrated in figure 34-2A, the probe is next moved slowly throughout the popliteal fossa to evaluate the popliteal vein. It is also necessary to observe for any cystic structures or masses, such as Baker's cyst (figure 34-2B). Moving the probe more distally, the examiner evaluates the gastrocnemius veins or muscular branches, small saphenous, and trifurcation region (figure 34-2C). The anterior tibial veins are usually difficult to visualize in this approach: An anterolateral approach is necessary.

- The small saphenous–popliteal junction must be carefully evaluated because, again, thrombosis of the superficial system at or near the deep system may require more aggressive treatment than isolated thrombosis in the superficial system alone.

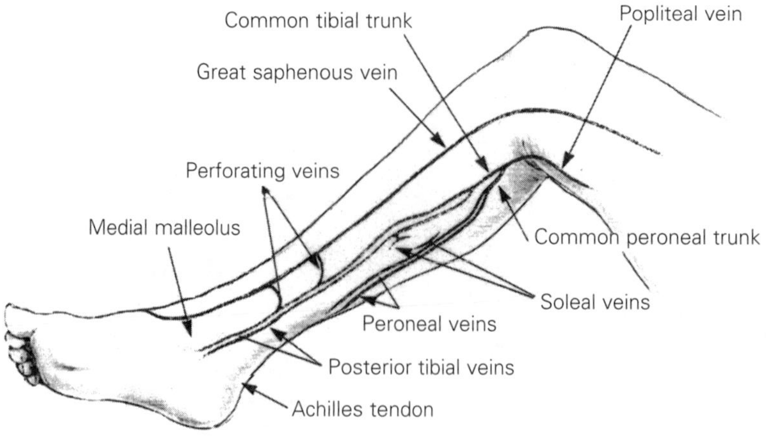

Figure 34-3. Lower extremity veins: paired posterior tibial and peroneal veins, great saphenous vein with perforators into the deep system, soleal veins off the peroneal and posterior tibial vessels, and the popliteal vein. From Talbot SR, Oliver MA: *Techniques of Venous Imaging.* Pasadena, Appleton Davies, 1992.

- Since the next step involves evaluating the calf vessels, see figure 34-3 for a review of the anatomy of the lower leg.
- The examiner moves the probe to the space between the medial malleolus and the Achilles tendon to locate the posterior tibial veins and artery (figure 34-4A). They are followed up the medial surface of the calf and evaluated for their compressibility and Doppler flow characteristics.
- The peroneal veins and artery are evident a few centimeters up the calf, deeper than the posterior tibial veins. The peroneal and posterior tibial veins can usually be followed up the calf at the same time (figures 34-4B and C).
- The examiner completes the evaluation of the calf vessels into the popliteal fossa (figures 34-4D and E).
- Although not a routine component of many testing protocols, the anterior tibial veins can also be evaluated. They are located running along the lateral surface of the tibia. Figure 34-4F shows those vessels between the tibia and fibula.

Peripheral Veins of the Lower Extremity—Chronic Venous Insufficiency (CVI)

Reversed Trendelenburg Position: Manual Technique

- While the vein of interest is being imaged (e.g., common femoral vein, popliteal vein), spectral analysis is activated. The patient is asked to perform the Valsalva maneuver or the technologist/sonographer applies proximal manual compression to determine the presence/absence of venous reflux.

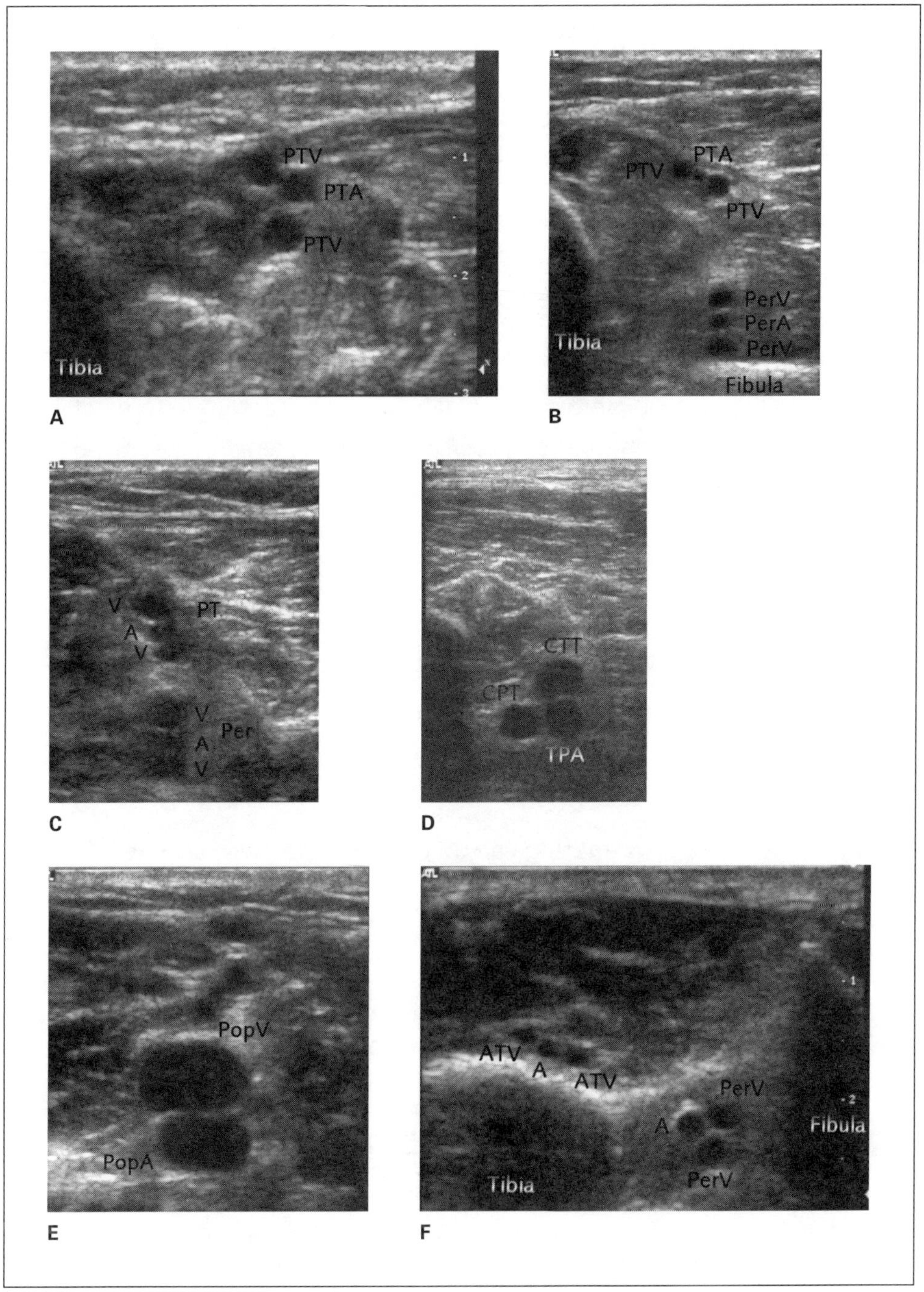

Figure 34-4. Transverse view of infrapopliteal veins from medial malleolus into popliteal fossa. **A** Posterior tibial veins (PTV) and artery (PTA) at ankle level. **B** The PT and peroneal vessels (Per) in the lower calf. **C** The PT and Per vessels in mid-calf. **D** The common tibial trunk (CTT), common peroneal trunk (CPT), and tibioperoneal artery (TPA) in lower popliteal fossa. **E** Popliteal vessels in popliteal fossa. **F** The anterior tibial veins (ATV) below the knee.

- Manual compression is next applied distal to the transducer as the examiner observes for the presence/absence of venous reflux following release of the distal compression.

- During these maneuvers, spectral analysis provides documentation of the presence/absence of reflux and allows the duration of the reflux to be measured.

- Doppler color flow imaging can also be utilized to note color changes during proximal compression or following distal compression maneuvers.
- If reflux is not evident in the reversed Trendelenburg position, the study must be repeated with the patient standing.

Standing Position: Manual Technique

- The leg to be examined is not to be bearing weight. When available, a two-step elevated stand or a one-step stool is used. For patient safety and comfort, the patient should have something to hold onto for balance.
- While the vein of interest is being imaged (e.g., common femoral vein, popliteal vein), spectral analysis is activated. The patient is asked to perform the Valsalva maneuver or the technologist/sonographer applies proximal manual compression to determine the presence/absence of venous reflux.
- Manual compression is then applied distal to the transducer as the examiner observes for the presence/absence of venous reflux following release of the distal compression.
- During these maneuvers, spectral analysis provides documentation of the presence/absence of reflux and allows the duration of the reflux to be measured.
- Doppler color flow imaging can be used to note color changes during proximal compression or following distal compression maneuvers.

Standing Position: Using an Automatic Cuff Inflator

- The leg to be examined is not to be bearing weight. When available, a two-step elevated stand or a one-step stool is used. In addition, the patient needs to hold onto something for balance.
- A 12 × 40 cm cuff is applied to the high-thigh portion of the leg.
- The duplex system is used to obtain an image and continuous Doppler spectral analysis of the common femoral vein. The transducer is positioned proximal to and within 5 cm of the cuff.
- Although difficult, it is crucial to maintain probe position over the site of interest so the signal is not lost during cuff inflation and deflation.
- Once the optimal Doppler signal is obtained, the examiner maintaining a 60 degree angle or slightly less if necessary, the cuff is inflated to 80 mmHg and maintained there for 1–2 seconds.
- The cuff is quickly deflated by the autoinflator/deflator or manually when faster cuff deflation is required, while the technologist/sonographer notes whether there is any reversal of venous flow after cuff deflation.

- If there is reversal of flow, the peak velocity (cm/sec) of the venous signal is measured, as well as the duration of the flow reversal (in seconds).
- The same steps above are applied to the great saphenous vein at the saphenofemoral junction.
- Although using different cuff inflation levels, the previous technical steps are also applied to the following vessels:

  To evaluate the popliteal vein and the great saphenous vein at the knee, the 12 × 40 cm cuff is placed at the calf level and inflated to 100 mmHg.

  To evaluate the posterior tibial vein, the 12 × 40 cm cuff is placed at the transmetatarsal (TM) level and inflated to 120 mmHg.

Peripheral Veins of the Upper Extremity

- In a transverse orientation, the examiner begins imaging the internal jugular vein along the lateral side of the neck (figure 34-5A), moving toward and under the clavicle, where the internal jugular vein joins the subclavian vein to form the innominate vein (figure 34-5B). Since compressibility is difficult to assess along the course of the subclavian vein, the examiner pays particular attention to confirming the presence of an anechoic lumen, normal Doppler signals, color filling, and partial to complete collapse of the vessel with deep inspiration. The vessel also will collapse at the onset of the Valsalva maneuver, although dilatation is evident as the patient continues the Valsalva.
- The subclavian vein may be evaluated using the supraclavicular and/or infraclavicular approach (figure 34-5C). It can usually be followed only to the outer border of the first rib, where the clavicle interferes with further visualization. A small section of the subclavian and axillary veins may be evident using an infraclavicular approach and angling the probe into the deltopectoral triangle. As the subclavian vein is followed laterally, the junction of the cephalic and axillary veins becomes evident. The cephalic vein can be followed as it courses toward the skin line and over the shoulder region, traveling down the lateral aspect of the arm.
- The examiner instructs the patient to raise the arm so that the probe can be placed in the axilla to continue evaluating the axillary vein (figure 34-5D).
- Moving through the axilla, the confluence of the basilic and paired brachial veins is evident with the basilic following a medial course fairly

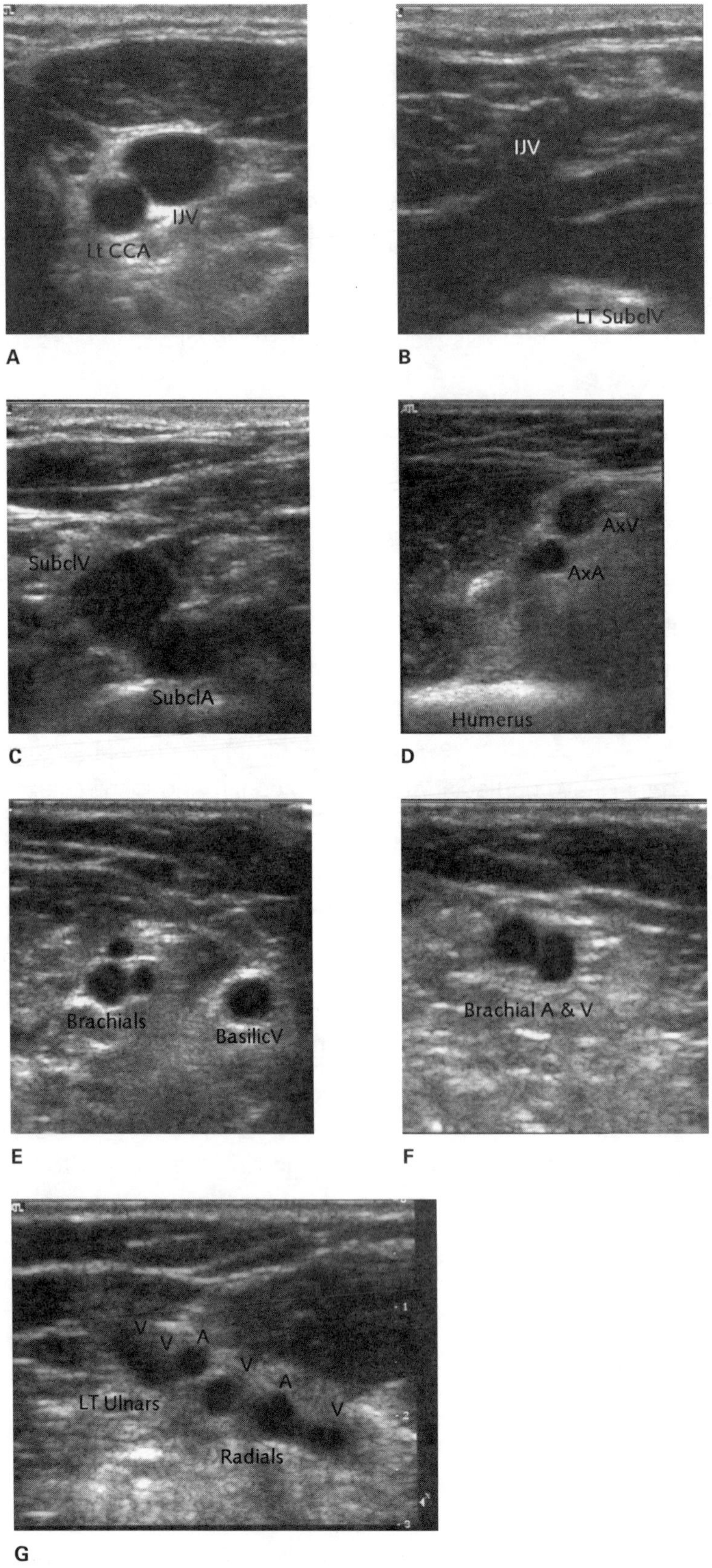

Figure 34-5. Transverse view of upper extremity veins from neck to antecubital fossa (bend of elbow). **A** Common carotid artery (CCA) and internal jugular vein (IJV). **B** Confluence of IJV and subclavian vein (SubclV) forming the innominate vein. A branch is also evident at that juncture. **C** The SubclV and artery from supraclavicular approach. **D** Axillary vessels. **E** The paired brachial veins and basilic vein in upper arm. **F** The common brachial vein in the antecubital fossa just after the confluence of the radial and ulnar veins. **G** The paired radial veins and paired ulnar veins in the more distal antecubital fossa.

close to the brachial veins (figure 34-5E). The examiner assesses the veins for compressibility and normal Doppler signals at frequent intervals. Figure 34-5F represents the brachial vein at the antecubital fossa.

- Traveling toward the hand, the brachial veins are evident, formed by the confluence of the paired radial and ulnar veins (figure 34-5G).

- The cephalic vein runs down the lateral aspect of the arm, while the basilic courses medially.

Abdominal and Pelvic Veins

- The evaluation of the inferior vena cava and pelvic veins begins in the transverse view at the level of the umbilicus.

- Evaluation of the other abdominal vessels begins in the transverse view at the xiphoid process.

- Accurate identification of the vessels depends on the examiner's careful attention to anatomic landmarks, e.g., pulsating aorta or aortoiliac bifurcation. See figure 34-6.

6 Interpretation:

Normal Peripheral Veins of the Lower Extremity

- Veins must be completely compressible to be considered normal (figure 34-7). Slightly more than usual pressure or changes in probe position may be necessary to completely compress the common femoral vein at the inguinal ligament and the femoral vein at the adductor hiatus.

- Normal Doppler venous signals are described beginning on page 309.

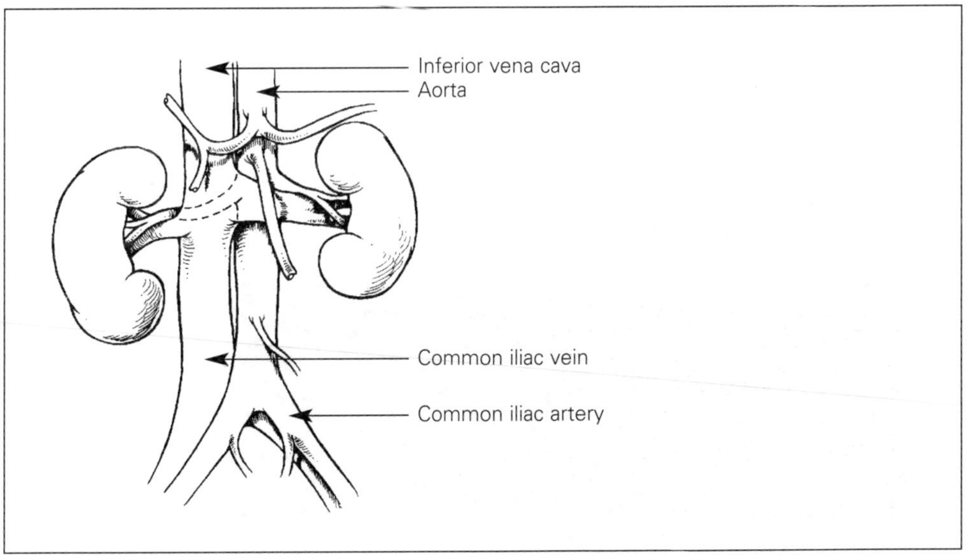

Figure 34-6. Abdominal vessels. The schematic drawing clarifies the relationships of structures. From Ridgway DP: *Introduction to Vascular Scanning,* 3rd edition. Pasadena, CA, Davies Publishing, 2004.

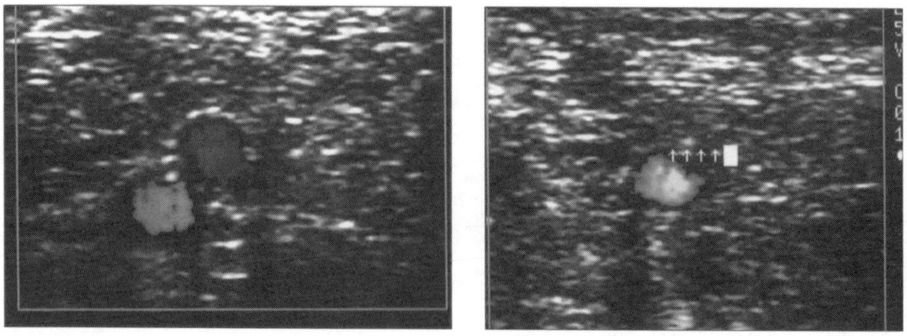

A B

Figure 34-7. Transverse view of the popliteal artery and vein. **A** Both vessels are open with complete color filling. **B** The popliteal vein completely compresses with probe pressure.

Normal Peripheral Veins of the Upper Extremity

- Veins covered by bones or other structures—part of the axillary, subclavian, proximal internal jugular, and innominate veins, for example—may not be completely compressible. One option during evaluation of the subclavian vein is to have the patient quickly take in a deep breath through pursed lips, as if sucking through a straw. This respiratory maneuver may collapse the vein and thereby ensure that it is thrombus free.
- Pulsatile venous Doppler signals in the internal jugular, subclavian, and innominate veins are normal because of their close proximity to the heart (figure 34-8).
- Augmentation with distal compression may normally be reduced or not evident in the subclavian vein and proximal axillary vein.
- If the patient has a patent dialysis access graft in the extremity being evaluated, expect the following changes in the outflow vein:

    Increased flow velocity and volume flow

    Pulsatile flow

    No response to distal compression

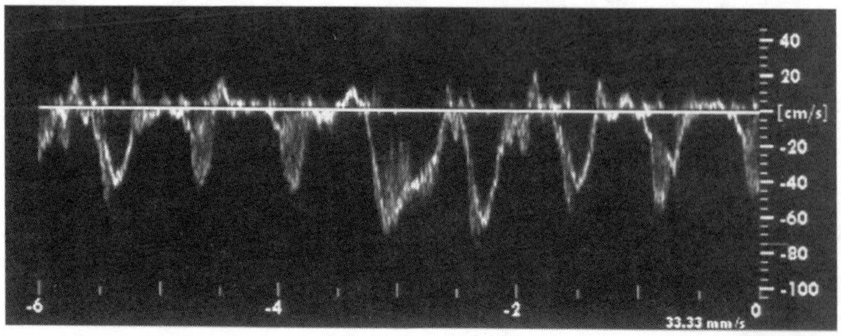

Figure 34-8. Pulsatile venous flow in this subclavian vein is an expected finding because of its close proximity to the heart.

Incompressible vessel

Collateral channels evident

Normal Abdominal and Pelvic Veins

- Because compressibility of the abdominal and pelvic veins is not possible, dilatation with deep inspiration should be evident in the inferior vena cava and iliac veins.
- Venous Doppler signals should be spontaneous.
- The portal vein, formed by the superior mesenteric and the splenic veins, carries blood *into* the sinusoids of the liver. Flow into the liver is called *hepatopetal*. Portal vein flow is characterized by flow patterns that exhibit minimally phasic, almost continuous Doppler signal as seen in figure 34-9. In the adult patient, flow in the portal vein is essentially not affected by normal respirations. A pulsatile waveform may be evident in cases of congestive heart failure and/or fluid overload. On B-mode imaging, the walls of the portal vein are hyperechoic.

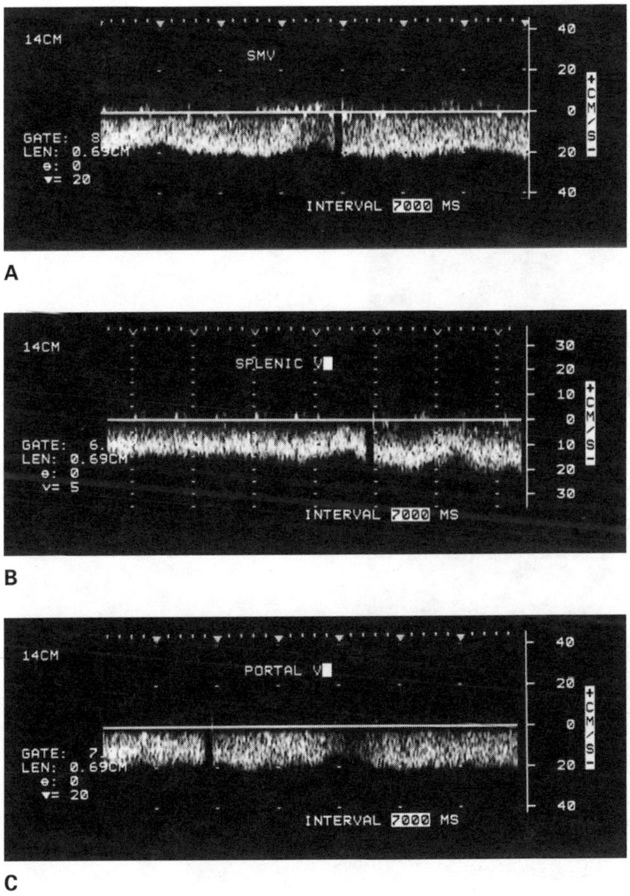

Figure 34-9. Doppler spectra showing hepatopetal flow, including the phasic, continuous signal typical of portal vein flow. **A** Superior mesenteric vein. **B** Splenic vein. **C** Portal vein.

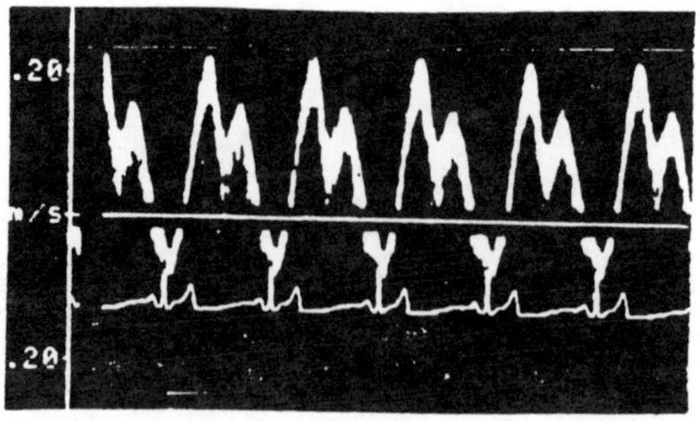

Figure 34-10. Doppler spectra showing hepatofugal flow in the hepatic vein, which is characteristically bidirectional (pulsatile).

- The hepatic veins carry blood *from* the liver into the inferior vena cava. Flow away from the liver is called *hepatofugal.* Hepatic vein flow is characterized by flow patterns that exhibit a minimally phasic, bidirectional, or pulsatile-appearing Doppler signal as seen in figure 34-10. On B-mode imaging, the walls of the hepatic veins are not clearly visible, unlike the bright walls of the portal vein.

- Figure 34-11 demonstrates differences between portal and hepatic vein flows.

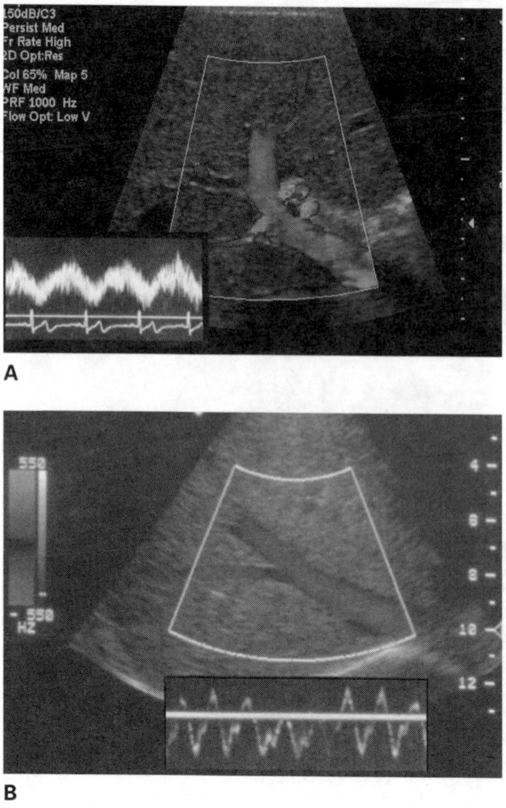

Figure 34-11. **A** Sagittal view of the portal vein documenting normal flow characteristics. Hepatopetal flow (moving into the liver) is considered to be unidirectional and more continuous in quality compared to hepatic vein flow. **B** Sagittal view of the hepatic vein documenting normal flow characteristics. Hepatofugal flow (away from the liver) is considered to be more pulsatile (because it flows into the IVC) compared to the more continuous hepatopetal flow of the portal vein.

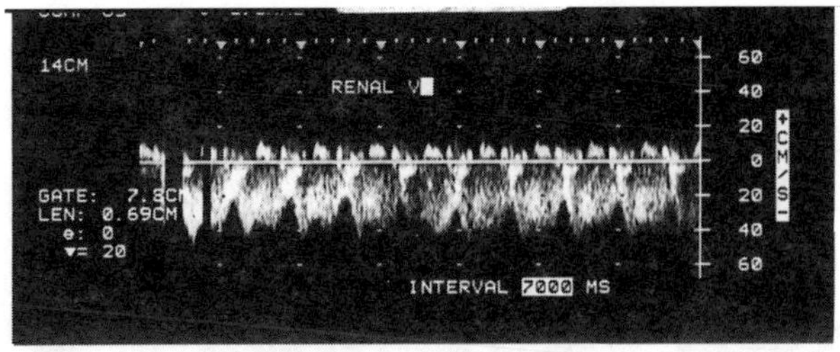

Figure 34.12.
Renal vein flow is similar to that in the hepatic vein.

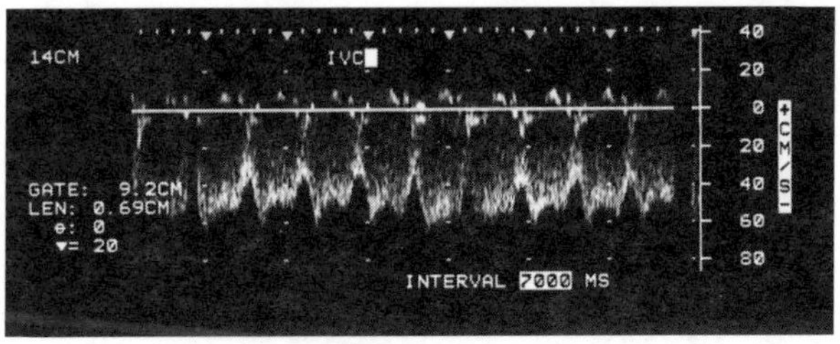

Figure 34-13.
Doppler spectra from the inferior vena cava.

- Because the renal veins carry blood into the inferior vena cava, the Doppler venous signal will be similar to the minimally phasic, bidirectional, or pulsatile-appearing Doppler signal as seen in the hepatic veins and displayed in figure 34-12.
- The Doppler signal of the inferior vena cava normally exhibits phasic, bidirectional/pulsatile Doppler signals (figure 34-13).

Acute Thrombosis

- In cases of acute venous thrombosis the peripheral veins are not completely compressible, and the examiner may observe very low-level echoes within the venous lumen (figure 34-14). The inferior vena cava and iliac veins, on the other hand, are too deep to be compressible.

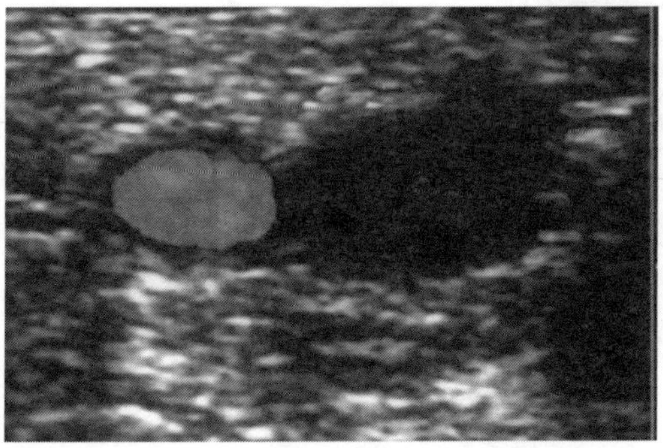

Figure 34-14.
Transverse view of the common femoral vein documenting anechoic filling of an incompressible vein, which is more characteristic of acute thrombosis of the deep veins.

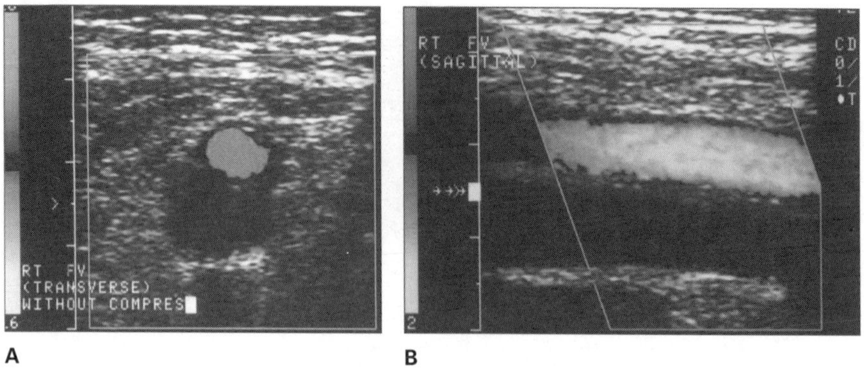

Figure 34-15. Hypoechoic filling of an incompressible vein is consistent with an acute thrombosis. **A** Transverse view of the femoral vein and artery. **B** Sagittal view of the same vessels.

- Visible thrombus may be evident.
- The vessel is usually dilated compared to its accompanying artery (figure 34-15). Figure 34-16 (see color plate 18) documents partial thrombosis dilating the portal vein. Although there is no accompanying artery for comparison, dilatation is evident by evaluating the contour of the vein wall.
- Abnormal Doppler signals are usually evident (see also chapter 33 for details on venous Doppler signals):

  If flow is not spontaneous at the common femoral, femoral, and/or popliteal veins, there may be an obstruction (DVT or extrinsic compression) distal to or at that site. Spontaneous flow is not expected in the tibial vessels, where flow is normally reduced.

  If flow is not phasic but, rather, continuous in the common femoral, femoral, and/or popliteal veins, the possibility of a proximal obstruction should be considered.

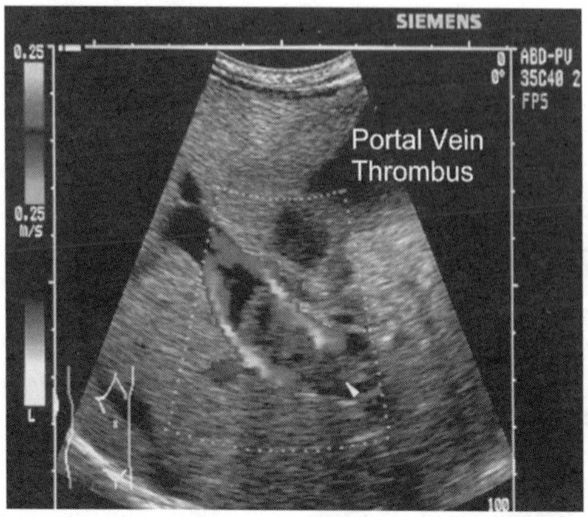

Figure 34-16. Acute, partial thrombosis of this portal vein (as seen in this long view) is evident by the lack of color filling and dilatation of the vein wall. Although there is no accompanying artery for comparison, dilatation is evident by evaluating the contour of the vein wall in the segment visualized.

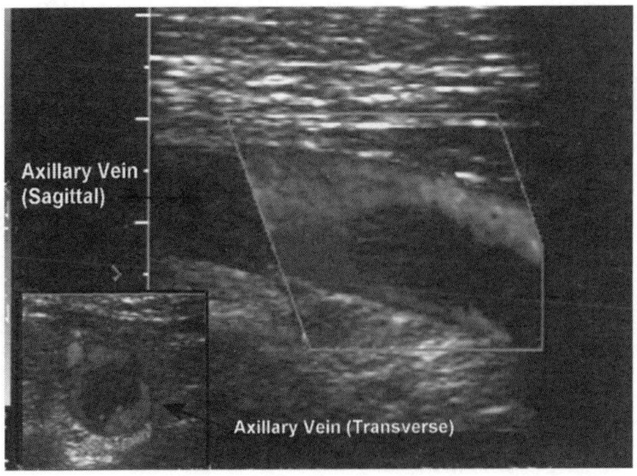

Figure 34-17. Longitudinal and transverse (inset) view of the axillary vein documenting acute thrombosis. The vein is dilated with very low-level echoes evident in sagittal view. Color flow imaging reveals partial filling of the vessel consistent with partial deep venous thrombosis.

If there is no augmentation with distal compression, there may be an obstruction between where you are compressing and where you are listening, or something slightly more proximally.

If there is no augmentation with proximal release, the possibility of a more proximal obstruction should be considered.

If flow increases during proximal compression, there may be venous reflux.

- Color flow imaging reveals no filling or only partial filling of the lumen.
- The thrombus may be poorly attached to the wall (figure 34-17; see also color plate 19).
- The thrombus may have a spongy texture.
- *Rouleau,* from a French word meaning "roll," is used to describe the appearance of red blood cells arranged like rolls of coins. Rouleau formation represents very sluggish flow which is seen with B-mode imaging as heterogeneous material moving with respirations and augmentation maneuvers. A compressible vessel with evidence of rouleau formation suggests very slow flow due to a proximal obstruction, increased proximal venous pressure, or other processes such as an increase in plasma immunoglobulin.

Chronic Changes following Acute DVT

- As thrombus ages, it becomes more echogenic, with very bright echoes eventually displayed as those in figure 34-18.
- Visible collateralization or recanalization may be evident. Linear, echogenic intraluminal striations called *synechiae,* may also be evident as the thrombus is slowly converted to a fibrous band over time (figure 34-19).

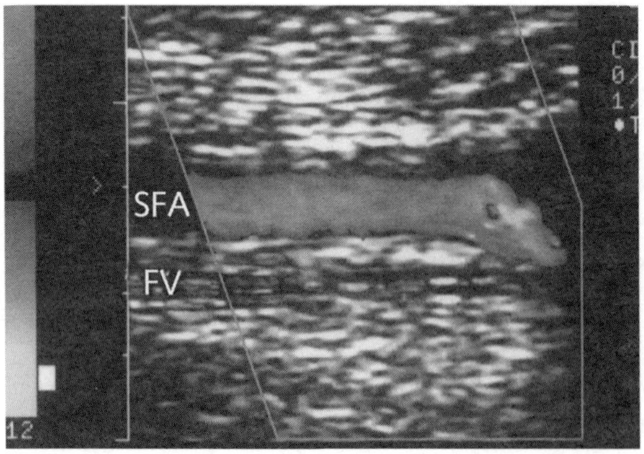

Figure 34-18. Sagittal view of the femoral vein and artery. The bright echoes are consistent with chronic changes.

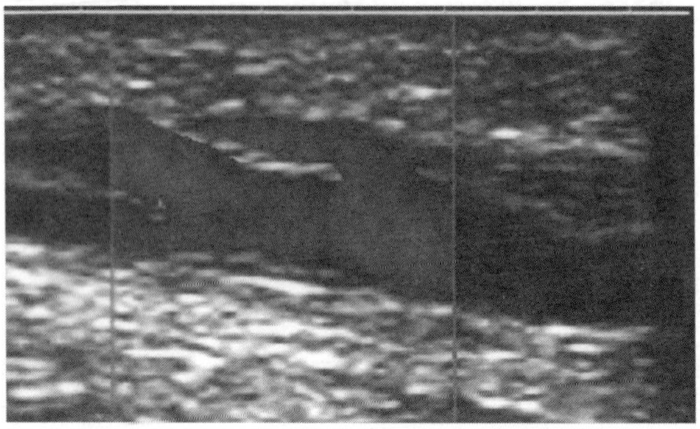

Figure 34-19. Sagittal view of vein showing linear, echogenic striations called synechiae, which may be evident as the thrombus is slowly converted to a fibrous band.

- The vessel is not dilated, and the aging thrombus retracts over time, leaving a thickened wall in some cases following acute deep venous thrombosis (figure 34-20).
- Abnormal Doppler venous signals may be evident.

Chronic Venous Insufficiency (CVI)

- No reflux should be observed with Doppler spectral analysis during proximal manual compression or the Valsalva maneuver, nor should there be reflux following distal compression. Nevertheless, a very short duration of flow in the opposite direction may be seen. If less than 0.5 seconds, this is often considered to be normal valve closure time.
- Venous reflux lasting > 0.5 or > 1 second, depending on the literature source (figure 34-21), is considered abnormal.
- Color flow imaging reveals venous reflux as a directional shift in color from flow away from the probe to flow toward the probe during the Valsalva maneuver and/or following distal compression (figure 34-22).

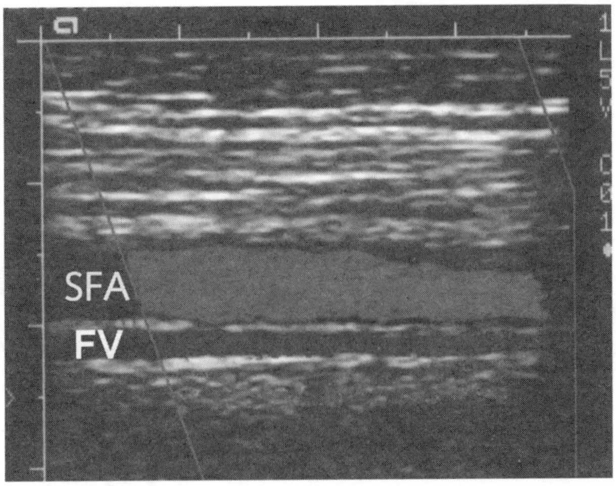

Figure 34-20. Longitudinal view of femoral vein. Sonographically, the vein with chronic changes usually appears to be retracted (smaller than normal diameter) with hyperechoic walls. Doppler flow characteristics may be normal or abnormal depending on the proximal outflow veins.

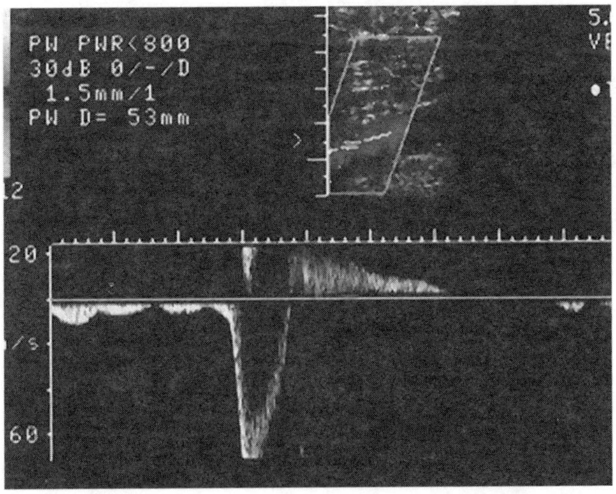

Figure 34-21. Sagittal view of the femoral vein. Flow characteristics confirm retrograde flow for > 2 sec in the vein following distal augmentation. This finding is consistent with venous reflux.

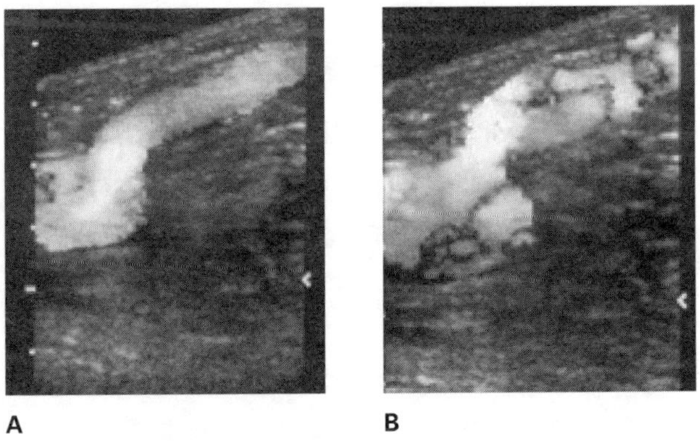

Figure 34-22. Sagittal view of proximal great saphenous vein. **A** With patient at rest, blood is moving in its normal direction, i.e., cephalad. **B** During the Valsalva maneuver, which is comparable to a proximal compression, reversed flow is evident. Flow reversal may also be seen following release of distal compression and is consistent with venous reflux. See color plates 20 and 21.

Table 34-1. General guidelines for assessing an avascular mass of the lower extremity.

Structure	***Usual Location***	***Typical Sonographic Presentation***	***Notes***
Abscess	Anywhere.	Various appearances from well- to ill-defined; heterogeneous.	Must correlate with signs/symptoms (area of warmth, redness, tenderness, pain).
Baker's cyst	Medial popliteal fossa; a ruptured Baker's cyst can extend along pre-tibial area.	Well-defined structure filled with synovial fluid. Usually anechoic. Connection ("neck") to joint space must be evident. If ruptured, fluid leaking into tissue has appearance of "rat-tail."	May become infected. Ortho surgery may be necessary. Usually associated with arthritis, but also with trauma, e.g., medial meniscus tear from a sports injury. A ruptured cyst is more common in geriatric and osteoarthritic patients.
Cyst	Anywhere.	A simple cyst is a well-defined fluid collection that is anechoic, with posterior enhancement.	Not related to trauma or injury.
Effusion	Near joint; usually midline.	Anechoic.	Signs/symptoms near or at joint.
Hematoma	Site of trauma or injury.	Various appearances from well- to ill-defined. Usually heterogeneous with low-level echoes initially; may become anechoic over time (months). Little or no through-transmission/posterior enhancement.	May be caused by vigorous exercise or a complication of anticoagulation therapy.
"Loose bodies" in synovial fluid	Recesses of knee.	Hyperechoic with shadowing.	Most often associated with calcium (usually piece of cartilage).

Miscellaneous Findings in the Extremities

During duplex evaluations, a nonvascular condition may be observed. Documentation of these findings could assist the physician in determining the cause of the patient's signs and symptoms. Although examples of nonvascular conditions (e.g., masses, structures) follow, it is critical that the vascular technologist/sonographer understand that the sonographic appearance of various types of structures may be similar. The information derived from the duplex scan must be correlated with the patient's history, risk factors, and signs and symptoms before a conclusion can be

Structure	Usual Location	Typical Sonographic Presentation	Notes
Lymph node	Neck, axilla, abdomen, groin, popliteal fossa.	Hyperechoic oval-shaped center (hilum) encircled by a hypoechoic halo (parenchyma). Blood flow usually seen in center of node. Flow around perimeter may indicate neoplasm.	Enlargement can be related to an inflammatory process or neoplasms, e.g., lymphoma.
Mass superficial to deep veins	Variable.	Variable.	Consider: neuroma (neoplasm from cells of the nervous system), sarcoma (neoplasm of connective tissue), schwannoma (tumor of nerve tissue).
Meniscal cyst	Lateral knee area.	Heterogeneous.	Mixed echoes (synovium in tissue).
Muscle tear	Site of injury.	Size and shape related to amount of bleeding. Initially hyperechoic with decreasing echogenicity over time (weeks).	Should correlate with patient's history.
Nerve	Travels adjacent to vessels.	Heterogeneous with striations seen in sagittal view.	
Tendon	Does not travel adjacent to vessels.	Hyperechoic. Echogenicity changes with angulation of transducer (tendon brightens and darkens with angulation—called *amasotrophy*).	Typically circular in transverse view (short axis) and tapered in sagittal view (long axis) when followed from muscle to bone.

reached. Table 34-1 provides some general guidelines for assessing an avascular mass in the lower extremities.

- *Edema:* Figure 34-23 documents B-mode findings consistent with edema, an excessive amount of fluid in body tissue.

- *Lymph node:* Figure 34-24A illustrates the typical appearance of a lymph node with brighter echoes in the center surrounded by low-level echoes. Figure 34-24B (see also color plate 22) is from a patient with lymphoma.

Figure 34-23. Edema, an excessive amount of fluid in the tissue, is evident sonographically as streaks or pockets of very low-level echoes.

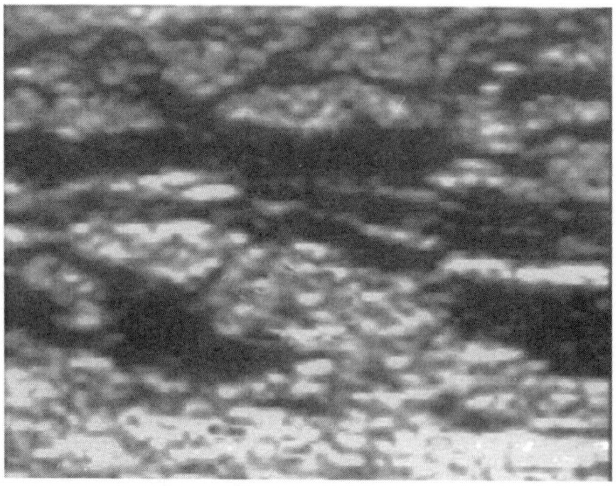

Figure 34-24. Lymph nodes, structures that can be found along the course of the lymphatic channels, are frequently seen during duplex exams. These images were obtained in the infrainguinal ligament area. **A** The hyperechoic center surrounded by low-level echoes is consistent with a lymph node. **B** The B-mode image confirms a mass. However, color flow imaging revealed a vascularized lymph node consistent with this patient's history of lymphoma.

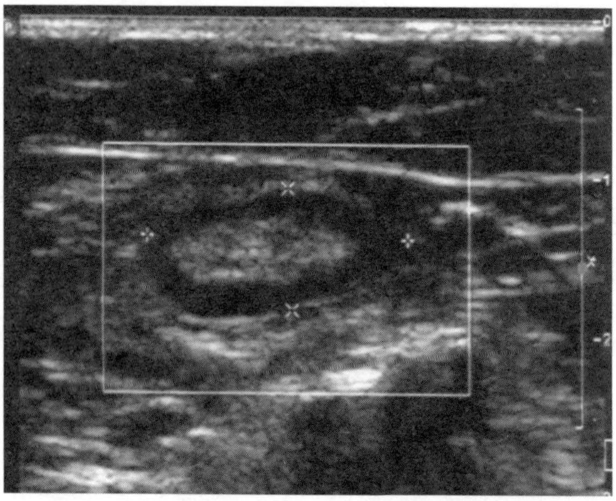

A

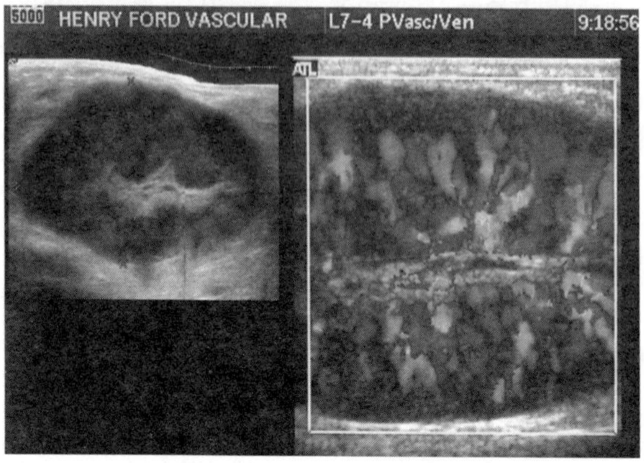

B

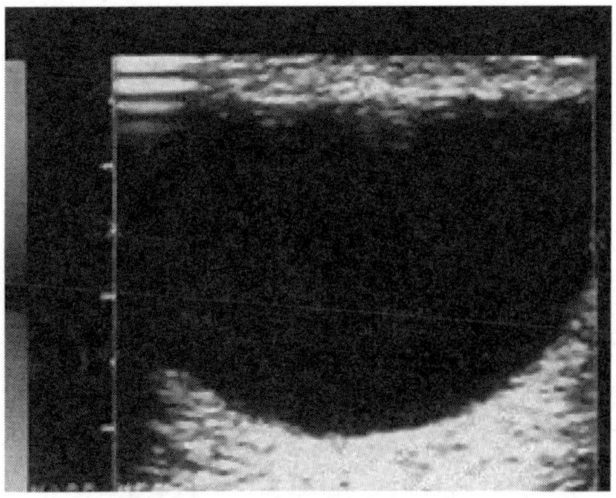

Figure 34-25. The low-level echoes of this well-defined avascular mass is consistent with an acute hematoma, in this case following a surgical procedure. Echo characteristics of hematomas vary, as well as how well-defined their borders are.

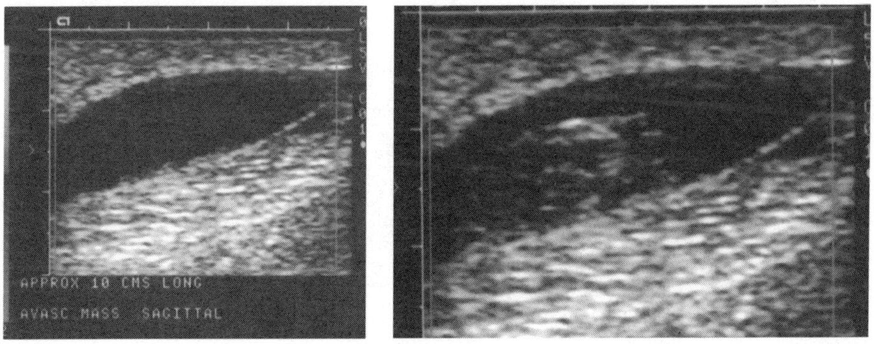

A B

Figure 34-26. These well-defined structures were found in the popliteal fossa posterior and medial to the knee joint. **A** Very low-level echoes are evident in a fluid-filled structure, and the location is consistent with a Baker's cyst. **B** The length of the structure suggests a ruptured Baker's cyst.

- *Hematoma:* Figure 34-25 demonstrates the localized collection of blood.
- *Baker's cyst:* Figure 34-26A documents a cystic structure in the popliteal fossa consistent with a Baker's cyst. The structure, composed primarily of synovial fluid, is found posterior and medial to the knee joint and may account for the patient's pain and swelling. Figure 34-26B shows a ruptured Baker's cyst with a heterogeneous appearance.
- *Muscle tear:* Figure 34-27A reveals a muscle tear producing pain and swelling. Figure 34-27B demonstrates normal anatomy.
- *Nerve:* Figure 34-28A (see color plate 23) documents a nerve that was mistaken for venous thrombosis. This illustrates the importance of evaluating a structure in transverse and sagittal views as well as following it proximally and distally to ensure accurate interpretation of the findings (figure 34-28B).

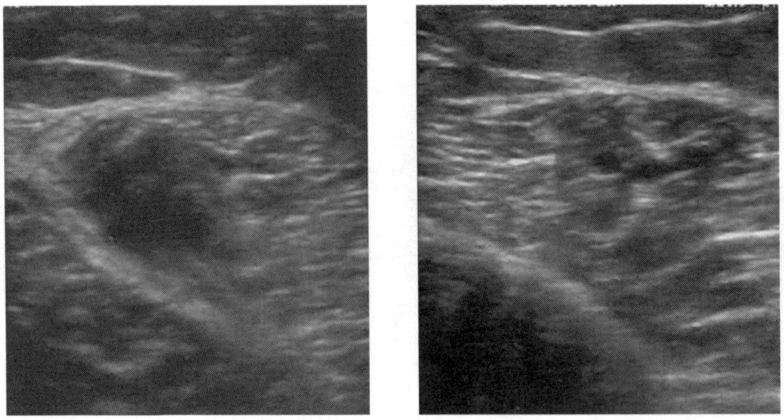

A B

Figure 34-27. Sonographic evaluation of the gastrocnemius muscle. **A** The heterogeneous material found within muscle tissue is consistent with a muscle tear and with the patient's complaints of focal pain in the area after abrupt movement. **B** An image of the contralateral calf demonstrates normal tissue.

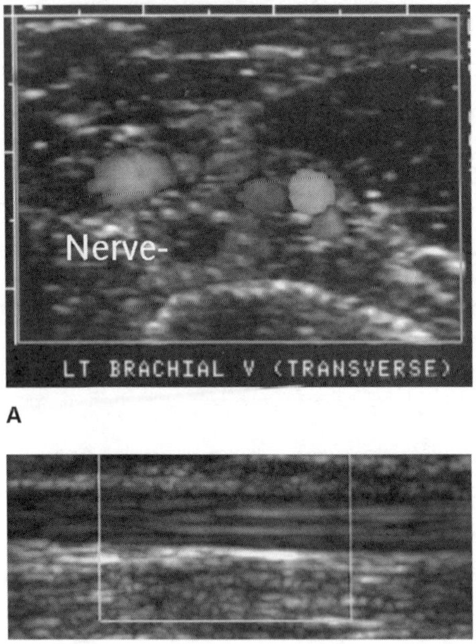

A

B

Figure 34-28. **A** Transverse view of the brachial vessels, the basilic vein, as well as a circular, incompressible structure. **B** The striations throughout the structure in sagittal view confirm that it is not a vessel. The unchanging size and course of the structure when scanned proximally and distally is consistent with a nerve.

- *Focal dilatation as a vein:* Figure 34-29 documents an unusual dilatation of the vein that may be consistent with a venous aneurysm.

Miscellaneous Findings in the Abdomen

- A vena cava interruption device is usually placed below the level of the renal veins and may appear as bright echogenic lines.
- Persistent and widespread venous dilation is consistent with systemic venous hypertension.

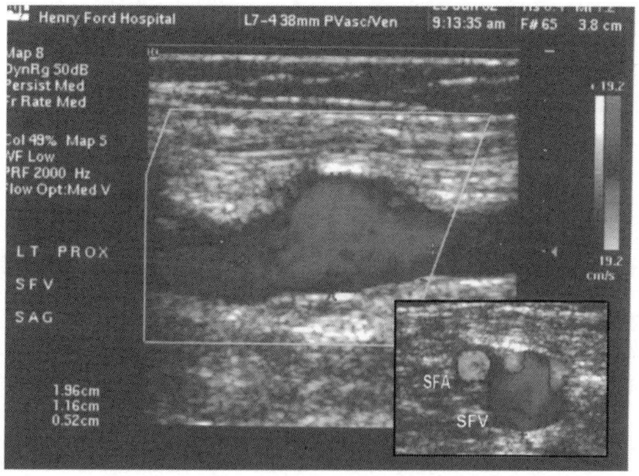

Figure 34-29. Sagittal scan of the femoral vein documenting focal dilatation of the femoral vein consistent with a venous aneurysm. The dilatation is not at a valve site and valve cusps are not visualized. **Inset:** Transverse view confirming this unusual finding.

- Portal hypertension is increased blood pressure within the portal vein usually resulting from increased resistance to blood flow.

  Increased resistance can result from pathology of the portal vein, small intrahepatic portal vein radicles, the hepatic parenchyma, or the hepatic veins. The pathology varies and includes cirrhosis, cancer, pancreatitis, thrombosis, and trauma.

  B-mode finding may include evidence of portosystemic venous collaterals.

  Because flow patterns are related to pressure gradients, the increased resistance can result in hepatofugal flow (flow away from the liver) rather than the normally expected hepatopetal flow (flow into the liver).

- Budd-Chiari syndrome results from hepatic vein occlusion. Some abnormal clinical findings include hepatomegaly, abdominal pain, and sudden onset of ascites. The causes of this syndrome are many and vary with regard to the primary site of obstruction (hepatic vein level, sinusoids, or the inferior vena cava).

Invasive Tests and Therapeutic Intervention

CHAPTER 35

D-dimer

1 Capabilites

- D-dimer is a measurable product of the thrombotic process detected on a quantitative blood assay.
- A positive result (increased level) is consistent with lysis, or breakdown, of thrombus.
- A negative result (normal or less-than-normal level) implies the absence of a thrombotic process.

2 Limitations

- Many different methods/products are available to detect D-dimer in the blood with tremendous variation in the results obtained.
- D-dimer may be elevated with pregnancy, liver disease, renal disease, cancer, or any thrombotic process, including recent surgery.
- A positive finding is not highly sensitive for deep venous thrombosis (DVT).

- D-dimer should be used with caution on inpatients/hospitalized patients.
- D-dimer should not be used on patients already being treated with anticoagulation therapy.

3 Application

These are some observations as to how physicians may utilize the results of D-dimer:

- Outpatients for whom there is a low clinical likelihood of acute deep venous thrombosis and a negative D-dimer test will likely not be treated for DVT or further evaluated by duplex ultrasonography on an emergent or urgent basis, if at all.
- Outpatients for whom there is a moderate clinical likelihood of DVT and a negative D-dimer test may undergo urgent venous duplex study. If the study is not immediately available, the patient may be given a subcutaneous injection of low-molecular weight heparin (LMWH) and asked to report to the vascular laboratory/ultrasound department the next day. If there is a contraindication to LMWH, the patient may be admitted to a clinical decision unit until venous duplex is available.
- Outpatients for whom there is a high likelihood of DVT and a negative D-dimer test may be managed according to the previous step or receive intravenous unfractionated heparin until the duplex study is available.
- Outpatients for whom there is a high likelihood of DVT and a positive D-dimer test are considered to have DVT until proven otherwise.

Contrast Venography

For quick reference to contrast venograms from actual cases, see plates 14–17 on pages 425–427, illustrating both normal and abnormal conditions including thrombosis and collateralization.

1 Capabilities:

Although contrast venography may still be considered the gold standard by which all other venous tests are compared, its use has markedly decreased because duplex scanning is so accurate and in fact provides information that venography cannot. Two types of venograms are performed:

- Ascending venography, used to evaluate acute deep venous thrombosis, congenital venous disease, and/or anomalies, and in the evaluation of chronic venous changes.

- Descending venography, used primarily to detect and quantify reversed flow from incompetent venous valves.

2 Limitations:

- Venography is highly technical in technique and interpretation.
- It is relatively expensive.
- It may be uncomfortable for the patient.
- It can produce adverse effects, such as allergic reactions or extravasation of contrast medium, and it may be contraindicated in patients with severe allergies to iodine or with severe peripheral vascular occlusive disease.

3 Technique:

- A radiopaque contrast material is injected into the foot veins to visualize the venous anatomy.
- Ascending venography requires that the contrast agent is injected into a vein on the dorsum of the foot.
- Descending venography usually has the contrast agent injected into the common femoral vein.
- Serial x-rays are taken as the contrast material passes through some of the deep veins. An example of ascending venography is seen in figure 35-1 below.

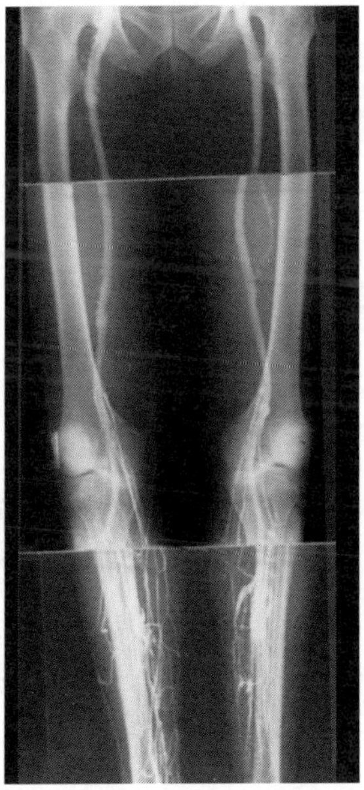

Figure 35-1. Normal venogram (phlebogram) in which a radiopaque contrast agent fills the venous system. The tibial, popliteal, femoral, and common femoral veins are shown. During venography, serial x-rays are taken.

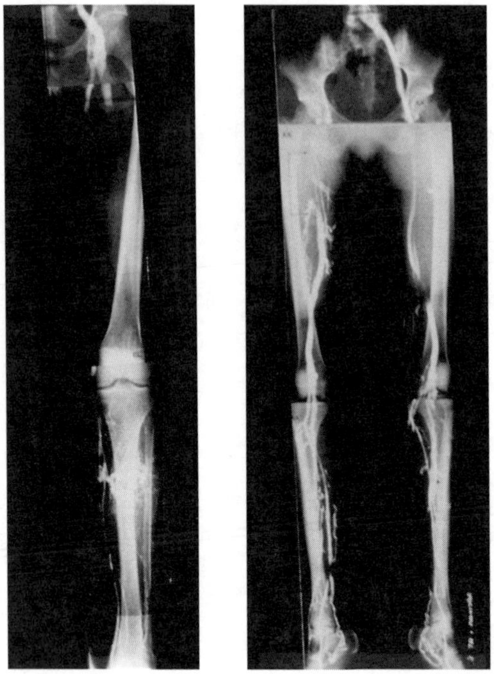

A B

Figure 35-2. Venograms showing filling defects that suggest the displacement of contrast material by thrombus. In image **A**, there is only minimal filling of the contrast agent in the tibial veins, with none seen until the confluence of the femoral vein and deep femoral vein into the common femoral vein. The significant filling defect probably represents deep venous thrombosis. In image **B**, contrast agent fills many more vessels, but filling defects indicating thrombosis appear in the right tibial veins, proximal femoral, and in confluence of the common femoral and external iliac veins. On the left, there is minimal filling of the tibial veins with filling defect in the femoral vein at mid thigh.

- Any deviation from normal—i.e., a filling defect indicating the displacement of contrast material by thrombus—is evidence of obstruction as seen on ascending venography (figure 35-2).

Lung Perfusion Scan

1 Capabilities:

A *VQ* scan (lung ventilation/perfusion scan) is a screening test for the detection of perfusion defects of the lungs, most commonly attributed to a pulmonary embolism (foreign material, usually originating from a deep venous thrombosis, that eventually lodges in the small vessels of the lungs and prevents adequate perfusion).

2 Limitations:

Disorders other than microembolism can cause perfusion defects, including emphysema, asthma, pneumonia, cancer of the bronchus, congestive heart failure, liver cirrhosis, radiotherapy, multiple blood transfusions, and postoperative phenomena. Pulmonary angiography has been considered the definitive diagnostic tool—the gold standard—in the diagnosis of a pulmonary embolus. Nevertheless, advances in computed tomography angiography (CTA) of the chest have shown CTA to be more sensitive in some cases, as well as being more readily available.

3 Technique:

- A radioactive contrast medium is injected, usually into an arm vein.
- Images of lung perfusion are taken with the patient in a variety of positions.
- Scans are interpreted as representing high, moderate, or low probability of pulmonary embolism, or they may be considered indeterminate.

Therapeutic Intervention

1 Medical therapy:

The prevention and treatment of venous thrombosis and its sequelae—pulmonary embolism (PE) and chronic venous insufficiency—include:

Controlling Risk Factors

All risk factors fall into one of the three causes of venous thrombosis as defined by Virchow's triad: venous stasis, trauma/endothelial damage, or hypercoagulability.

- Decrease venous stasis by limiting long periods of inactivity or bed rest and by promoting venous drainage when inactive. Various methods of promoting venous drainage include wearing support hose or elastic stockings, elevating legs, using intermittent pneumatic calf compression during and after surgery, and reducing weight.
- Prevent endothelial damage from injury or infection to the extremities.
- Be aware of hypercoagulability states and factors and follow prescribed treatment plan.

Anticoagulant Therapy for Prophylaxis

- Low-dose unfractionated heparin, e.g., 5000 units subcutaneously every 12 hours before and after surgery, decreases the postoperative risk of deep venous thrombosis.
- Low-molecular-weight heparin, e.g., Lovenox, administered subcutaneously, may be utilized to provide a bridging mechanism, that is, to provide anticoagulation when Coumadin must be discontinued for an invasive procedure.

Anticoagulant Therapy for Acute DVT and/or PE

- Although protocols vary, a loading dose of 10,000 units of heparin followed by continuous intravenous infusion for 5–10 days may be recommended for the treatment of acute deep venous thrombosis.

Patients receiving intravenous heparin, i.e., unfractionated heparin, are placed on strict bed rest to decrease the risk of an embolic process secondary to muscle contraction.

- Heparin interferes with the formation of a blood clot by slowing the conversion of prothrombin to thrombin, increasing the effect of antithrombin III, and decreasing platelet adhesiveness. It does not dissolve (lyse) an existing thrombus, but helps to prevent propagation of the clot. Since the body has its own mechanism for clot lysis, lytic therapy is usually only offered when the patient has an acute iliofemoral thrombosis and/or is in danger of limb loss.

- Dosage is regulated to ensure that the patient's partial thromboplastin time (PTT)—the time needed for a fibrin clot to form—is 1.5–2 times normal.

- Low-molecular-weight heparin administered subcutaneously may be utilized to provide outpatient anticoagulation. Current research indicates that within 30 minutes of administration it is safe for these patients to ambulate.

- Oral anticoagulation (e.g., Coumadin) is also administered. The dosage is regulated to ensure that the patient's prothrombin time (PT) is 1.5–2 times normal.

- Heparin can be discontinued after 5–10 days if there has been sufficient overlap with the Coumadin (i.e., a minimum of 4 days), if the PT and PTT remain at therapeutic levels, and if there are no signs of active thrombosis. The patient remains on oral anticoagulation for 3–6 months.

2 Surgical and endovascular therapy:

Acute DVT or PE

- Vena caval interruption procedures may be indicated in patients with acute deep venous thrombosis who cannot be anticoagulated to prevent pulmonary embolism. Using fluoroscopy, an interruption device such as a Greenfield or bird's nest filter may be placed in the inferior vena cava via the jugular or femoral vein. An external caval clip may also be placed around the vena cava during abdominal surgery to decrease the risk of pulmonary embolism. Although inferior vena caval (IVC) filters have traditionally been considered permanent devices, an IVC filter may be placed temporarily for certain surgical procedures, such as gastric bypass.

- An iliofemoral venous thrombectomy may be performed in a patient with impending limb loss (e.g., because of phlegmasia cerulea dolens) if thrombolytic therapy (e.g., streptokinase or urokinase) does not dissolve the clot.

Acute Phlebitis at Junction of Deep and Superficial Systems

- Thrombosis of the great saphenous vein at or near the saphenofemoral junction may require more aggressive treatment than an isolated superficial system thrombosis, i.e., ligation of the great saphenous vein, to prevent propagation into the deep system. Some protocols call for serial scans to determine extension of the thrombus into the deep system.

- Although thrombosis of the small saphenous vein at or near the saphenopopliteal junction is rarely seen, it also may require more aggressive treatment than an isolated superficial system thrombosis, i.e., ligation of the small saphenous vein, to prevent propagation into the deep system. Some protocols call for serial scans to determine extension of the thrombus into the deep system.

Chronic Venous Insufficiency

- Nonhealing venous ulcers may be treated in a variety of ways, including these:

  Medicated wrap/dressing, e.g., Unna boot, Profore dressings

  Hyperbaric oxygen therapy

  Ligation of incompetent perforators

- Valvular reconstruction or valve transplantation procedures are infrequently performed.

Varicose Veins

- Saphenous vein removal or local excision of varicosities (phlebectomy) is commonly performed.

- Sclerotherapy can be used to treat small varicosities. A sclerosing agent such as sodium tetradecyl sulfate is injected into the varix followed by application of a compression dressing.

- Venous ablation procedures may be used to close down the varicose vein by radio-frequency or laser techniques. Using ultrasound, the caliper of the vein and depth of the vessel are determined, as well as important landmarks, e.g., the superficial epigastric vein as it joins the great saphenous vein in the groin. The size of the vessel determines whether the procedure can be performed; the depth of the vessel determines whether precautions must be taken to avoid burning the skin. Duplex imaging is also helpful for proper administration of tumescent anesthesia: a large volume of a local anesthetic infiltrates the tissue around the vein to provide anesthesia. The engorged tissue helps to compress the vessel and also provides a cushion between the vein and the skin to prevent skin burning and necrosis. Under ultrasound guidance, a catheter is advanced to the proximal portion of the vein being treated. The energy emitted by the device heats the vein, causing the walls to permanently coapt. Follow-up duplex scans are performed, e.g., 48–72 hours postprocedure, to ascertain that the vein has been ablated and that other vessels have not been affected by the procedure. Subsequent venous duplex studies may also be scheduled 3 and 6 months after venous ablation and yearly thereafter. In addition to ablation of the great saphenous vein (GSV), some sites are using the technique to treat incompetent perforators. Radio-frequency or laser ablation of the GSV below the knee appears to have a higher incidence of postprocedure neuralgia since the nerve and vein are in close proximity. This usually resolves slowly over time.

Portal Hypertension

- Transjugular intrahepatic portosystemic shunt (TIPS or TIPSS): a percutaneous procedure to create a shunt/communication between the portal and hepatic veins to decompress the portal vein.
- TIPSS is accomplished by the following:

  Insert a catheter in the right internal jugular vein.

  Under fluoroscopic guidance, advance the catheter into the right hepatic vein.

  Create a bridge into the portal vein, e.g., advance the catheter from the hepatic vein into the portal vein.

  Support the bridge with an endoprosthesis, i.e., a stent (figure 35-3).

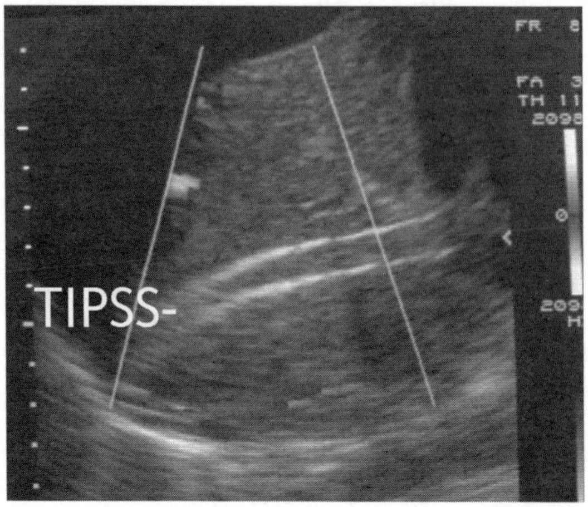

Figure 35-3. The hyperechoic lines represent the walls of the endoprosthesis that was placed to decompress the portal vein. The lack of color filling confirms thrombosis of the stented track.

Test Validation, Statistics, and Patient Safety

PART IV

STATISTICAL PROFILE AND CORRELATION

PATIENT SAFETY AND INFECTION CONTROL

Statistical Profile and Correlation

CHAPTER 36

QA Statistics

It is important to evaluate the value of a noninvasive study by comparing it to a reference test that is assumed to be The Truth: a gold standard. This comparison not only provides a certain level of confidence in the results of the noninvasive test, but also identifies the limitations of a particular noninvasive study, technologist, and/or reading physician. Ideally, the results of a noninvasive study will match those of the gold standard 100% of the time. Realistically, noninvasive studies that by all measures correlate with the gold standard 90% or more of the time are good. There are five measures by which such comparisons are analyzed:

- Sensitivity
- Specificity
- Positive predictive value
- Negative predictive value
- Accuracy

Sensitivity

1. Definition: The ability of a test to detect disease. The abnormal non-invasive test result is confirmed or supported by an abnormal result of the gold standard. In other words, how many noninvasive results were correctly called positives on the basis of gold standard results?
2. Calculation: Number of true positive noninvasive diagnoses/number of all positive diagnoses by the gold standard.

Specificity

1. Definition: The ability of a test to identify normality. The normal test result is confirmed or supported by a normal gold standard result. In other words, of the gold standard negatives, how many did you correctly call negative on the basis of the noninvasive results?
2. Calculation: Number of true negative noninvasive diagnoses/number of all negative diagnoses by the gold standard.

Positive Predictive Value

1. Definition: The percentage of noninvasive test results that accurately predict abnormality. In other words, of your positive noninvasive studies, what percentage correctly predicted disease as supported by the gold standard?
2. Calculation: Number of true positive noninvasive tests/number of all positive noninvasive studies (i.e., true + false positives).

Negative Predictive Value

1. Definition: The percentage of noninvasive test results that accurately predict normality. In other words, of your negative noninvasive studies, what percentage correctly predicted the absence of disease as supported by the gold standard?
2. Calculation: Number of true negative noninvasive tests/number of all negative noninvasive studies (i.e., true + false negatives).

Accuracy

1. Definition: Percentage of correct noninvasive diagnoses. In other words, how well does the noninvasive test both detect and rule out disease?

2 Calculation: Total number of correct tests/total number of all studies.

3 **Note:** The value for accuracy falls between those for sensitivity and specificity and between the positive and negative predictive values.

Comparing Noninvasive and Gold Standard Results

1 The 2 × 2 factorial table (table 36-1) provides a well-accepted method of calculating statistics appropriate for the vascular laboratory.

- *True positives (TP), A:* You (the vascular laboratory) said abnormal, and they (the gold standard) said abnormal. You both agree that the test results are positive (+).
- *False positives (FP), B:* You (the vascular laboratory) said abnormal, but they (the gold standard) said normal. The noninvasive study is positive, but it is wrong—a false positive. Because you and they disagree, the table indicates your positive result (+) and their negative result (−).
- *False negatives (FN), C:* You (the vascular laboratory) said normal, but they (the gold standard) said abnormal. The noninvasive study is negative, but it is wrong—a false negative. Because you and they disagree, the table indicates your negative result (−) and their positive result (+).
- *True negatives (TN), D:* You (the vascular laboratory) said normal, and they (the gold standard) said normal. You both agree that the test results are negative (−).

Table 36-1.

		Gold Standard	
		+	−
TEST	+	A (True Positives)	B (False Positives)
	−	C (False Negatives)	D (True Negatives)

Table 36-2.

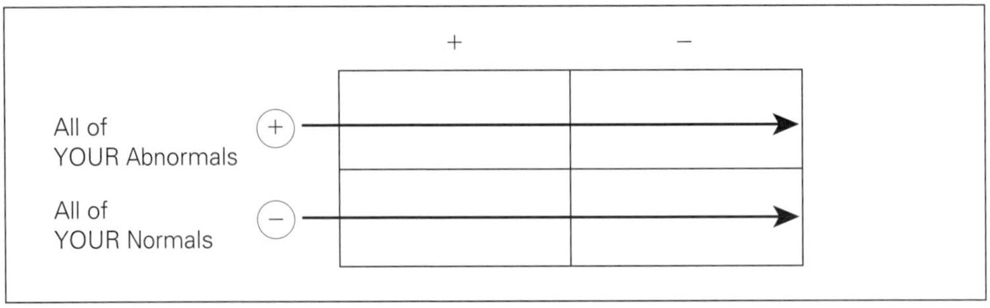

Table 36-3.

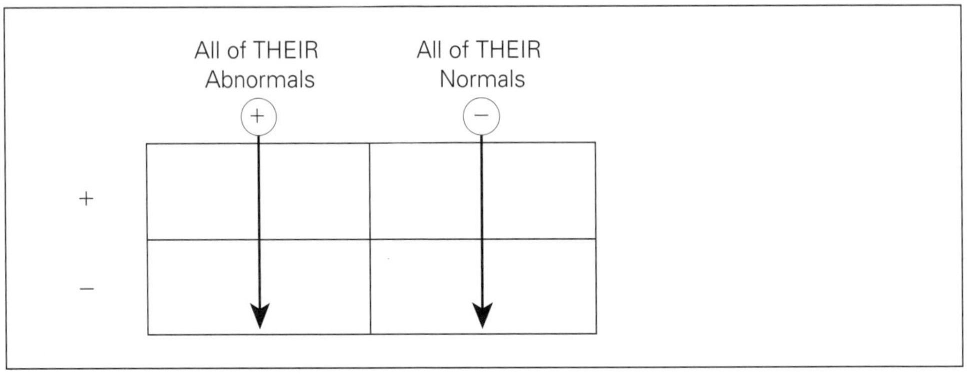

2 Tables 36-2 and 36-3 illustrate another way of understanding the process:

- After comparing the results of noninvasive and gold standard studies and placing the numbers in the correct boxes, the five standard values are calculated: sensitivity, specificity, positive predictive value (PPV), negative predictive value (NPV), and accuracy. The word "group" may appear before each term, e.g., "group positive predictive value (GPPV)."
- The calculations can be made using the following rules:

Sensitivity = TP/(TP + FN)
Specificity = TN/(TN + FP)
Positive Predictive Value = TP/(TP + FP)
Negative Predictive Value = TN/(TN + FN)
Accuracy = (TP + TN)/(TP + FP + FN + TN)

- Table 36-4 illustrates another way of calculating sensitivity and specificity:

Sensitivity = A/(A + C)
Specificity = D/(B + D)

- Table 36-5 illustrates another way of calculating positive predictive and negative predictive values:

Positive Predictive Value = A/(A + B)
Negative Predictive Value = D/(C + D)

- Table 36-6 illustrates another way of calculating overall accuracy:

Accuracy = (A + D)/(A + B + C + D)

Table 36-4.

Gold Standard

TEST		+	−
	+	A (True Positives)	B (False Positives)
	−	C (False Negatives)	D (True Negatives)

Sensitivity Specificity

Table 36-5.

Gold Standard

TEST		+	−	
	+	A (True Positives)	B (False Positives)	Positive predictive value
	−	C (False Negatives)	D (True Negatives)	Negative predictive value

Table 36-6.

Gold Standard

TEST		+	−
	+	A (True Positives)	B (False Positives)
	−	C (False Negatives)	D (True Negatives)

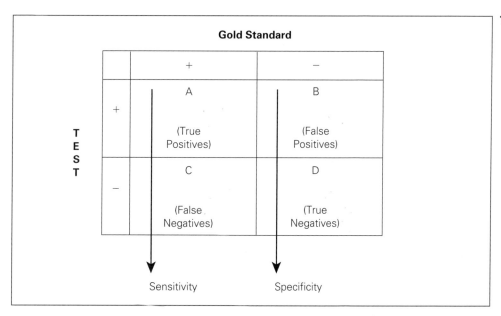

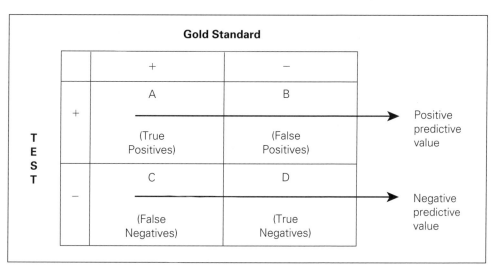

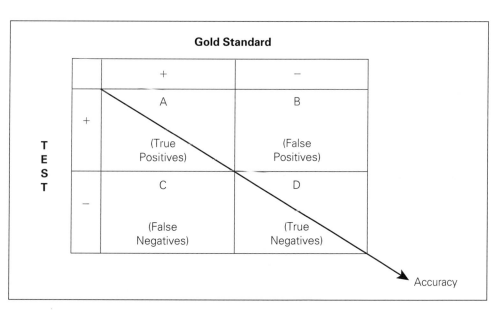

Accuracy

Example of Statistical Calculations

One hundred twenty (120) patients had 4-vessel cerebral angiography performed following duplex evaluation. One hundred thirty-two (132) of the noninvasive and invasive studies documented a hemodynamically significant stenosis of the internal carotid artery (ICA). Ninety-eight (98) ICAs showed very mild occlusive disease by both methods. However, there was disagreement between the invasive and noninvasive findings in 10 vessels: 7 ICAs were thought to have significant stenosis by duplex, but the angiogram showed mild disease; 3 vessels thought to be within normal limits by duplex had significant disease on angiography.

- Step 1: Evaluate this information and determine values for true positive, true negative, false positive, and false negative. (The number of true positives is 132: those abnormal by both methods. The number of true negatives is 98: those considered normal by both methods. In 7 vessels, the duplex was abnormal, but it was wrong: those are false positive. In 3 vessels, the duplex was normal, but it was wrong: those are false negatives.)
- Step 2: Place the values in the appropriate cells of the 2 × 2 table. (See table 36-7 for the correct placement of the information.)
- Step 3: Complete the calculations.

  Sensitivity = 132/135 = 98%

  Specificity = 98/105 = 93%

  Positive predictive value = 132/139 = 95%

  Negative predictive value = 98/101 = 97%

  Accuracy = 230/240 = 96%

Table 36-7.

		Gold Standard	
		+	−
TEST	+	132 (True Positives)	7 (False Positives)
	−	3 (False Negatives)	98 (True Negatives)

Miscellaneous

- If the columns and rows were switched (as might happen in the registry exam)—i.e., the rows now containing the positive and negative numbers by the gold standard and the columns displaying the positive and negative numbers by the noninvasive test (see table 36-8)—it is recommended that you redraw the 2 × 2 table as you learned it. *Then* make the calculations.

- The positive/negative rows or columns may also be switched. Again, it is recommended that you redraw the 2 × 2 table as you learned it, and then make the calculations.

- Remember that accuracy falls somewhere between sensitivity and specificity *as well as* between positive predictive value and negative predictive value.

- If the number of true positive studies changes and all of the other variables remain the same, the following calculations will be affected: sensitivity, positive predictive value, and accuracy. Likewise, if the number of true negative studies changes and all of the other variables remain the same, then specificity, negative predictive value, and accuracy will be altered.

Table 36-8.

GOLD STANDARD		TEST +	TEST −
	+	132 (True Positives)	3 (False Positives)
	−	7 (False Negatives)	98 (True Negatives)

Patient Safety and Infection Control

CHAPTER 37

Universal Precautions and Disinfection

- All personnel should follow good hand-washing techniques before and after direct patient contact, after handling bodily fluid/secretions, and after using the bathroom.
- Standard/universal precautions must be maintained. This means that all bodily fluids are considered infectious. Personal protection equipment must be available (e.g., gloves, gowns, masks) and used by personnel when appropriate.
- Disinfect external ultrasound probes following the manufacturer's specific recommendations. General guidelines:

  First, remove visible residue with a soft cloth lightly dampened with mild soap, then wipe with water-dampened cloth. Although sterile saline may be used to remove visible residue on the transducer, it is *not* a disinfectant agent.

  Commonly used high-level disinfectants include gluteraldehyde (e.g., Cidex) and hydrogen peroxide (≥6%).

  Commonly used low-level disinfectants include quarternary ammonium, N-alkyl (e.g., T Spray II, Sani-Cloth, or hydrogen peroxide [3%]).

If an isopropyl alcohol solution is used, do not let the alcohol air dry.

Do not use undiluted bleach, abrasive cleaners, or solvents such as paint thinner or benzene.

Environmental Safety

- Patients must be provided with a safe environment. This includes being escorted to and from examination rooms and being assisted on and off examination tables as needed. If the patient is on a stretcher, the side rails must be up before and after the study is completed, as well as any time the technologist/sonographer is not right next to the stretcher.
- All electric/electronic equipment and devices must meet the safety code requirements recommended by the manufacturer and the healthcare facility.

Special Considerations

- For hospitalized patients, additional care is given or observations made with regard to special equipment and healthcare issues, e.g., oxygen tanks, IV medications.
- A test procedure should not be carried out if the technologist/sonographer determines that the patient cannot safely tolerate or cooperate with the protocol.

Medical Emergencies

- All personnel providing direct patient care should be trained in basic cardiac life support.
- An emergency response system must be available. This could include an in-house emergency team or a community emergency number such as 911.

Case Studies for Self-Assessment

APPENDIX A

Arterial Evaluation

1 What would be the significance of finding this waveform (figure A-1) in a peripheral artery?

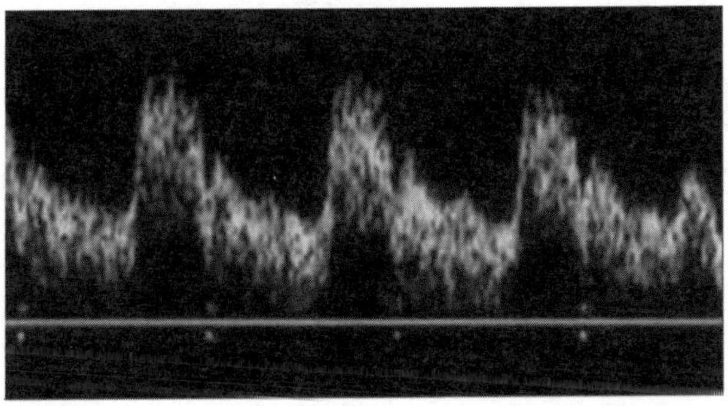

Figure A-1. This peripheral arterial waveform applies to question 1.

A. Blood is flowing into a low-resistance vascular bed rather than the expected high-resistance bed.

B. The arterioles of the distal vascular bed are vasoconstricted.

C. This waveform is consistent with a distal occlusion.

D. This flow pattern suggests a proximal arteriovenous fistula.

2 Your department includes auscultation as part of your limited physical exam. Which of the following is NOT true about that technique?

A. Auscultation means to listen with a stethoscope.

B. You can auscultate the aorta, femoral, and popliteal arteries.

C. The absence of a bruit excludes disease.

D. Bruits can be graded from 1–3.

E. A bruit would indicate turbulent flow.

3 Ms. Morris presents to your vascular lab for arterial testing because she has pain in her legs when she walks. If she has vascular disease, what is the most likely cause?

A. Atherosclerosis.

B. Embolism.

C. Arteritis.

D. Thromboangiitis obliterans.

E. Raynaud's phenomenon.

4 What do these CFA waveforms (figure A-2) represent?

Figure A-2. These CW Doppler waveforms apply to questions 4 and 5.

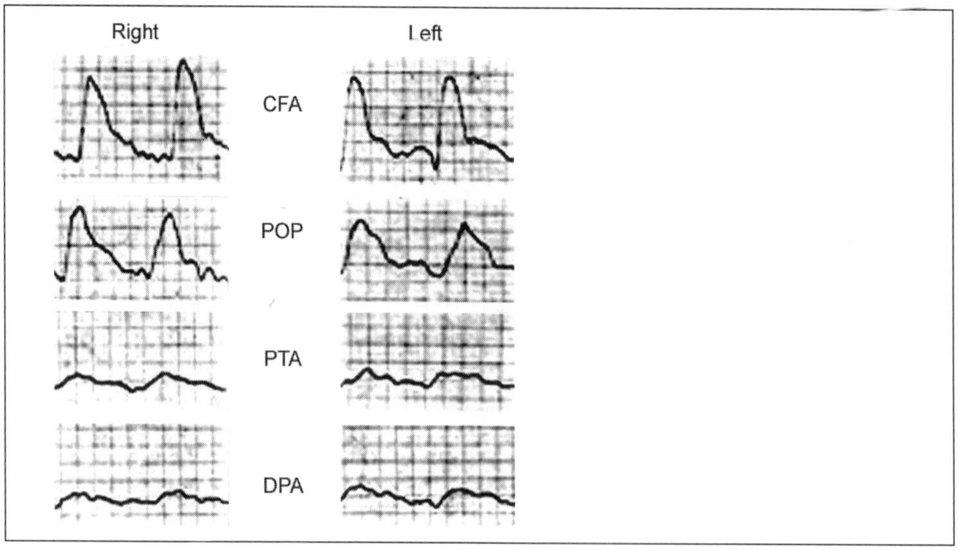

A. You suspect the patient has an acute occlusion of her aorta.

B. The waveforms are within normal limits and exclude proximal disease.

C. You believe that the patient has inflow disease.

D. The waveforms are consistent with a distal embolism.

E. You think your patient has Buerger's disease.

5 Your description of flow patterns in the tibial arteries (figure A-2) would include all of the following EXCEPT:

A. Waveforms are consistent with a low-flow state.

B. There is decreased acceleration time.

C. The flow is monophasic.

D. Waveforms are severely dampened.

6 All of the following are true about the four-cuff method EXCEPT:

A. High-thigh cuff measures proximal SFA pressure.

B. Above-the-knee cuff measures distal SFA pressure.

C. The cuff width size used on the thighs and calves is the same as that on the arms.

D. Cuff artifact is not a consideration with this method.

E. Cuff placement allows determination of the level of peripheral arterial disease.

7 With the three-cuff method, how should the thigh pressure compare to the brachial pressures in the following situations?

A. You would expect the thigh pressures to be lower than the brachial pressures to exclude proximal disease.

B. Thigh pressures are expected to be at least 30 mmHg higher than the highest brachial pressure to indicate the absence of in-flow disease.

C. Thigh pressures are expected to be similar to brachial pressures in the absence of aortoiliac occlusive disease.

D. The three-cuff method is not accurate in the evaluation of proximal disease when brachial pressures are high.

8 Which segmental pressure measurements are obtained last, if at all?

A. Brachial pressures.

B. Ankle level.

C. High thigh.

D. Above the knee/low thigh.

E. Below the knee/calf.

9 Which of the following guidelines should not be used when obtaining segmental pressures?

A. Inflate the cuff 20–30 mmHg beyond the last audible sound, then slowly deflate the cuff.

B. Inflate the cuff 20–30 mmHg higher than highest brachial pressure, then slowly deflate it.

C. Wait approximately 1 minute before repeating the pressure measurement.

D. Inflate the cuff until disappearance of the sound, and then slowly deflate it.

10 Your next patient is a 35-year-old male, Mr. Nichols, who complains of severe pain in his feet at night. Which of the following is the most significant of his risk factors?

A. He has high blood pressure.

B. He smokes 3 packs a day.

C. He has high cholesterol.

D. He had a myocardial infarction (MI) in the recent past.

11 If you suspect that Mr. Nichols has an acute arterial occlusion, which of the following symptoms would not likely be evident?

A. Pain and pallor.

B. Pulselessness and paresthesia.

C. Swelling and rubor.

D. Patches of purple discoloration.

12 Your next patient, Ms. Peters, presents to your department with a bluish discoloration of her right big toe. All pulses are palpable. She denies a history of pain in her legs with activity or pain that wakes her up at night. She has no risk factors associated with arterial disease. What do you suspect is the most likely cause of her problem (mechanism of disease)?

A. Atherosclerosis.

B. Aneurysmal disease.

C. Raynaud's phenomenon.

D. Coarctation of the aorta.

E. Thromboangiitis obliterans.

13 Calculate the ABIs using the pressure measurements in figure A-3.

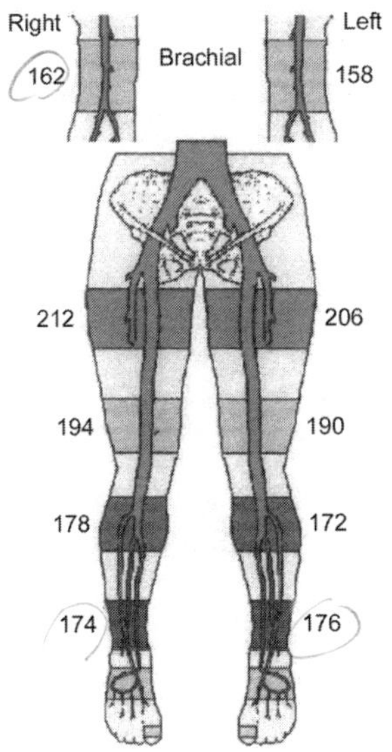

Figure A-3. These segmental pressure measurements were obtained from Ms. Peters and apply to questions 13, 14, and 15.

A. Rt: 174/162 and Lt: 176/158.

B. Rt: 174/158 and Lt: 176/162.

C. Rt: 174/162 and Lt: 176/162.

D. Rt: 178/158 and Lt: 172/162.

14 What toe pressures would you likely obtain based on Ms. Peter's current complaints?

A. 110 mmHg on the right and 124 on the left.

B. 90 mmHg on the great toes bilaterally.

C. 86 mmHg on the right great toe and 0 on the left.

D. 0 on the right great toe and 94 mmHg on the left.

E. 28 mmHg on the right toe and 34 on the left.

15 Based on your assessment of Ms. Peters, what additional noninvasive testing might provide an explanation for her signs and symptoms?

A. Pressures before and after treadmill.

B. Duplex imaging.

C. Tracings before and after immersing feet in cold water.

D. Reactive hyperemia.

E. Repeat tracings after wrapping toes in warm towels.

16 What is the significance of the thigh pressures in figure A-4?

Figure A-4.
These pressure data apply to questions 16 and 17.

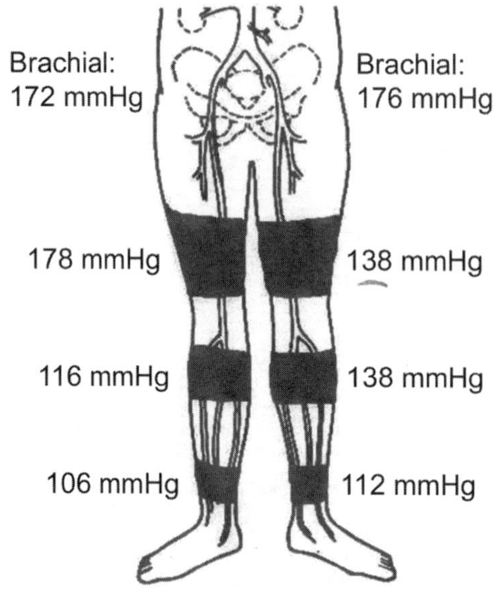

A. It indicates bilateral aortoiliac disease, left side greater than right.

B. The pressures are consistent with proximal disease on the left.

C. This study demonstrates bilateral inflow disease.

D. This patient has significant right common iliac disease.

E. You suspect the patient has incompressible vessels.

17 What information do the below-knee/calf pressure measurements in figure A-4 provide?

A. This patient has hemodynamically significant disease of the femoral-popliteal segment bilaterally.

B. You suspect an occlusion of the left tibial arteries.

C. Your patient has femoral-popliteal disease of the right lower extremity.

D. The measurements exclude arterial occlusive disease.

E. This patient has occluded popliteal arteries bilaterally.

18 Your next patient is complaining of pain in his left calf every time he climbs a flight of stairs. Because the ABIs are within normal limits, you will have the patient complete some form of exercise. Which of the following is not a contraindication to a constant load treadmill?

A. Recent history of severe cardiac arrhythmias.

B. History of hyperlipidemia.

C. A past total hip replacement resulting in a limp.

D. The patient walks with a cane.

E. 10-year history of severe hypertension.

19 After completing the exercise study, you believe the patient has single-level arterial disease based on which of the following conclusions?

A. The patient takes 2–6 minutes to return to his pre-exercise pressures.

B. Recovery time is 6–12 minutes.

C. Recovery time is > 12 minutes.

D. The patient never returns to his pre-exercise pressures.

20 If your patient cannot walk on the treadmill, your protocol is to complete reactive hyperemia. All of the following are true about that test EXCEPT:

A. It provides an alternative method for stressing peripheral circulation.

B. Both large thigh cuffs are inflated to 30 mmHg above the higher brachial pressure.

C. The high-thigh pressure is maintained for 1–3 minutes and then ABIs are obtained.

D. Normal limbs may show a temporary drop of ≤34%.

E. Multilevel disease is considered with a > 50% drop in pressure.

21 Your next patient is scheduled for a coronary artery bypass graft (CABG), and the surgeon wants to use her right radial artery for the bypass. Which of the following is NOT true about the test you will complete?

A. You will perform an Allen test.

B. The study determines the patency of the radial artery.

C. While compressing the radial artery at the wrist, you will ask the patient to make a tight fist.

D. While still compressing the radial artery, you have the patient relax her hand and you observe for a return to its normal color.

E. If color does not reappear, you suspect a radial artery occlusion or palmar arch obstruction.

22 As you evaluate the plethysmographic waveforms from a patient complaining of bilateral hip/thigh pain with activity, you notice that there is no notch (reflection) on the downslope of the waveforms bilaterally. What is the significance of that observation?

A. This is an expected finding.

B. The lack of reflection rules out vascular disease.

C. The finding means the vessels are incompressible.

D. This finding is likely based on the result of collateral arterial branches.

23 A patient presents with left lower extremity claudication. The right lower extremity has normal plethysmographic waveforms and Doppler pressures. The exam results on the left show rounded, broad plethysmographic waveforms at the thigh level, which are similar in quality at the calf and ankle levels. The Doppler pressures are bilateral brachial = 150, left thigh = 100, calf = 90, ankle = 90. These findings suggest:

A. Multilevel disease of the left lower extremity.

B. Significant lesion of the aortoiliac system.

C. Suspected left ileo-femoral arterial obstructive disease.

D. Suspected left tibial vessel arterial stenosis.

24 35-year-old Mrs. Emerson presents with worsening bilateral finger pain, numbness, and coldness, which she has experienced intermittently for the past 15 years. Bilateral digital PPG pressures and waveforms are obtained both before and after cold stress. Before cold, the waveforms have a sharp upstroke, a narrow peak, and a reflection on the downslope. After cold, the waveforms change to a slow upstroke and downslope, a severely rounded peak, and disappearance of reflection on the downslope. These findings remain for as long as 5 minutes after cold immersion. The postcold waveforms are most consistent with:

A. Embolic process.

B. Thromboangiitis obliterans.

C. Primary Raynaud's.

D. Phlegmasia cerulea dolens.

E. Atherosclerosis.

25 The following techniques/instruments can be used as "end point detectors" EXCEPT:

A. Photoplethysmography.

B. Displacement plethysmography.

C. Stethoscope.

D. Volume plethysmography.

E. CW Doppler.

26 A 71-year-old male presents to the vascular laboratory for postoperative evaluation of his right femoral–posterior tibial reverse saphenous vein bypass graft. At rest, the ABI is 0.92. The biphasic Doppler velocities are 85 cm/sec in the native common femoral artery, 100 cm/sec in the proximal anastomosis, and 50 cm/sec at the knee level, with a range of 48–52 cm/sec at the distal anastomosis and native posterior tibial artery. No poststenotic turbulence is observed. What is the most likely cause of the decrease in velocities?

A. Proximal obstruction.

B. Collateralization.

C. Arteriovenous fistula.

D. Increase in graft diameter.

27 If the PSV in the region of stenosis is measured at 224 cm/sec, what would the prestenotic PSVs have to be for this stenosis to be considered ≥75% diameter reduction? (See figure A-5.)

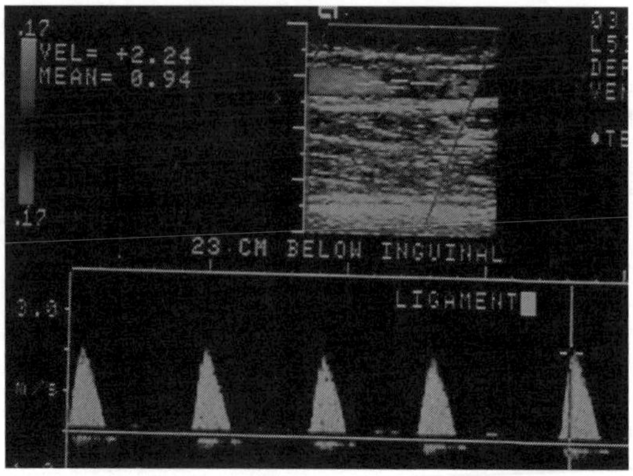

Figure A-5.
This color duplex image applies to question 27.

A. 112 cm/sec.

B. 448 cm/sec.

C. 56 cm/sec.

D. 150 cm/sec.

28 What would your preliminary report include if your next patient has the following upper extremity pressure measurements?

Right		***Left***
132	Brachial	126
108	Radial	128
106	Ulnar	130

A. The pressures are consistent with a right subclavian/axillary stenosis.

B. You suspect the patient may have a stenosis in the right distal brachial artery.

C. This patient may have some type of obstructive process in the right radial and ulnar arteries.

D. B and C.

29 Your next patient, Mr. Richards, is being evaluated for impotency. Your noninvasive study includes all of the following EXCEPT:

A. Obtain Doppler waveforms of the CFA, popliteal artery, PTA, and DPA bilaterally.

B. Apply blood pressure cuffs to the ankles and base of penis.

C. Use the Doppler to obtain ABIs.

D. Obtain penile pressures with either PPG or Doppler.

E. Calculate ABI and PBI.

30 Mr. Richards has ABIs of 0.8 bilaterally using both the posterior tibial artery and dorsalis pedis artery, with a PBI of 0.38. What is your impression of these findings?

A. Mr. Richards has some large-vessel disease, but his PBI is within normal limits.

B. Mr. Richards may be impotent due to vascular disease of his aortoiliac arteries.

C. You suspect blockage of the vessels feeding the penis but can exclude more proximal disease.

D. The PBI rules out poor blood flow to the penis as a source of the problem.

E. Mr. Richards has a vasculogenic impotence from internal iliac artery occlusive disease.

31 All of the following statements about popliteal entrapment are true EXCEPT:

A. It is compression of the popliteal artery by the head of the soleus muscle.

B. The repeated trauma can cause thrombosis and/or emboli.

C. Young men are affected more often than young women.

D. Patients complain of pain in their lower extremity with jogging and/or very fast walking.

E. Bilateral involvement is seen in one-third of cases.

32 What pathology is evident in this film (figure A-6)?

Figure A-6.

This image applies to questions 32 and 33.

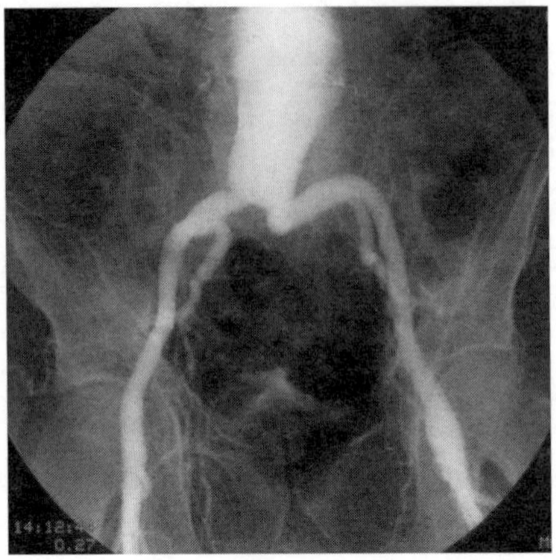

A. Aortoiliac occlusive disease (AIOD).

B. Infrarenal abdominal aneurysm.

C. Coarctation of the aorta.

D. Fibromuscular dysplasia (FMD).

E. Pseudoaneurysm of distal aorta.

33 Of the following choices, what is a frequent complication from the disease process represented in figure A-6?

A. Vasospasm.

B. Rupture.

C. Acute occlusion.

D. Compartment syndrome.

34 All of the following statements about endovascular stents are true EXCEPT:

A. They provide intraluminal support, acting as a kind of scaffold.

B. Insertion is usually through the CFA.

C. Over time the stent can migrate.

D. Restenosis can cause problems.

E. Postoperative surveillance requires angiography.

35 A referring physician asks you to document just the flow velocities in the dialysis graft, rather than performing a complete evaluation. What is the significance of a PSV of 85 cm/sec in this case?

A. This is normal flow for a dialysis graft.

B. You suspect the patient has an arterial inflow problem.

C. This is accelerated flow and is consistent with graft stenosis.

D. You are concerned that the graft is too large and shunting too much blood from the patient's arterial system into his venous system.

36 If PSVs are > 400 cm/sec in the superficial femoral artery, the probable diameter reduction is:

A. <50%. B. 50%.

C. ≥75%. D. ≥90%.

37 As you evaluate a reversed saphenous vein bypass graft, all of the following information is abnormal EXCEPT:

A. ABIs are reduced by < 0.15 compared to previous exam.

B. Doppler flow patterns are multiphasic.

C. Wall defect is observed.

D. Decreased PSVs are evident at the smallest graft diameter.

38 Your next patient is a thin, frail looking 90-year-old female scheduled for an abdominal duplex. As you begin her evaluation, which transducer will you most likely choose to start the exam?

A. 10 MHz CW probe.

B. 7 MHz linear array.

C. 5 MHz curved array.

D. 3 MHz curved array.

39 All of the following transverse measurements are consistent with an abdominal aortic aneurysm EXCEPT:

A. 2.2 × 2.2 cm.

B. 3.4 × 3.6 cm.

C. 4.2 × 4.4 cm.

D. 5.0 × 5.2 cm.

E. 5.4 × 5.8 cm.

40 Low-resistance flow patterns are expected in all of the following arteries EXCEPT:

A. Preprandial superior mesenteric artery.

B. Postprandial celiac artery.

C. Renal arteries.

D. Preprandial hepatic artery.

E. Splenic artery.

41 What would be the significance of finding PSVs of 80 cm/sec in the aorta and 240 cm/sec in the left renal artery?

A. Consistent with a 50% diameter reduction of the left renal artery.

B. Within normal limits.

C. Documents a ≥60% diameter reduction of the left renal artery.

D. Consistent with ≥75% diameter reduction of the left renal artery.

E. Suggests a 30% diameter reduction of the left renal artery.

42 As you evaluate the cause for the flow acceleration in the celiac artery, which of the following will most likely help you figure out what's going on?

A. Comparing PSVs with inspiration and exhalation.

B. Determining if there are flow changes postprandially.

C. Determining if the flow is different when the patient fasts.

D. Comparing PSVs with the patient in a slightly different position.

E. Helping the patient relax and then retaking the velocity measurements.

43 A 69-year-old patient presents with a one-year history of uncontrollable hypertension. Duplex findings include elevated, low-resistance PSVs (160–180 cm/sec) throughout the right renal artery without poststenotic turbulence, as well as cystic right kidney, absence of flow in the left renal artery, and a small left kidney with PSVs of 8 cm/sec. What is the most likely explanation for the elevated velocities in the right renal artery?

A. Accessory renal artery.

B. Contralateral renal artery occlusion.

C. Hydronephrosis.

D. Contralateral nephrosclerosis.

44 A 65-year-old patient presents to the vascular laboratory complaining that for the past several months she has experienced crampy abdominal pain after eating. During this time she has lost 15 pounds. A pre/postprandial duplex study yields the following findings, all of which are considered abnormal EXCEPT:

A. The SMA's end diastolic velocities are approximately 2 cm/sec pre/postprandially.

B. The celiac artery's PSVs are > 280 cm/sec with spectral broadening.

C. The SMA and celiac artery branch off of a common trunk.

D. The IMA's PSVs are approximately 280 cm/sec followed by turbulence.

45 An angiogram reveals significant aortoiliac occlusive disease in a 55-year-old male. Vascular laboratory findings show resting ankle/brachial indices of 0.4 on the right and 0.3 on the left. All of the following situations could be possible EXCEPT:

A. The patient complains of impotence.

B. The patient has 1/2 block claudication.

C. The patient has palpable pedal pulses.

D. The patient experiences ischemic rest pain.

46 As you evaluate flow patterns in a patient, you document a PSV of 104 cm/sec at the distal aorta, 218 cm/sec at the take-off of the left common iliac artery (CIA), and 226 cm/sec at the right proximal CIA. What would your preliminary report include?

A. This patient has a ≥75% diameter reduction of the distal aorta.

B. These findings are consistent with ≥75% diameter reduction of the right CIA.

C. Your patient has >50% diameter reduction of the CIAs bilaterally.

D. You have documented a >50% diameter reduction of the right CIA and >75% diameter reduction of the left.

E. These flows are within normal limits.

47 Your patient says he has been recently diagnosed with congestive heart failure (CHF). How might CHF be associated with his dialysis access?

A. The two are not related at all.

B. You suspect that the access is close to the heart and a large amount of blood is being shunted into the venous circulation, stressing the heart.

C. You're thinking that the access is too small and the blood going around the dialysis access graft through collateral channels is stressing the circulatory system.

D. CHF is an expected, long-term result of a dialysis access site anywhere in the body.

E. You know that CHF is a temporary finding when a dialysis access site is created.

48 All of the following statements about interpreting flow in a native artery of the upper extremity are true EXCEPT:

A. PSVs can vary widely secondary to changes in temperature.

B. There is no criteria consistent with a >50% diameter reduction.

C. The stenosis profile is not evident in upper extremity disease.

D. Absence of Doppler signals is most likely consistent with an occlusion.

E. Distal emboli may be associated with a subclavian aneurysm.

49 A 64-year-old male presents to the vascular laboratory with a history of left extremity pain (calf claudication) after 150 feet, ulceration on the great toe, and ankle Doppler systolic pressures of 300 mmHg (brachial pressures are 140 mmHg bilaterally). What are these findings most consistent with?

A. Small-vessel disease.

B. Elevated pressure due to the large cuff size.

C. Erroneous pressures due to probable arterial calcification.

D. Severe hypertension.

E. Signs/symptoms are not related to vascular disease.

50 What would your preliminary report include if the aortic PSV were 70 cm/sec and the right renal artery had peak velocities of 280 cm/sec?

A. These findings are within normal limits.

B. This patient has ≥60% diameter reduction of the right renal artery.

C. This is consistent with a 50% diameter reduction.

D. You suspect the patient has a 75% diameter reduction.

E. These findings suggest an occlusion of the right renal artery.

51 As you begin your next physiologic study, which will include CW Doppler waveforms and segmental pressure measurements, you understand that the size of the cuffs affects the accuracy of the systolic pressure measurements. Which is the LEAST accurate statement about the selection of blood pressure cuffs?

A. The length of the cuff is a critical factor in obtaining accurate blood pressure measurements.

B. A narrow cuff can result in artifactually high pressures.

C. The width of the cuff should be 20% of limb diameter.

D. The most frequently used cuff for brachial blood pressure measurements is 12 cm × 40 cm.

E. A wide cuff can result in artifactually low pressures.

52 Your department happens to have a piece of physiologic equipment that can obtain four blood pressure measurements simultaneously. This method can be utilized in all but which of the following scenarios?

A. Right finger, left arm, penis, right ankle.

B. Left calf, right arm, left toe, right thigh.

C. Penis, left arm, left ankle, right arm.

D. Right forearm, right calf, left arm, left ankle.

53 You are completing an abdominal ultrasound evaluation on a patient following endovascular repair of an abdominal aortic aneurysm. If you were to detect flow in the aneurysmal sac at the distal/inferior attachment site, what type of problem would you have identified?

A. Type I endoleak

B. This is not a problem, but an expected finding.

C. Type II endoleak.

D. Flow through the IMA.

E. This finding is unexpected but not considered an endoleak.

54 You are concerned when you see a well-defined area of very low-level echoes within the aneurysmal sac. Although you are unable to detect flow in that area with color flow Doppler, you suspect something is going on. Which of the following best describes your finding?

A. This finding is consistent with a type III endoleak.

B. This is not a problem, but an expected finding.

C. You know the endograft is now at risk of rupture.

D. You suspect something called endotension.

E. It is consistent with blood leaking through the graft material.

55 All of the following are points to remember when scanning the abdomen EXCEPT:

A. Deeper vessels require a lower-frequency transducer.

B. You may need to use a variety of approaches.

C. More pressure on the transducer may be required to see areas of interest.

D. You will make large pivoting movements with abdominal imaging.

E. Color flow Doppler assists with vessel identification.

Cerebrovascular Evaluation

56 What type of flow patterns do you normally expect from vessel B in figure A-7?

Figure A-7. This color flow image applies to questions 56 and 57.

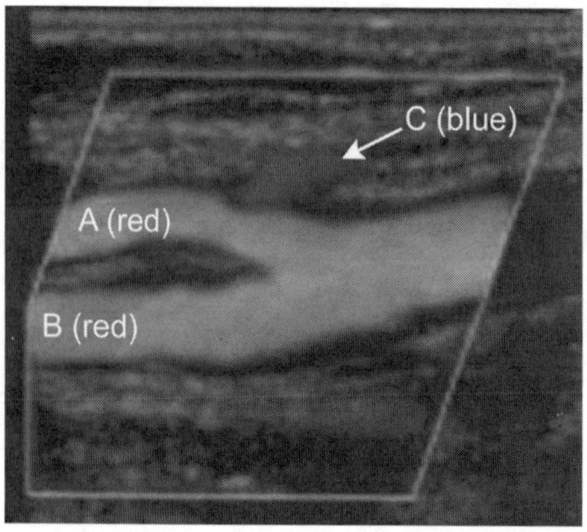

A. More flow in end-diastole than in vessel A.

B. Caudal flow.

C. Turbulent flow.

D. High-resistance.

E. Increased acceleration time.

57 Which of the following is most likely true about vessel C in figure A-7?

A. It is the occipital artery with abnormal flow.

B. It is the superior thyroid artery with normal flow.

C. It represents a collateral vessel with normal flow.

D. It is the superficial temporal artery with abnormal flow.

E. It is the facial artery with normal flow.

58 Your next patient complains of increasing episodes of unilateral facial and arm weakness rather than leg problems, and behavioral changes. Which vessel do you most likely suspect has an obstruction?

A. Anterior cerebral artery.

B. Middle cerebral artery.

C. Vertebrobasilar system.

D. Posterior cerebral artery.

E. These complaints are not localized to any specific vessel.

59 A previously healthy 74-year-old female is undergoing a carotid duplex study because of a pulsatile mass in her neck. Which one of the following is the most likely explanation for a pulsatile mass in the cervical carotid artery?

A. You are most likely to find a localized dilatation of the common carotid artery (CCA) on ultrasound.

B. You anticipate findings of a carotid body tumor.

C. The physical findings are most likely due to a tortuous CCA.

D. The pulsatile mass is the result of trauma, e.g., carotid endarterectomy.

E. Atherosclerosis can be a causative agent.

60 Mr. Ernst is coming to your department for his 12-month scan following right carotid endarterectomy. You notice that the incision is healing nicely and the patient denies any cerebrovascular signs/symptoms. The Doppler peak systolic velocities are approximately 130 cm/sec throughout the internal carotid artery. Your image (figure A-8) documents all of the following findings EXCEPT:

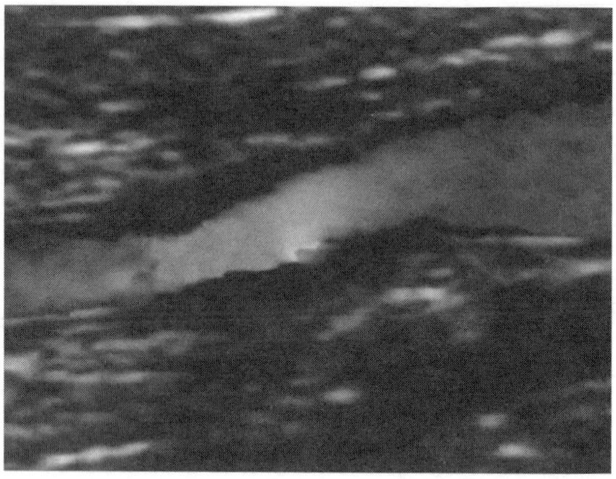

Figure A-8. This image of your patient's right distal common carotid artery (CCA)–proximal internal carotid artery applies to questions 60 and 61.

A. Patent vessels.

B. Hyperechoic region near proximal ICA.

C. Evidence of wall irregularity.

D. Absence of poststenotic turbulence.

E. No hemodynamically significant wall irregularities in distal CCA.

61 Your preliminary report to Mr. Ernst's vascular surgeon would most likely be based on which of the following pieces of information (see figure A-8)?

A. The patient has a preocclusive lesion in the proximal ICA.

B. The postoperative carotid duplex examination is within normal limits.

C. You suspect the patient has had a rapid production of smooth muscle cells.

D. Mr. Ernst has developed fibromuscular dysplasia.

E. This postoperative scan reveals that another plaque formation has developed.

62 As you start the carotid duplex examination, what do you suspect based on this initial duplex scan of the left side (figure A-9)?

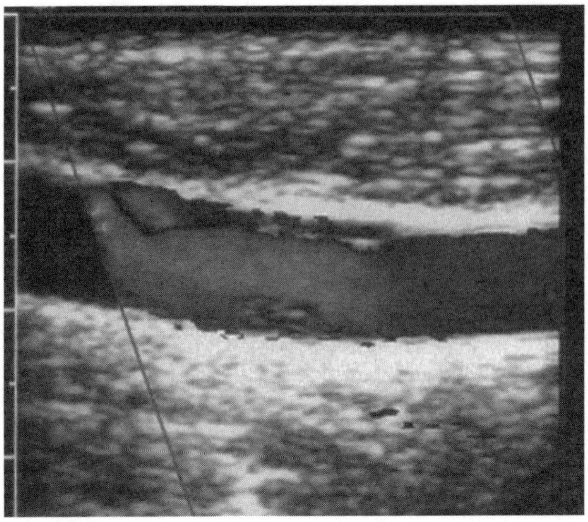

Figure A-9. This scan of a 64-year-old woman who was recently in a car accident applies to questions 62 and 63. A carotid duplex examination was ordered because the ER physician heard a bruit on the left side of her neck. Color flow Doppler displays predominantly red color with an area of blue at the superficial wall edge.

A. A hemodynamically significant stenosis.

B. A carotid body tumor.

C. An arterial dissection.

D. A possible pseudoaneurysm.

E. A form of arteritis.

63 You think about the pre-exam history and physical that you complete before every noninvasive study and recall that your patient's bilateral brachial blood pressures were 134/82 mmHg on the right and 130/80 mmHg on the left. You also heard the bruit in the left cervical carotid area. Although you auscultated the bruit, what would you call it if you palpated it?

A It is still referred to as a bruit.

B. It would be called a thrill.

C. Most people consider it a bounding pulse.

D. Some of your coworkers document it as being phasic.

64 You begin a carotid duplex scan of your next patient and measure PSVs of 185 cm/sec in her right internal carotid artery. Although this corresponds to a hemodynamically significant lesion, you do not see any plaque formation. You understand that accelerated flow, which can lead to overestimation of disease, can result from all of the following EXCEPT:

A. Inappropriate Doppler angle.

B. Stenosis at an area of dilatation.

C. Increased cardiac output.

D. A tortuous vessel.

E. Vessel providing a collateral pathway.

65 A patient arrives at your facility for a carotid duplex evaluation. Based on the nonlateralizing signs/symptoms of dizziness and an episode of blurred vision bilaterally, you suspect she may have which of the following problems?

A. An MCA infarction.

B. A stroke related to a PCA lesion.

C. VBI.

D. Bilateral ICA stenosis.

E. An ACA infarction.

66 All of the following statements about figure A-10 are true EXCEPT:

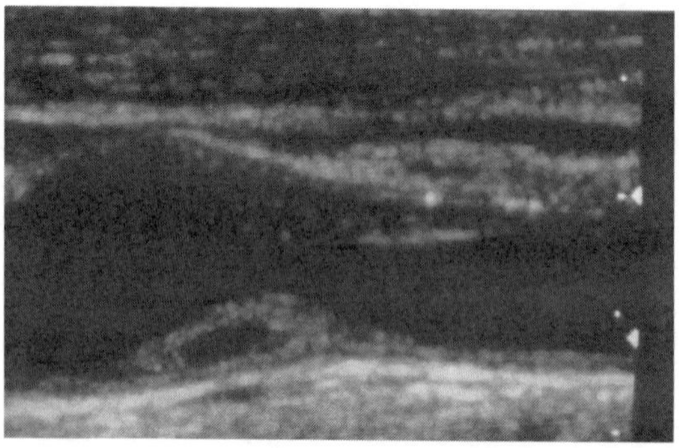

Figure A-10. This image, from the patient in question 65, applies to question 66.

A. The patient has a heterogeneous plaque formation.

B. This lesion accounts for the patient's symptomatology.

C. You would be concerned that this structure may be unstable.

D. This scan is suggestive of an area of intraplaque hemorrhage.

67 Which of the following statements is true about this image (figure A-11) from Mr. Kallen, a previously healthy 54-year-old with no history of vascular disease?

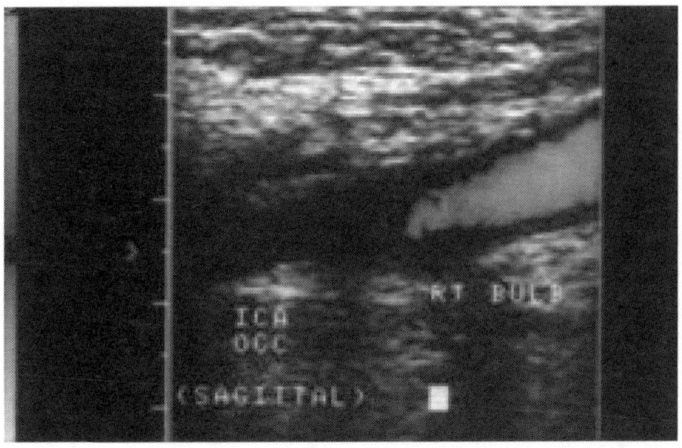

Figure A-11. This image of an internal carotid artery applies to question 67.

A. The patient has a calcific plaque formation resulting in an occluded internal carotid artery (ICA).

B. Since the blockage is on the right side, Mr. Kallen is at risk for a neurologic deficit related to his left hemisphere.

C. The B-mode findings are consistent with thrombosis of the ICA.

D. You suspect Mr. Kallen has fibromuscular dysplasia.

68 What is the most likely cause of Mr. Kallen's pathology?

A. It may be secondary to a carotid dissection.

B. Mr. Kallen has intraplaque hemorrhage.

C. The occlusion is the result of heterogeneous plaque.

D. You suspect intimal hyperplasia.

E. It may be related to temporal arteritis.

69 Oculopneumoplethysmography (OPG-Gee) is ordered on Mr. Kallen to assess flow to the ophthalmic arteries, an indicator of possible collateralization. As you start the study you understand that the contraindications to OPG-Gee include all of the following EXCEPT:

A. Conjunctivitis.

B. Acute or unstable glaucoma.

C. A past retinal detachment.

D. Eye surgery within the last 6 months.

70 Before starting the OPG-Gee, you obtain the following brachial pressures on Mr. Kallen: 162/90 on the right and 158/88 on the left. As you follow normal testing protocol, you instill a local anesthetic to the eyes and apply 500 mmHg pressure to the eye cups. You immediately document pulsations on the recording. What is your next step in the process?

A. You would repeat the process but apply 300 mmHg pressure instead of the 500 mmHg.

B. The study cannot be completed because of the patient's hypertension.

C. You reapply the local anesthetic and apply 500 mmHg pressure to the eye cups.

D. Pulsations that are immediately evident on the recording is an expected finding. You would release the eye cups and calculate the ophthalmic systolic pressures.

71 Your next patient, Ms. Lewis, is being evaluated for acute left hemiplegia and aphasia. As you start evaluating her right cervical carotid arteries, you document the flow in her right common carotid artery (figure A-12). This flow pattern most likely suggests which of the following?

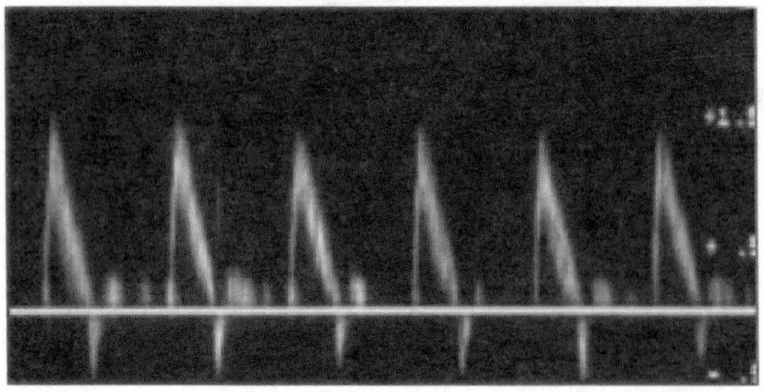

Figure A-12. This waveform from a right common carotid artery applies to question 71.

A. If unilateral, it would suggest a distal occlusion.

B. It represents normal flow in the right CCA.

C. It suggests a proximal lesion.

D. You think this signifies collateral flow secondary to contralateral occlusion.

E. You are concerned about decreased cardiac output.

72 Your next patient is having a carotid duplex examination because she wants a second opinion. Her original study, performed elsewhere, is consistent with a high-grade stenosis of her left carotid artery. This waveform of her left ICA (figure A-13) is consistent with all of the following EXCEPT:

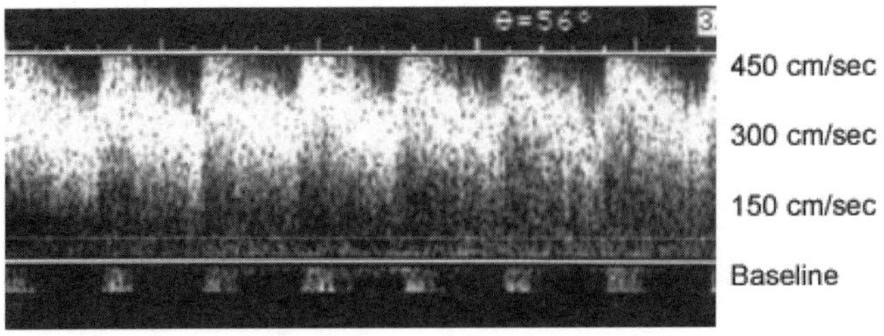

Figure A-13. This waveform from a left internal carotid artery applies to question 72.

A. It confirms the findings of her original study.

B. This signal is greater than one-half the PRF.

C. The signal exceeds the Nyquist limit.

D. It results any time the Doppler angle is < 60 degrees.

E. The PSV cannot be accurately obtained.

73 A local neurologist refers a patient for a TCD examination. The nonimaging technique uses a 2 MHz pulsed Doppler for which of the following reasons?

A. The vessels being evaluated are shallow.

B. The technique requires range resolution.

C. The location of the signal helps to confirm which vessel is being evaluated.

D. A and B.

E. B and C.

74 What angle of insonation is assumed when doing TCD?

A. 0 degrees.

B. 45–60 degrees.

C. 60 degrees.

D. 45 degrees.

E. 90 degrees.

75 As you begin the TCD study, you know that various windows are used to insonate specific intracranial vessels. Which of the following vessels is least likely to be evaluated through the transtemporal window?

A. ACA.

B. Basilar artery.

C. MCA.

D. PCA.

E. Distal ICA.

76 All of the following information accurately describes MCA findings EXCEPT:

A. Antegrade flow is expected.

B. Normal velocity is 55 +/− 12 cm/sec.

C. The signal is obtained at a depth of 60–70 mm.

D. The probe is angled anterior and superior.

E. Bidirectional flow at its origin is an important landmark.

77 As you continue your evaluation you detect antegrade flow in the ipsilateral ACA. What is the significance of that finding?

A. It represents crossover collateralization.

B. It is an expected finding and within normal limits.

C. You anticipate a contralateral ICA occlusion.

D. It represents compensatory posterior-to-anterior flow.

E. You are concerned about a vasospasm.

78 What findings are documented in figure A-14?

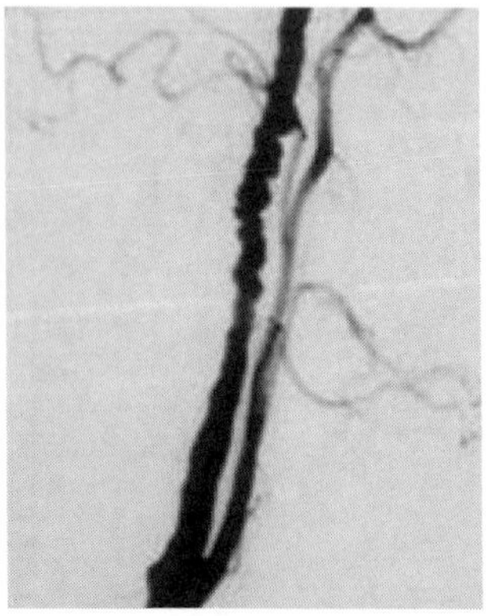

Figure A-14. This angiogram applies to question 78.

A. Fibromuscular dysplasia.

B. Intimal hyperplasia.

C. Carotid body tumor.

D. Atherosclerotic plaque formation.

E. Carotid dissection.

79 How would this angiogram (figure A-15) be interpreted?

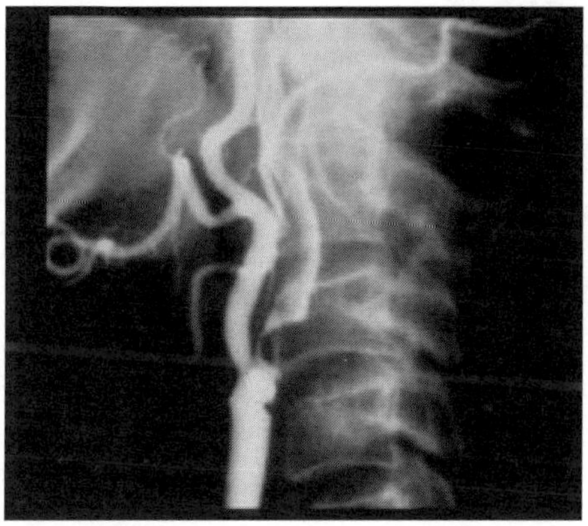

Figure A-15. This angiogram applies to questions 79 and 80.

A. This patient has a hemodynamically significant lesion of the ECA.

B. This film documents a preocclusive lesion of the ICA.

C. This study shows a severe problem with the distal CCA.

D. You suspect an occluded ICA.

E. An absent ECA is evident in this film.

80 If the true lumen measured 8 mm and the residual lumen measured 2 mm, what is the diameter reduction (figure A-15)?

A. 25%

B. 33%

C. 50%

D. 66%

E. 75%

Venous Evaluation

81 Based on your knowledge of anatomy, you know that vessels A and B (figure A-16) are which vessels?

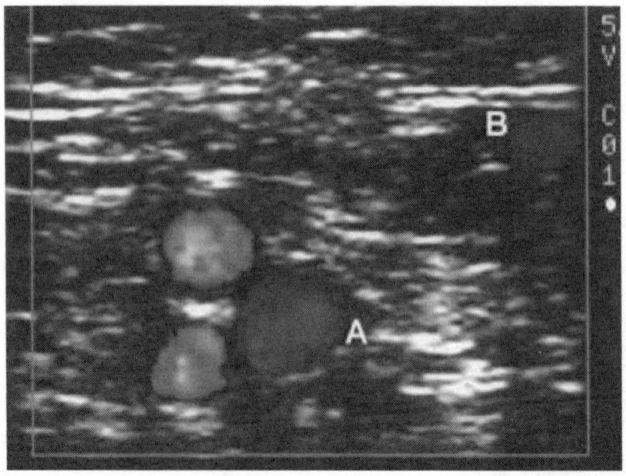

Figure A-16. Questions 81 and 82 relate to this duplex image obtained just below the inguinal ligament of the right lower extremity. Color flow Doppler displays vessels A and B in blue and the vessels adjacent to A in red.

A. A is the common femoral vein and B is the femoral vein.

B. A is the profunda femoris vein and B is the femoral vein.

C. A is the common femoral vein and B is the great saphenous vein.

D. A is the femoral vein and B is the great saphenous vein.

82 Name the superficial and deep vessels to the left of vessel A represented in red (figure A-16):

A. Common femoral and superficial femoral arteries.

B. Superficial femoral and profunda femoris arteries.

C. Common femoral and profunda femoris arteries.

D. Superficial femoral and deep superior epigastric arteries.

83 Even though there are many risk factors for DVT, all of them fall into which three basic categories known as Virchow's triad?

A. Trauma, stasis, hypercoagulability.

B. Stasis, immobility, extrinsic compression.

C. Intrinsic trauma, heart disease, immobility.

D. Cancer, myeloproliferative disorders, heart disease.

84 The duplex image in figure A-17 documents all of the following flow characteristics EXCEPT:

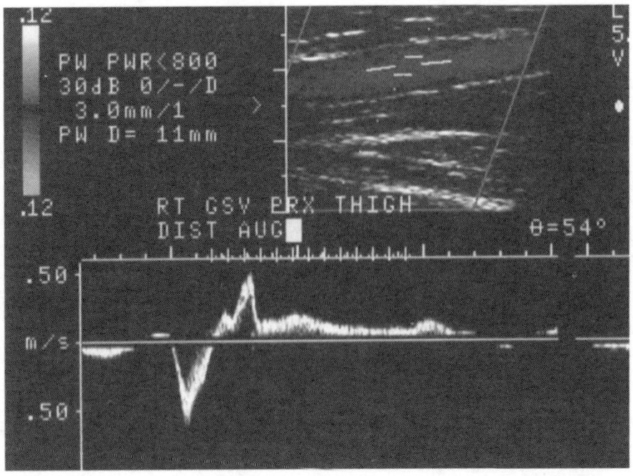

Figure A-17. This duplex image applies to questions 84 and 85.

A. A patent vessel.

B. Presence of spontaneous venous flow.

C. Augmentation with distal compression.

D. Competent valves.

85 As you ask your patient to do a Valsalva maneuver, which flow characteristic do you normally expect to see (see figure A-17)?

A. Cessation of flow during the maneuver.

B. Augmented flow when the patient relaxes and breathes normally.

C. Increased flow as the patient bears down.

D. A and B.

86 What is the most likely cause for the discoloration in the patient's legs (figure A-18)?

Figure A-18.
Questions 86, 87, and 88 are based on this photograph.

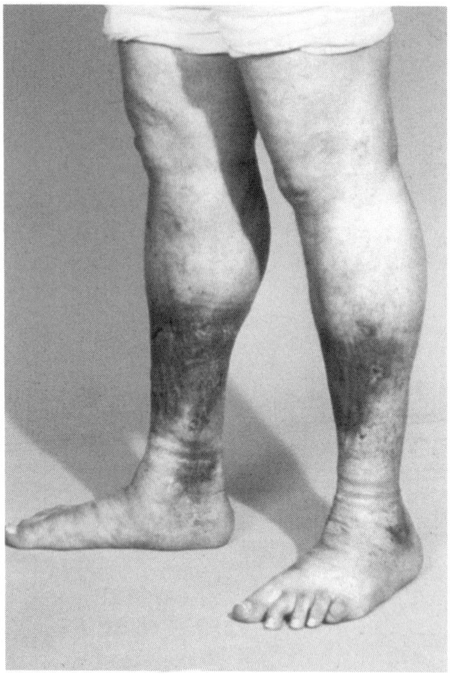

A. Acute peripheral arterial disease.

B. Acute deep venous thrombosis.

C. Inflammatory process.

D. Chronic arterial insufficiency.

E. Chronic venous insufficiency.

87 The discoloration illustrated in figure A-18 is related to alterations in blood flow that can result from all of the following conditions EXCEPT:

A. Increased capillary pressure.

B. Reduced venous pressure.

C. Fluid moving into the tissue, causing edema.

D. Other blood components, e.g., red blood cells, moving into the tissue.

88 In completing a physiologic study (e.g., PPG) on the patient appearing in figure A-18 for complaints of increased swelling of the lower extremities, which of the following findings would be LEAST likely?

A. An initial venous refill time (VRT) of 34 seconds, with a VRT of 26 seconds with application of a tourniquet.

B. An initial venous reactive time (also referred to as VRT) of 14 seconds with a VRT of 12 seconds with tourniquet application.

C. A VRT of 10 seconds with and without tourniquet.

D. An initial VRT of 16 seconds, decreasing to 9 seconds following tourniquet application.

E. A VRT of 14 seconds with and without use of a tourniquet.

89 The duplex image in figure A-19 documents all of the following flow characteristics EXCEPT:

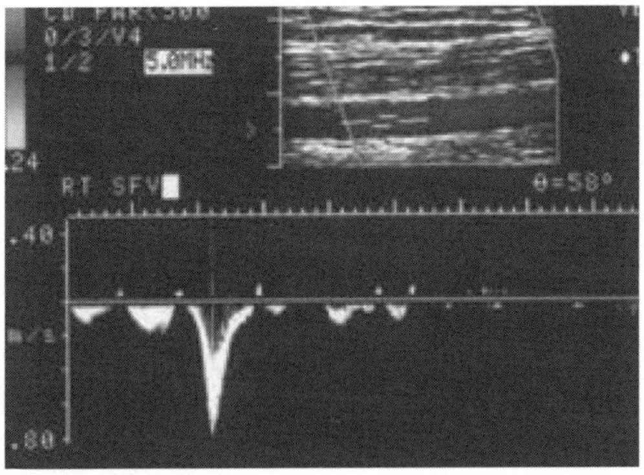

Figure A-19. This duplex image applies to question 89.

A. Spontaneous flow.

B. Phasic flow.

C. Augmentation with distal augmentation.

D. Valvular incompetence.

90 Based on your knowledge of venous anatomy, what vessel is being compressed (figure A-20)?

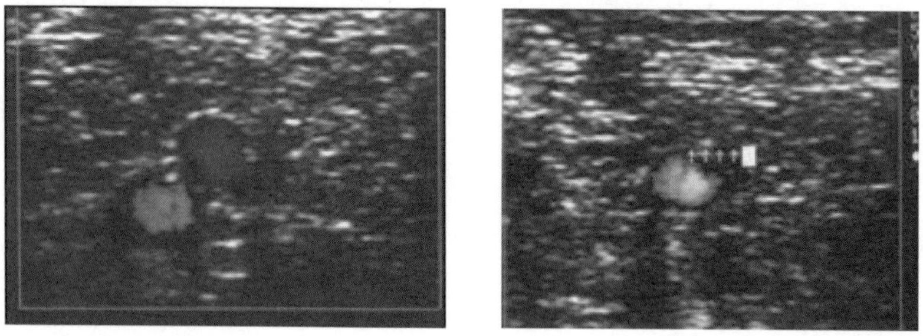

A

B

Figure A-20. Question 90 is based on these two images.

A. Common femoral vein.

B. Femoral vein.

C. Popliteal vein.

D. Posterior tibial vein.

91 What does this venous duplex image of the right lower extremity (figure A-21) most likely document, with color flow Doppler showing red in the vessel on the left and no color in the vessel on the right?

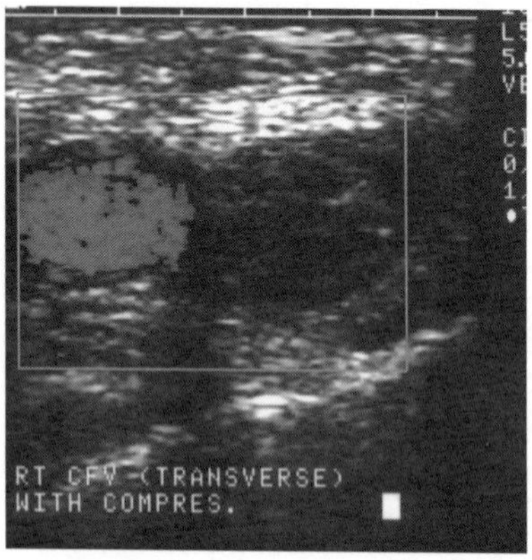

Figure A-21. Questions 91 and 92 are based on this venous color flow image obtained during compression.

A. Absence of a deep venous thrombosis.

B. Presence of chronic venous insufficiency.

C. The study is within normal limits.

D. An acute deep venous thrombosis.

E. Chronic venous thrombosis.

92 Which of the following terms would NOT accurately describe the structure on the right in figure A-21?

A. Predominantly hypoechoic/low-level echoes.

B. Dilated.

C. Patent.

D. Walls do not coapt.

93 The internal jugular vein carries blood into which vessel?

A. External jugular vein.

B. Subclavian vein.

C. Superior vena cava.

D. Innominate vein.

E. Cephalic vein.

94 Where does blood flow from the cephalic vein?

A. Brachial vein.

B. Subclavian vein.

C. Axillary vein.

D. Internal jugular vein.

E. Basilic vein.

95 The superior vena cava is formed by the confluence of which two vessels?

A. Jugular and subclavian veins.

B. Innominate and subclavian veins.

C. Cephalic and axillary veins.

D. Bilateral innominate veins.

E. Subclavian and basilic veins.

96 Your next patient is scheduled for vein mapping and the doctor is asking that you map his upper extremity veins because the superficial veins of the lower extremity have either been harvested or are not suitable as a bypass conduit. All of the following are true about upper extremity vein mapping EXCEPT:

A. The cephalic vein is lateral to the basilic vein.

B. The basilic vein is medial to the brachial veins.

C. The radial veins are medial to the ulnar veins.

D. The cephalic vein is lateral to the radial veins.

E. The basilic vein is medial to the ulnar veins.

97 If the lower and upper extremity superficial veins are not suitable as an arterial conduit for coronary artery bypass surgery, which vessel will most likely be considered for this procedure?

A. Brachial vein.

B. Radial artery.

C. Ulnar vein.

D. Ulnar artery.

E. Profunda femoris vein.

98 You are performing an upper extremity venous duplex study on Ms. Granger. As you move the transducer distally from her left axilla to her upper arm, you document this image (figure A-22). Color flow Doppler shows vessel A as blue, B without color, and C as red. Vessel A most likely depicts which vessel?

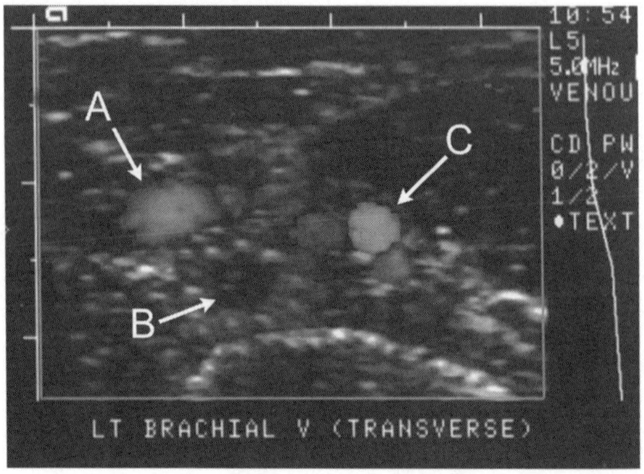

Figure A-22. Questions 98 and 99 are based on this duplex image. Note mixed echoes in structure B.

A. Cephalic vein.

B. Basilic vein.

C. One of the brachial veins.

D. Axillary vein.

E. Radial vein.

99 As you evaluate the vessels of interest as well as the surrounding soft tissue, you are a little concerned about structure B (see figure A-22). What does it most likely represent?

A. Thrombosed brachial vein.

B. Occluded radial artery.

C. A nerve.

D. A calcified accessory artery.

100 Your next patient is scheduled for an ultrasound evaluation of the portal vein. What are normal flow characteristics of that vessel?

A. Hepatofugal: into the liver; bidirectional and phasic.

B. Hepatopetal: into the liver; minimally phasic and almost continuous.

C. Hepatofugal: out of the liver; bidirectional and phasic.

D. Hepatopetal: into the liver; bidirectional and phasic.

E. Hepatopetal or hepatofugal depending on the normal metabolic demands of the liver.

101 You detect abnormal blood flow in the portal vein and suspect that your patient has portal hypertension and that the radiologist may consider doing a transjugular intrahepatic portal-systemic shunt (TIPSS). What flow changes would you expect to see if the TIPSS procedure is successful?

A. The pressure in the portal vein would decrease.

B. Blood would be shunted from the portal vein into the superior mesenteric vein.

C. Decreased flow in the hepatic vein would be evident.

D. The pressure in the hepatic vein would decrease.

102 Because you have been so helpful in the past, a physician is now calling to ask if you can perform photoplethysmography (PPG) on one of his patients. What is the role of PPG in venous evaluations?

A. It can detect acute deep venous thrombosis (DVT).

B. PPG is an accurate way to assess collateral development.

C. The study documents the presence/absence of venous insufficiency.

D. It evaluates the status of the calf muscle pump.

E. PPG can differentiate chronic from acute thrombus.

103 As you assess your patient before the PPG study you expect to see which of the following symptoms?

A. A swollen, painful lower extremity.

B. A hot, red leg.

C. Brawny discoloration around the gaiter zone.

D. A painful mass behind the knee.

E. Cyanosis of the lower extremity.

104 Your next patient, Ms. Montez, is being evaluated for a two-day history of a painful, swollen right lower extremity. Because your diagnostic facility does not offer 24-hour service, the patient is sent for a venogram. Where is the contrast agent injected to provide the images in figure A-23?

Figure A-23. Questions 104 and 105 are based on the following venogram.

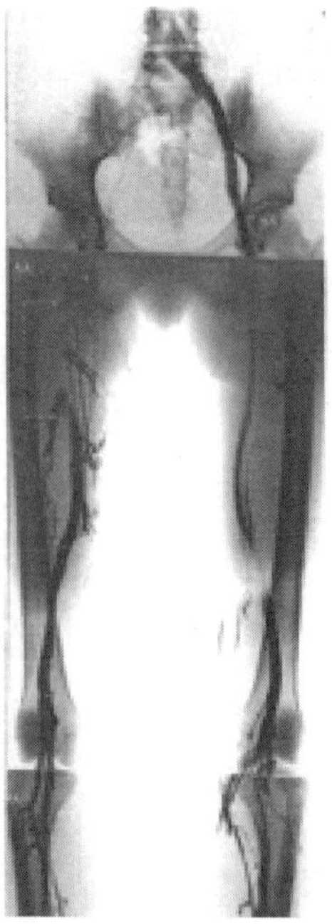

A. A superficial vein on the dorsum of the foot.

B. The anterior tibial veins.

C. The popliteal vein.

D. The common femoral vein.

E. A superficial vein near the groin.

105 What pathology is documented on the film in figure A-23?

A. Chronic thrombosis of the popliteal vein is evident.

B. A proximal acute deep venous thrombosis is documented.

C. Venous reflux is present.

D. A Baker's cyst is seen.

E. This patient has superficial varicosities.

106 As you perform venous imaging on your next patient, you document this finding in the patient's popliteal fossa (figure A-24). The structure on the left of the image is LEAST likely to represent which of the following?

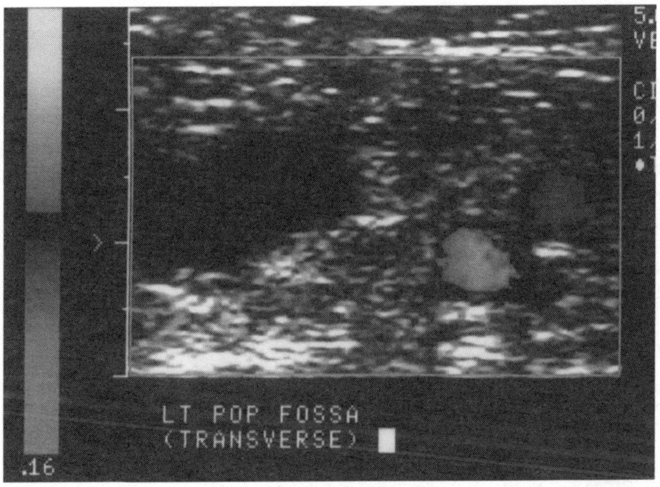

Figure A-24. Questions 106, 107, and 108 are related to the following duplex image of the left popliteal fossa. The structure on the left has no color filling. On the right, the superficial structure is blue and the deeper structure is red by color flow Doppler.

A. Hematoma.

B. Cyst.

C. Sarcoma.

D. Acute deep venous thrombosis.

E. Ruptured Baker's cyst.

107 Regarding figure A-24, which of the following information related to the patient's history and physical exam is LEAST important to you as you put together your preliminary report?

A. The patient has a history of arthritis.

B. The knee area has been painful over the last four days.

C. A color change in the toes/foot is occasionally experienced.

D. The patient complains of discomfort in the proximal calf.

108 You contact the referring physician with your preliminary findings, and the physician tells you he thinks the patient has a Baker's cyst (figure A-24). What is the cyst composed of?

A. Synovial fluid.

B. Blood.

C. Lymph fluid.

D. Water.

E. Pus.

109 With regard to figure A-25, what is the LEAST likely rationale for the flow characteristics in the subclavian vein?

Figure A-25.
Question 109 is based on this upper extremity venous duplex study.

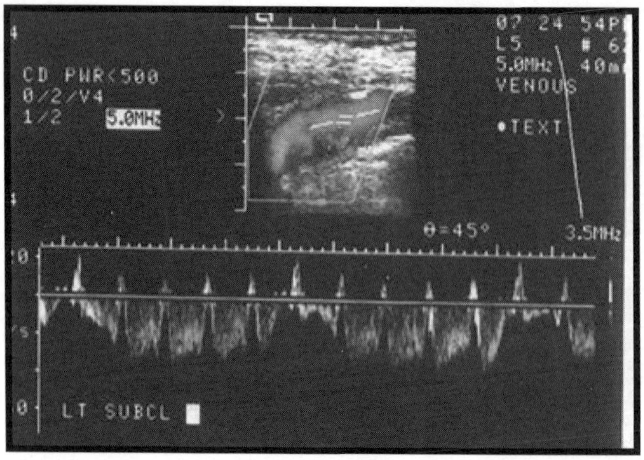

A. These flow patterns are expected because of the close proximity to the heart.

B. The pulsatile flow may be secondary to fluid overload.

C. Flow patterns suggest an obstructive process.

D. It may indicate tricuspid insufficiency.

110 As you evaluate your patient for chronic swelling of the left lower extremity, what is the most significant finding in figure A-26?

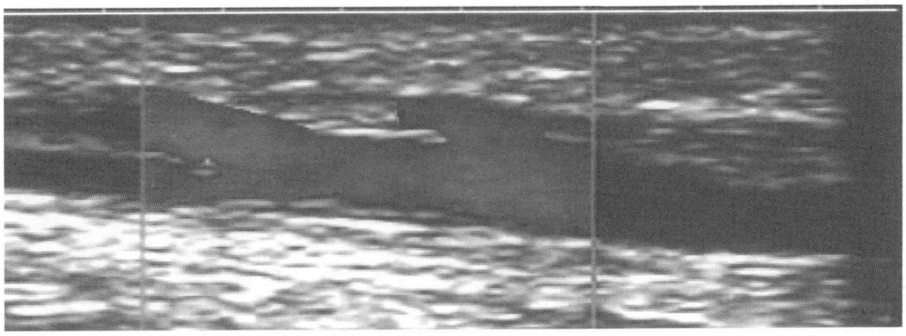

Figure A-26. Question 110 refers to this B-mode image of the left great saphenous vein.

A. Vessel is patent.

B. Wall irregularities present.

C. Chronic thrombus documented.

D. Acute thrombosis evident.

Test Validation, Statistics, and Patient Safety

111 As you provide care to your patients, you are very conscientious about avoiding cross-contamination. What is the most effective infection control practice that you engage in?

A. Hand washing.

B. Isolation practices.

C. Wearing gloves.

D. Having hair covered or pulled back.

112 Upon completion of an ultrasound examination, you always prepare the room for the next patient by discarding soiled items, changing the sheet/covering on the exam table, and cleaning the transducer. If you feel the transducer needs to be disinfected, which of the following products would not be appropriate for that purpose?

A. Hydrogen peroxide.

B. Sterile saline.

C. Gluteraldehyde.

D. Quarternary ammonium.

113 If sensitivity is 98%, specificity 93%, positive predictive value (PPV) 97%, and negative predictive value (NPV) 91%, what might overall accuracy be?

A. 99%.

B. 98%.

C. 96%.

D. 92%.

E. 91%.

114 An accrediting body wants to see documentation of how good your studies are in detecting disease when disease is present as well as excluding disease when the circulation is considered normal. What calculation would best answer their question?

A. Specificity.

B. Sensitivity.

C. Accuracy.

D. Negative predictive value.

E. Positive predictive value.

115 A new referring physician wants to know how often your positive studies are correct. Which of the following calculations would best answer her question?

A. Negative predictive value.

B. Positive predictive value.

C. Sensitivity.

D. Specificity.

E. Accuracy.

Box A-1.
Questions 116–120 are based on this scenario.

During the last two quarters, your department has compared 86 internal carotid arteries evaluated by carotid duplex with angiography. The results:

- 70 vessels had a hemodynamically significant lesion documented by the noninvasive and invasive studies.
- 12 showed no evidence of a hemodynamically significant lesion by either method.
- In 1 case, the noninvasive study was negative, but the angiogram documented a 70–99% diameter reduction.
- In 3 cases, the noninvasive studies were consistent with a greater than 50% diameter reduction, but the invasive study was read as normal.

116 What is your department's specificity in carotid duplex?

A. 82/86.

B. 70/71.

C. 70/73.

D. 12/15.

E. 12/13.

117 What is the positive predictive value of carotid duplex in your department?

A. 82/86.

B. 70/71.

C. 70/73.

D. 12/15.

E. 12/13.

118 How accurate is your diagnostic facility in performing carotid duplex?

A. 82/86.

B. 70/71.

C. 70/73.

D. 12/15.

E. 12/13.

119 What is the negative predictive value of carotid duplex in your department?

A. 82/86.

B. 70/71.

C. 70/73.

D. 12/15.

E. 12/13.

120 What is your department's sensitivity with carotid duplex?

A. 82/86.

B. 70/71.

C. 70/73.

D. 12/15.

E. 12/13.

Answers

1. A
2. C
3. A
4. C
5. B
6. D
7. C
8. C
9. D
10. B
11. C
12. B
13. C
14. D
15. B
16. B
17. C
18. B
19. A
20. C
21. E
22. D
23. C
24. C
25. B
26. D
27. C
28. D
29. A
30. B
31. A
32. B
33. B
34. E
35. B
36. C
37. B
38. C
39. A
40. A
41. B
42. A
43. B
44. C
45. C
46. C
47. B
48. C
49. C
50. B
51. A
52. B
53. A
54. D
55. D
56. A
57. B
58. B
59. C
60. B
61. C
62. C
63. B
64. B
65. C
66. B
67. C
68. A
69. A
70. B
71. A
72. D
73. E
74. A
75. B
76. C
77. A
78. A
79. B
80. E
81. C
82. B
83. A
84. D
85. D
86. E
87. B
88. A
89. D
90. C
91. D
92. C
93. D
94. B
95. D
96. C
97. B
98. B
99. C
100. B
101. A
102. C
103. C
104. A
105. B
106. D
107. C
108. A
109. C
110. C
111. A
112. B
113. C
114. C
115. B
116. D
117. C
118. A
119. E
120. B

Application for CME Credit

APPENDIX B

Introduction

Vascular Technology: An Illustrated Review is a continuing medical educational (CME) activity approved for 12 hours of credit by the Society of Diagnostic Medical Sonography and may be used by more than one person (see *Note* on following page).

Who May Apply for CME Credit

This credit may be applied as follows:

1. Sonographers and technologists may apply these hours toward the CME requirements of the ARDMS, ARRT, and/or CCI, as well as to the CME requirements of ICAVL and AIUM for technologists and sonographers in facilities accredited by those organizations.
2. Physicians may apply a certain maximum number of SDMS-approved credit hours toward the CME requirements of the ICAVL and AIUM for accreditation of diagnostic facilities. (Be sure to confirm current requirements with the pertinent organizations.) Physicians who are registered

sonographers or technologists may apply all of these hours toward the CME requirements of the ARDMS, ARRT, and/or CCI. SDMS-approved credit is not applicable toward the AMA Physician's Recognition Award.

If you have any questions whatsoever about CME requirements that affect you, please contact the responsible organization directly for current information. CME requirements can and sometimes do change.

NOTE

The original purchaser of this CME activity is entitled to submit this CME application for an administrative fee of $39.50. Please enclose a check payable to Davies Publishing Inc. with your application or a 16-digit credit card number and expiration date. Others may also submit applications for CME credit by completing the activity as explained above and enclosing an administrative fee of $49.50. The CME administrative fee helps to defray the cost of processing, evaluating, and maintaining a record of your application and the credit you earn. Fees may change without notice. For the current fee, call us at 626-792-3046, e-mail us at **cme@daviespublishing.com**, or write to us at the address on page 405. We will be happy to help!

Objectives of this Activity

Upon completion of this educational activity, you will be able to:

1. Identify the gross anatomy of the central and peripheral arterial and venous systems.
2. Describe the physiology and fluid dynamics of the central and peripheral circulation.
3. Describe how, when, and why imaging and nonimaging techniques are applied to the noninvasive diagnosis of vascular disease.
4. Describe the invasive tests and therapeutic interventions used in vascular disease.
5. List the diagnostic criteria for both imaging and nonimaging tests used in the noninvasive diagnosis of vascular disease.
6. Apply statistical techniques to the analysis of the sensitivity, specificity, accuracy, and negative and positive predictive values of a diagnostic test.

How to Obtain CME Credit

To apply for credit, please do all of the following:

1. Read and study the book and complete the interactive exercises it contains.

2. Photocopy and complete the applicant information form, CME answer sheet, and evaluation form.

3. Make copies of the completed forms for your records and then return the forms together with payment of the applicable administrative and processing fee (see *Note* on page 404) to the following address:

   CME Coordinator
   Davies Publishing, Inc.
   32 South Raymond Avenue, Suite 4
   Pasadena, California 91105-1935

   or fax to 626-792-5308

   Please allow 15 working days for processing. Questions? Please call us at 626-792-3046.

4. If more than one person will be applying for credit, be sure to photocopy the applicant information, evaluation form, and CME quiz so that you always have the original on hand for use.

APPLICANT INFORMATION

Name ______________________________ Date of Birth ______________

Your degrees and credentials ______________________________

Street address ______________________________

City/State/Zip ______________________________

Telephone ______________ Fax ______________ eMail ______________

ARDMS # ______________ ARRT # ______________ SDMS # ______________ CCI # ______________

Check enclosed ❑ or credit card # ______________________________ Exp. date ______________

Signature and date certifying your completion of the activity ______________________________

Answer Sheet

Circle the correct answer below and return this sheet to Davies Publishing Inc.

1. A B C D	11. A B C D	21. A B C D
2. A B C D	12. A B C D	22. A B C D
3. A B C D	13. A B C D	23. A B C D
4. A B C D	14. A B C D	24. A B C D
5. A B C D	15. A B C D	25. A B C D
6. A B C D	16. A B C D	26. A B C D
7. A B C D	17. A B C D	27. A B C D
8. A B C D	18. A B C D	28. A B C D
9. A B C D	19. A B C D	29. A B C D
10. A B C D	20. A B C D	30. A B C D

Evaluation—You Grade Us!

Please let us know what you think of this book. Participating in this quality survey is a requirement for CME applicants, and it benefits future readers by ensuring that current readers are satisfied and, if not, that their comments and opinions are heard and taken into account.

1 Why did you purchase this book? (Circle your primary reason.)

Registry review   Course text   Clinical reference   CME activity

2 Have you used the book for other reasons, too? (Circle all that apply.)

Registry review   Course text   Clinical reference   CME activity

3 To what extent did this book meet its stated objectives and your needs? (Circle one.)

Greatly   Moderately   Minimally   Insignificantly

4 The content of this book was (circle one):

Just right   Too basic   Too advanced

5 The quality of the exercises, illustrations, and case examples was mainly (circle one):

Excellent   Good   Fair   Poor

6 The manner in which the book presents the material is mainly (circle one):

Excellent   Good   Fair   Poor

7 If you used this book to prepare for the registry exam, did you also use other materials or take any exam-preparation courses?

No Yes (please specify what materials and courses)

____

____

8 If you used this book for a course, please cite the course, the instructor's name, the name of the school or program, and any other textbooks you may have used:

Course/Instructor/School or program ______________________

Other textbooks ______________________________

9 What did you like best about this book?

____

____

10 What did you like least about this book?

____

____

11 If you used this book to prepare for your registry exam in vascular technology, did you pass?

Yes No Haven't yet taken it

12 May we quote any of your comments in our catalogs or promotional material?

Yes No Further comment . . .

____

____

CME QUIZ

Please answer the following questions after you have completed the CME activity. There is one best answer for each question. Circle your choice on the answer sheet on page 406. Be sure to make a copy of the answer sheet if more than one person will take this quiz.

1 The term *tunica adventitia* refers to which part of the arterial wall:

A. The middle layer

B. The inner layer

C. The outer lining

D. The circumferential muscle fibers

2 Which of the following is **NOT** true about pressure/volume relationships, according to Bernoulli?

A. Pressure is higher at the region of stenosis

B. Pressure is lower at the region of stenosis

C. Pressure gradients are set up

D. Velocity and pressure are inversely related

3 Laminar flow is most likely to become disturbed when the Reynolds number exceeds:

A. 1

B. 100

C. 1000

D. 2000

4 Use the following lab findings to calculate the right ABI: right brachial pressure = 170 mmHg, left brachial pressure = 120 mmHg, right ankle pressure = 100 mmHg, and left ankle pressure = 80 mmHg.

A. 0.66

B. 0.47

C. 0.58

D. 0.83

5 Venous return to the heart is facilitated by all of the following **EXCEPT:**

A. Calf muscle pump

B. Inspiration

C. Venous valves

D. Expiration

6 A patient with an ABI of 0.65 is most likely to be:

A. Asymptomatic

B. Having rest pain

C. Suffering from severe arterial disease

D. Experiencing claudication

7 A high-resistance signal can be described by all of the following **EXCEPT:**

A. Triphasic

B. Pulsatile

C. Biphasic

D. Continuous

8 Why is the Adson maneuver performed?

A. To test the palmar arch

B. To evaluate dialysis access grafts

C. To assess thoracic outlet syndrome

D. To rule out carpal tunnel syndrome

9 The complications of arteriography commonly include all of the following **EXCEPT:**

A. Hematoma at the puncture site

B. Local artery occlusion

C. Pseudoaneurysm

D. Nerve damage

10 Branches of the external carotid artery include all of the following **EXCEPT:**

A. Occipital

B. Basilar

C. Facial

D. Ascending pharyngeal

11 The factor having the single most dramatic effect on blood flow is, according to Poiseuille's law:

A. Radius of the vessel

B. Length of the vessel

C. Elasticity of the vessel wall

D. Viscosity of the blood

12 The vessel that connects the left and right anterior cerebral arteries is the:

A. Middle cerebral artery

B. Basilar artery

C. Anterior communicating artery

D. Right and left posterior communicating arteries

13 All of the following vessels are part of the circle of Willis **EXCEPT:**

A. Anterior communicating artery

B. Posterior cerebral arteries

C. Anterior choroidal arteries

D. Posterior communicating arteries

14 One cause of right hemispheric infarction is:

A. Stenosis of the right external carotid artery

B. Occlusion of the right internal carotid artery

C. Stenosis of the left external carotid artery

D. Occlusion of the left internal carotid artery

15 Spectral broadening usually represents:

A. Laminar flow

B. Turbulent flow

C. Accelerated flow

D. Dampened flow

16 In a carotid duplex exam, all of the following are common causes for underestimating the degree of stenosis **EXCEPT:**

A. Improper placement of the sample volume

B. Low-level echoes from fresh thrombus

C. Tangent lesions

D. Fibromuscular dysplasia

17 The following artery is **NOT** routinely evaluated in a TCD exam:

A. Middle cerebral

B. Anterior cerebral

C. Posterior cerebral

D. Posterior communicating

18 Angiography reveals a stenotic right internal carotid artery with a residual lumen of 3 mm. If the normal lumen is 9 mm, the diameter reduction is:

A. 33%

B. 50%

C. 67%

D. 78%

19 A 75% area reduction equates to a diameter reduction of:

A. 37.5%

B. 50%

C. 70%

D. 75%

20 In a TCD exam the normal direction of blood flow in the middle cerebral artery is:

A. Bidirectional

B. Antegrade

C. Retrograde

D. Antegrade in dominant hemisphere, retrograde in nondominant hemisphere

21 Which of the following is **NOT** a deep vein of the lower extremity?

A. Superficial femoral

B. Profunda femoris

C. Great saphenous

D. Anterior tibial

22 All of the following are deep veins of the upper extremity **EXCEPT:**

A. Radial

B. Ulnar

C. Cephalic

D. Subclavian

23 The most important criterion for the correct identification of the deep veins is the:

A. Course of the vessel

B. Size of the vessel

C. Depth of the vessel

D. Adjacent artery

24 Which of the following flow characteristics is **LEAST** important in diagnosing acute DVT by continuous wave Doppler?

A. Nonpulsatility

B. Spontaneity

C. Phasicity

D. Augmentation

25 All of the following may cause false-negative continuous wave venous Doppler results **EXCEPT:**

A. Collateralization

B. Extrinsic compression

C. Partial thrombus

D. Bifid venous system

26 IPG detects volume changes by the following method:

A. One-wire

B. Two-wire

C. Three-wire

D. Four-wire

27 In venous duplex scanning for DVT, all of the following characteristics are of diagnostic value **EXCEPT:**

A. Venous reflux

B. Augmentation with distal compression

C. Wall compressibility

D. Echogenicity of the lumen

28 The most common sequela of DVT is:

A. Pulmonary embolism

B. Valvular destruction

C. Venous gangrene

D. Phlegmasia cerulea dolens

29 In sagittal view, color flow imaging displays flow within the great saphenous vein as blue; during a Valsalva maneuver venous flow is displayed in red. This signifies:

A. Venous obstruction

B. Rouleau formation

C. Normal flow patterns

D. Valvular incompetence

30 If sensitivity is 92%, specificity 89%, positive predictive value 90%, and negative predictive value 88%, the overall accuracy might be:

A. 90%

B. 91%

C. 92%

D. 93%

Bibliography

Bernstein ED: *Noninvasive Diagnostic Techniques in Vascular Disease,* 4th edition. St. Louis, Mosby, 1993.

Cohen JR: *Vascular Surgery for the House Officer.* Baltimore, Williams & Wilkins, 1992.

Fahey VA: *Vascular Nursing.* Philadelphia, WB Saunders, 1988.

Hershey FB, Barnes RW, Sumner DS [eds]: *Noninvasive Diagnosis of Vascular Disease.* Pasadena, Davies Publishing, 1984.

Kempczinski RE, Yao JST [eds]: *Practical Noninvasive Vascular Diagnosis,* 2nd edition. Chicago, Year Book, 1987.

Neumyer MM: *Techniques of Abdominal Vascular Sonography* [DVD]. Pasadena, Davies Publishing, 2007.

Ridgway DP: *Introduction to Vascular Scanning,* 3rd edition. Pasadena, Davies Publishing, 2004.

Ridgway DP, Size GP: *Vascular Anatomy and Physiology.* Pasadena, Davies Publishing, 2009.

Ridgen J: *MacMillan Encyclopedia of Physics,* volume 2. NY, Simon & Schuster, 1996, pp 597–599.

Rumwell C, McPharlin M, Strandness DE, Grant EG (eds): *Vascular Laboratory Policies and Procedures Manual.* Pasadena, Davies Publishing, 2000.

Illustrated Manual of Nursing Practice. Springhouse, PA, Springhouse Corporation, 1991.

Talbot SR, Oliver MA: *Techniques of Venous Imaging* [book], 2nd edition. Pasadena, Davies Publishing, 2009 (in press).

Talbot SR, Oliver MA: *Venous Imaging Techniques* [DVD], 2nd edition. Pasadena, Davies Publishing, 2007.

Zwiebel WJ: *Introduction to Vascular Ultrasonography,* 5th edition. Philadelphia, Elsevier Saunders, 2005.

Angiograms

It is instructive to see angiograms (arteriograms, aortograms, and venograms) from actual cases in one place. Each image in this section is thoroughly explained in its caption, and each caption cites the page numbers where more information on the topic can be found. See *Arteriography* in chapters 17 and 25, and *Contrast Venography* in chapter 35 for further discussion of these images and topics.

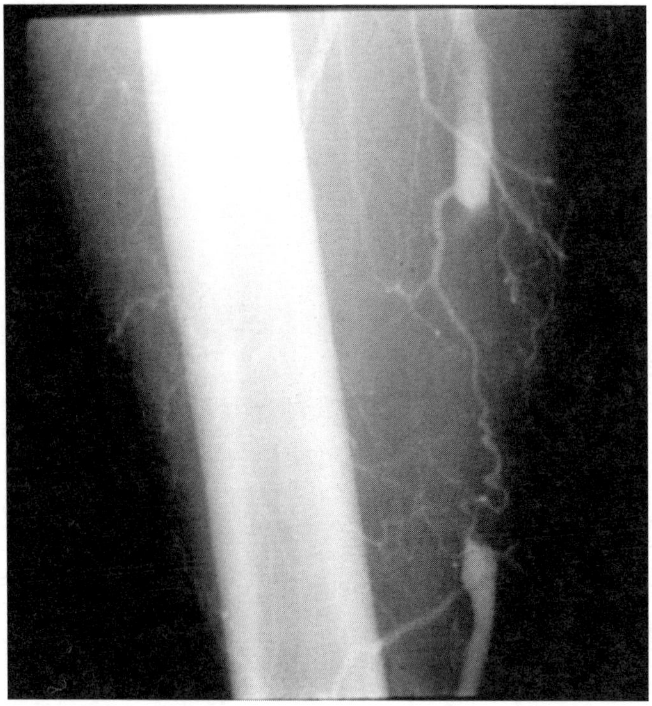

Plate 1.
Left femoral arteriogram revealing an occluded superficial femoral artery (SFA) with reconstitution at the distal SFA/proximal popliteal artery level via collateralization. See pages 173–178.

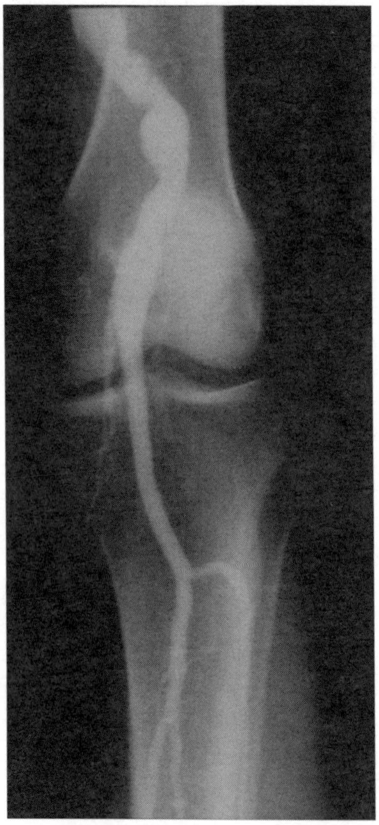

Plate 2.
Popliteal aneurysm. The thrombus evident in this type of peripheral aneurysm is often the source of distal embolization. It is not unusual for a person with one aneurysm (e.g., aortic) to have aneurysmal disease at other levels (e.g., femoral, popliteal). See pages 173–178.

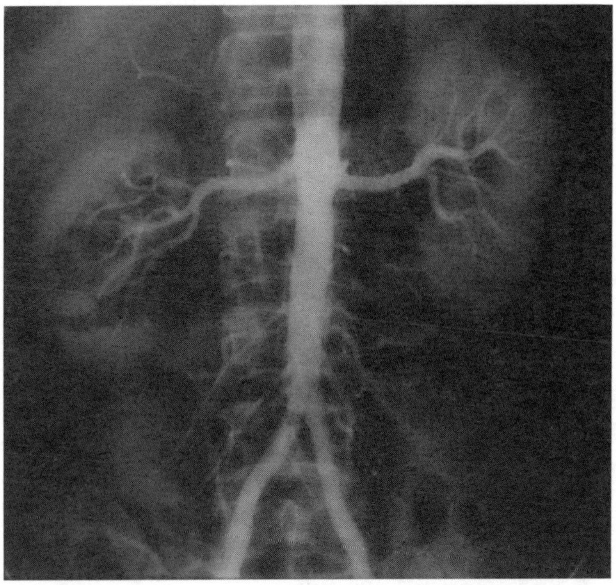

Plate 3.
Normal aortogram, including the renal artery branches. See pages 173–178.

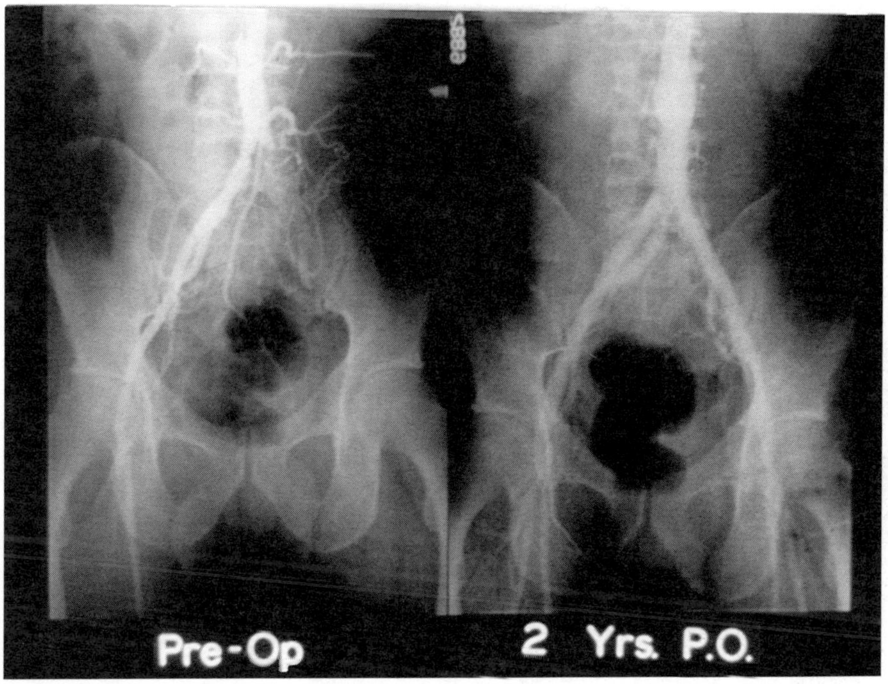

Plate 4.
Pre- and postoperative aortograms showing the abdominal aorta, proximal branches (including the renal arteries and the common iliac artery), and bifurcation vessels. **Preoperative:** Occluded left common iliac artery with minimal collateralization at the origin of the left common iliac artery. **Postoperative:** Normal aorta and right and left common iliac, external iliac, and internal iliac arteries. See pages 173–178.

Plate 5.

Aortic occlusion with collateralization. Several collateral branches provide flow to the lower extremities. As the body attempts to maintain perfusion, the distal capillary beds vasodilate. The normally high-resistance flow patterns assume the low-resistance flow qualities of that vasodilated vascular bed. See pages 173–178.

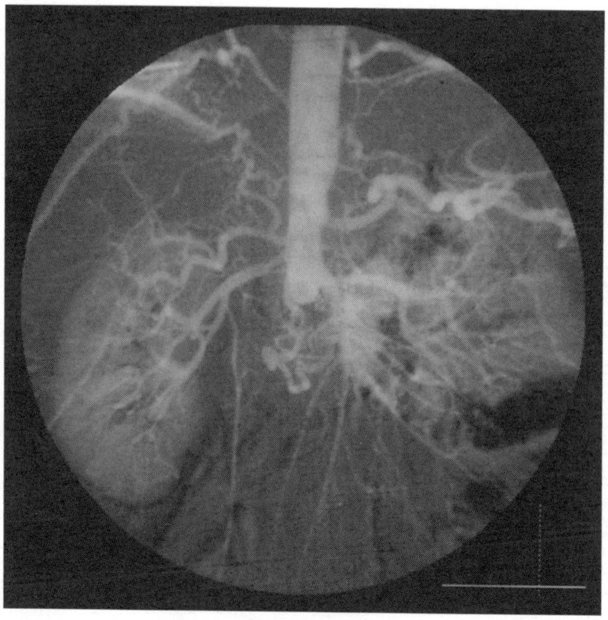

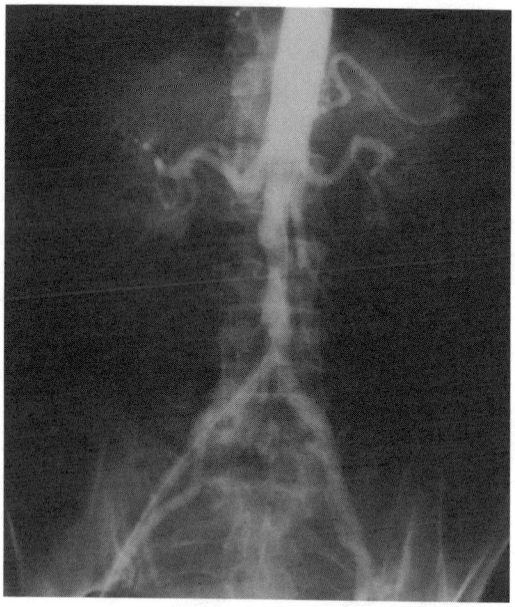

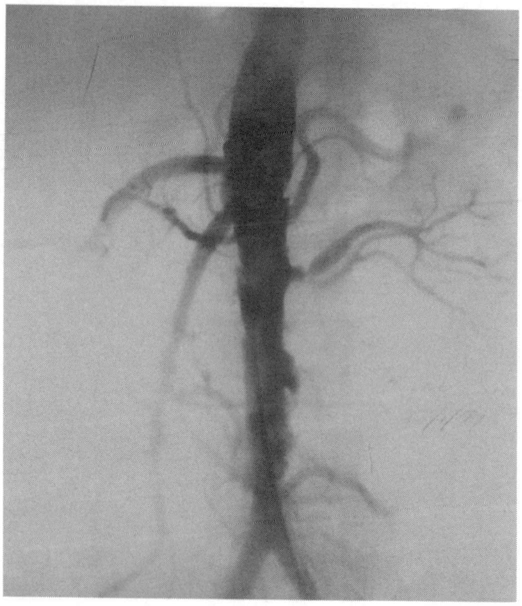

Plate 6.

Aortogram showing several areas of filling defects, although the renal arteries appear normally patent. See pages 173–178.

Plate 7.

Aortogram revealing multiple wall irregularities, a high-grade stenosis of the left renal artery, and poor to absent flow in the renal vasculature on the right. See pages 173–178.

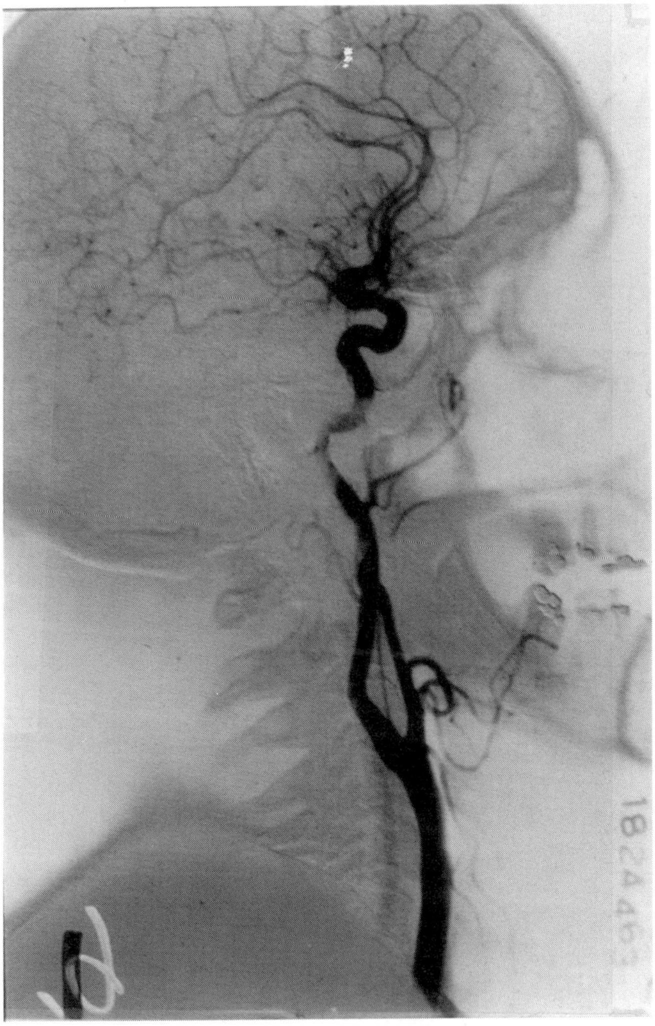

Plate 8.

Normal carotid angiogram in which contrast medium completely fills the vessels. See pages 251–256.

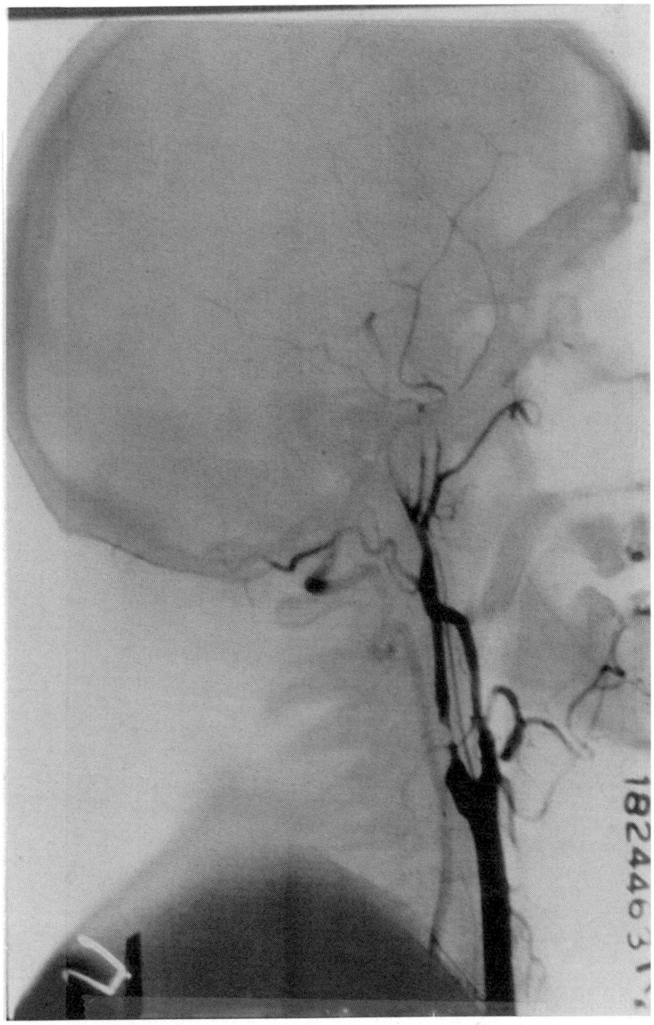

Plate 9.

Abnormal carotid angiogram with filling defect in the internal carotid artery. See pages 251–256.

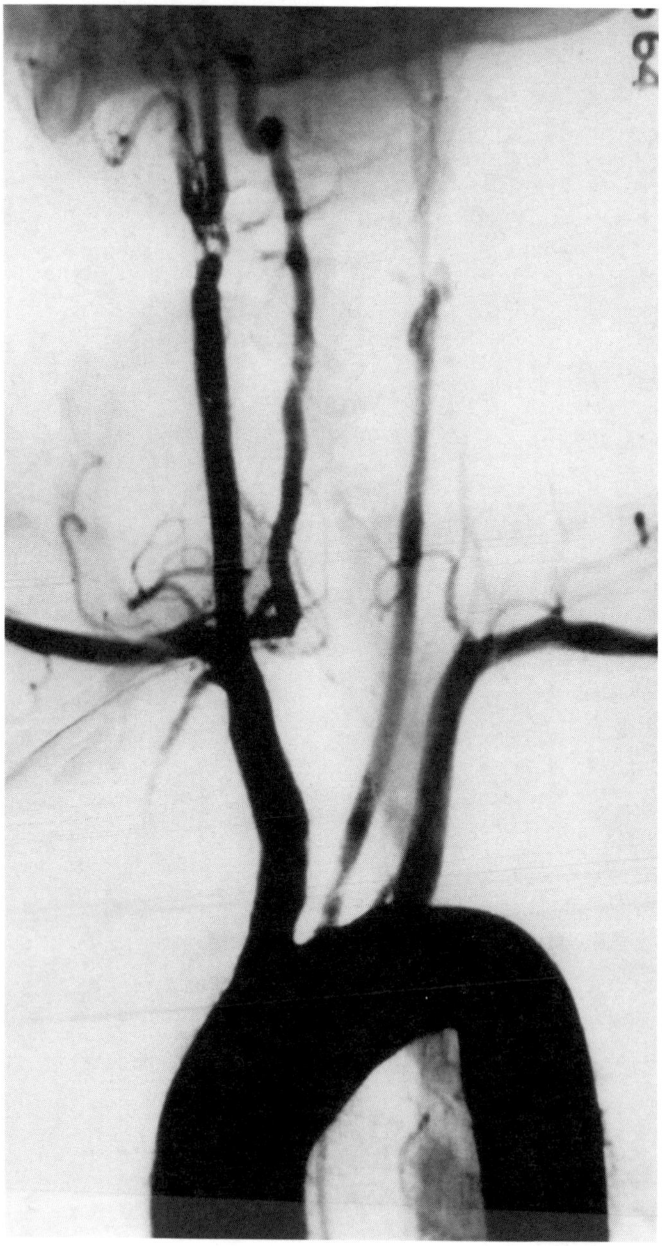

Plate 10.
Arch and carotid angiogram showing multiple disease in a patient admitted with acute CVA. **Right bifurcation vessels:** High-grade stenoses suggested by filling defects. **Left proximal common carotid artery:** High-grade stenosis with poor vessel filling throughout the remainder of the carotid system, but especially above the bifurcation. In this case, the proximal disease increases the time it takes for the contrast agent to move distally within the vessel. For better visualization, delayed films would be required. If the patient had undergone a carotid duplex scan, Doppler flow in the CCA distal to the stenosis would be of poor quality (reduced strength, rounded peak), but resistance would not change. On the other hand, if the flow were approaching a significant stenosis, such as on the right, resistance (pulsatility) would increase. **Left vertebral artery:** No flow noted, indicating occlusion or—another possibility—a congenitally absent left vertebral artery. See pages 251–256.

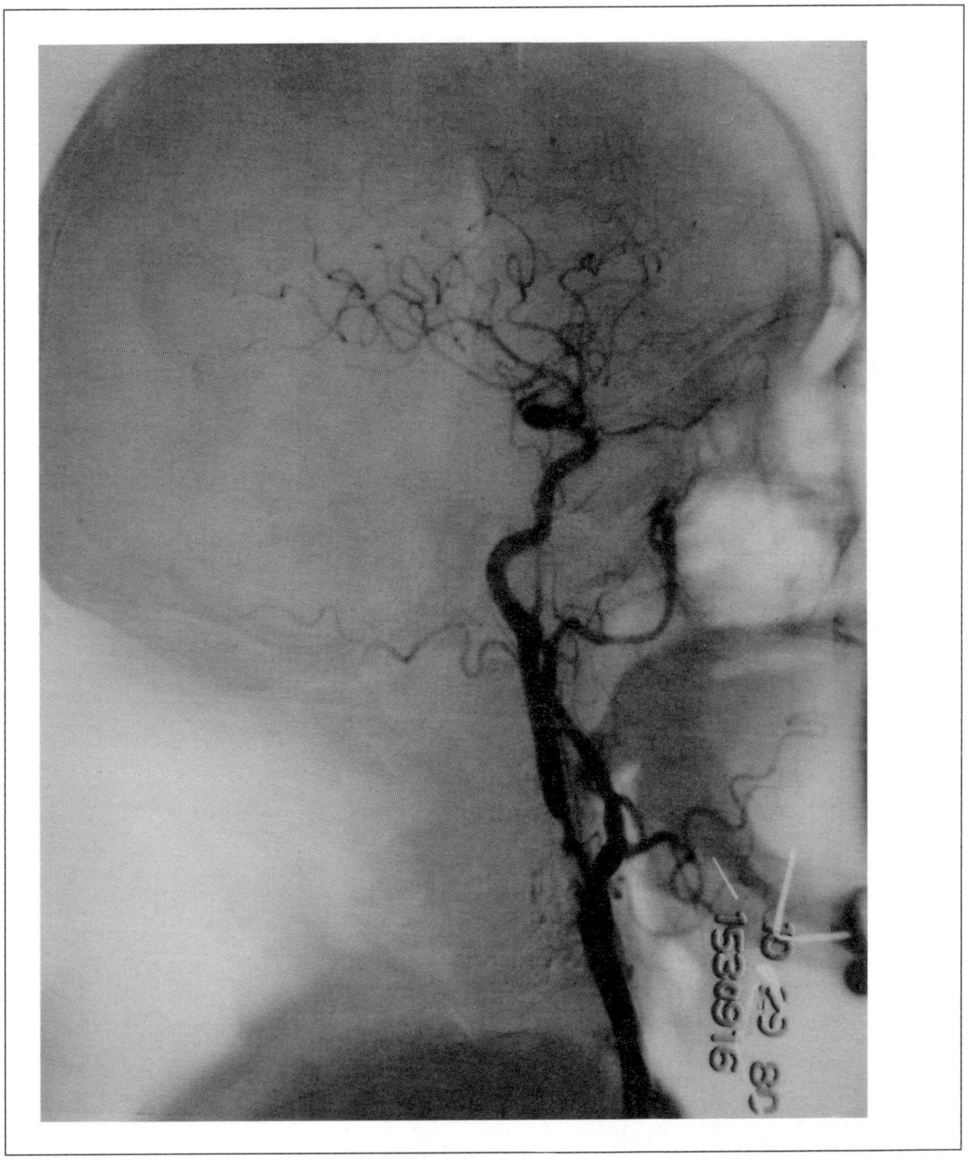

Plate 11.

Angiogram from patient with asymptomatic carotid bruit. In this image the absence of contrast agent (filling defect) reveals the irregularity of the plaque in the bulb and internal carotid artery. See pages 251–256.

Plate 12.
Angiogram showing internal carotid artery with fibromuscular dysplasia (FMD) of the distal portion. While most carotid disease is atherosclerotic and occurs at the bifurcation, FMD is usually located quite distal. On angiography, the vessel with FMD typically appears to be beaded, a characteristic referred to as a "string of beads." On duplex scanning, Doppler signals would show spectral broadening and increased velocities that vary with the location of the Doppler sample. See pages 251–256.

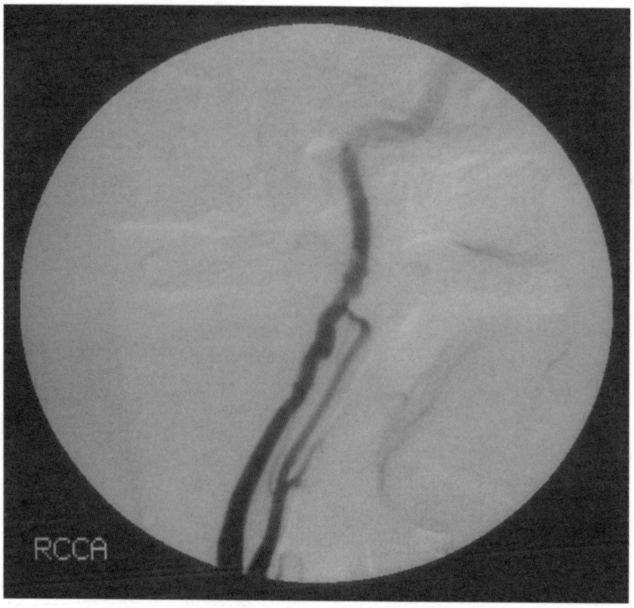

Plate 13.
Angiogram depicting a "string sign" in the carotid bulb. Doppler peak flow velocities are reduced because of the low flow state of this preocclusive lesion. Auscultation of the neck would probably fail to elicit a bruit because the disease is so severe. See pages 251–256.

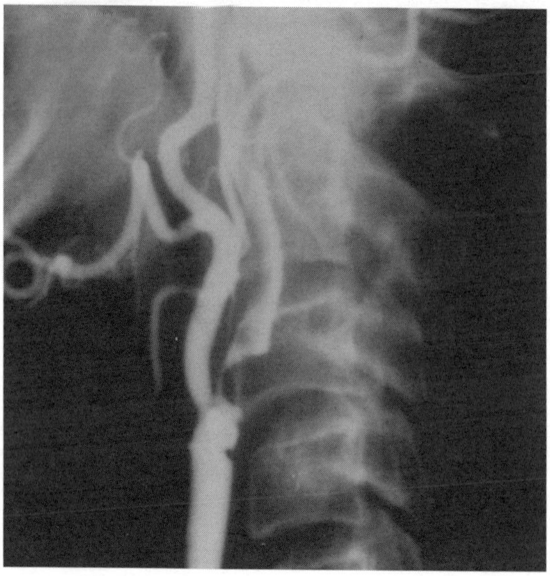

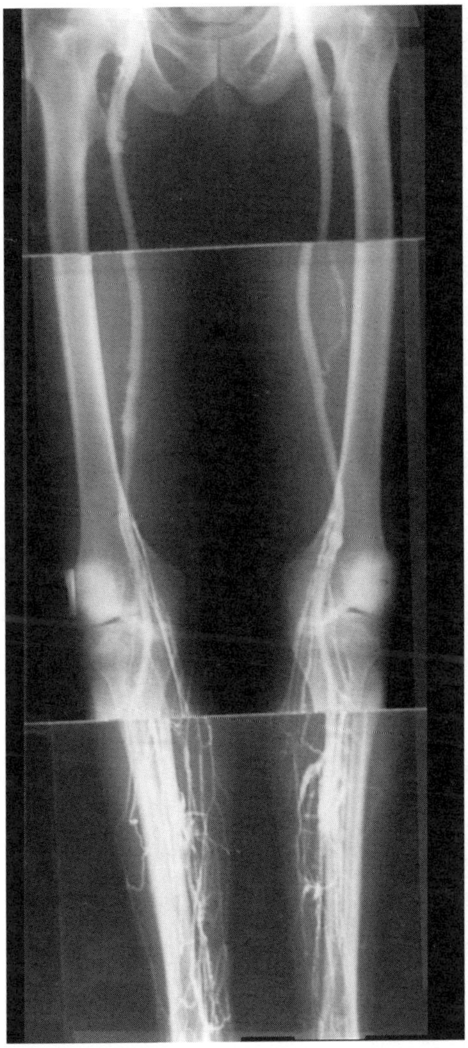

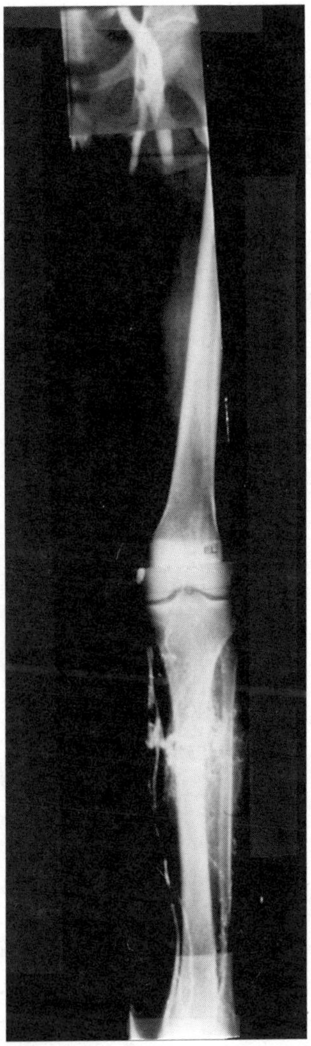

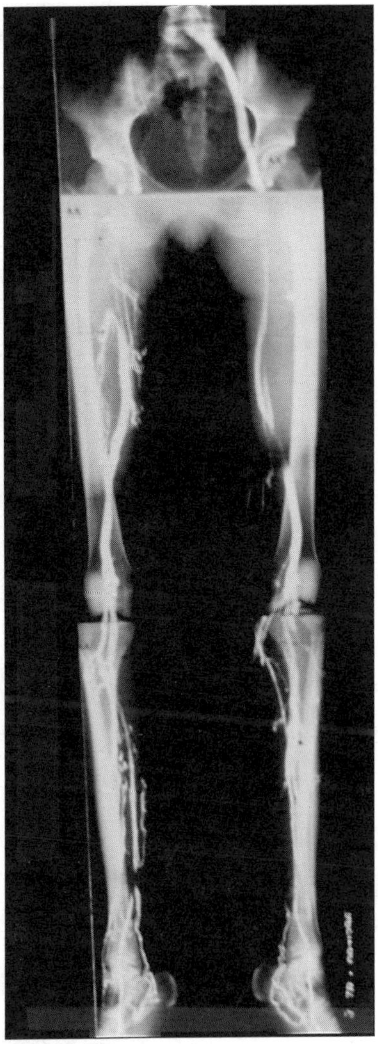

Plate 14.

Normal venogram (phlebogram) in which a radiopaque contrast agent fills the venous system. The tibial, popliteal, femoral, and common femoral veins are shown. During venography, serial x-rays are taken. See pages 340–342.

Plates 15 and 16.

Venograms showing filling defects that suggest the displacement of contrast material by thrombus. In plate 15, there is only minimal filling of the contrast agent in the tibial veins, with none seen until the confluence of the femoral vein and deep femoral vein into the common femoral vein. The significant filling defect probably represents deep venous thrombosis. In plate 16, contrast agent fills many more vessels, but filling defects indicating thrombosis appear in the right tibial veins, proximal femoral, and in the confluence of the common femoral and external iliac veins. On the left, there is minimal filling of the tibial veins with filling defect in the femoral vein at mid thigh. See pages 340–342.

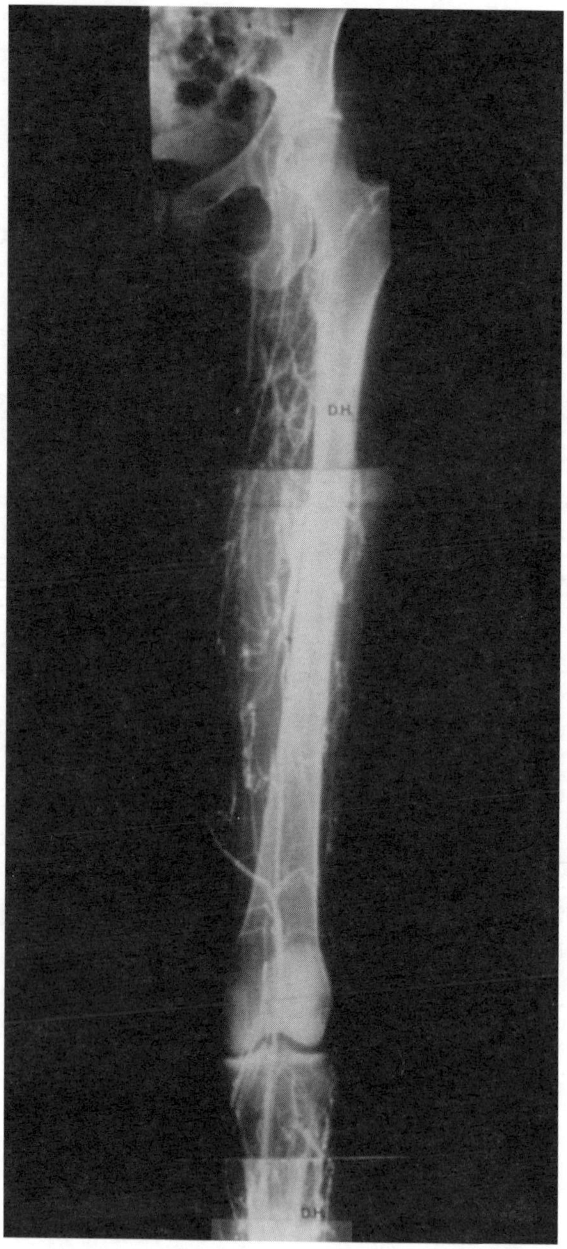

Plate 17.

Abnormal venogram showing the presence of multiple collateral veins. Only the popliteal vein appears to be patent, while the remainder of the proximal deep system is thrombosed. In the presence of the extensive network of collateral vein branches, continuous wave Doppler (without imaging) could elicit fairly normal venous signals even though the deep system is thrombosed. For this reason, duplex scanning is more accurate, since it images the veins and surrounding tissue. See pages 340–342.

Index

Page locators followed by an "f" indicate figures; those followed by a "t" indicate tables.